AF776922

The Downy Mildews - Biology, Mechanisms of Resistance and Population Ecology

Aleš Lebeda · Gerald J. Holmes · Brigitte Mauch-Mani · Michael J. Jeger
Editors

The Downy Mildews - Biology, Mechanisms of Resistance and Population Ecology

Previously published in *European Journal of Plant Pathology*
Volume 129, issue 2, 2011

Springer

Editors

Prof. Aleš Lebeda
Faculty of Science
Department of Botany
Palacký University in Olomouc
Slechtitelu 11, 783 71 Olomouc-Holice
Czech Republic
e-mail: ales.lebeda@upol.cz

Dr. Gerald J. Holmes
Valent USA Corporation
Pilot Mountain Court 2102
27502 Alex, North Carolina, USA
e-mail: gerald.holmes@valent.com

Prof. Brigitte Mauch-Mani
University of Neuchâtel
Laboratory of Cellular and Molecular Biology
Rue Emile Argand 11
CH-2009 Neuchâtel, Switzerland
e-mail: brigitte.mauch@unine.ch

Prof. Michael J. Jeger
Division of Biology
Imperial College London
Silwood Park Campus
SL5 7PY Ascot, United Kingdom
e-mail: m.jeger@ic.ac.uk

ISBN: 978-94-017-8448-1

Softcover re-print of the Hardcover 1st edition 2011

Cover illustration: Detail of sporangiophores and sporangia of *P. cubensis*

Printed on acid-free paper

Springer is part of Springer Science+Business Media (www.springer.com)

European Journal of Plant Pathology

Contents Volume 129 No. 2 February 2011

Special Issue
"The Downy Mildews - Biology, Mechanisms of Resistance, Population Ecology"
Guest Editors: A. Lebeda, G.J. Holmes, B. Mauch-Mani and M.J. Jeger

Eur J Plant Pathol (2011) 129:1
DOI 10.1007/s10658-010-9709-7

Foreword

"The downy mildews—Biology, mechanisms of resistance, population ecology"

Aleš Lebeda · Gerald J. Holmes · Brigitte Mauch-Mani · Michael J. Jeger

Published online: 17 December 2010

This book contains selected papers presented at the Downy Mildew Workshop during the International Congress of Plant Pathology (ICPP) held in August 2008 at Torino, Italy, and some additional relevant papers selected by the editors. The papers are focused on the biology, evolution, resistance mechanisms, host-parasite interactions, genetic variation, population ecology, epidemiology, chemical control and fungicide resistance of downy mildews. This volume represents a continuation of a previous volume entitled: The Downy Mildews—Genetics, Molecular Biology and Control (A. Lebeda, P.T.N. Spencer-Phillips and B.M. Cooke, Eds.; Springer, 2008, 206 pp., ISBN 978-1-4020-8972-5).

Knowledge of downy mildew pathogens, and the plant diseases they cause, has advanced substantially during last decade. Progress in some areas has been summarized in three consecutive volumes of "Advances in Downy Mildew Research" (2002, 2004 and 2007). However, the last Downy Mildew Workshop provided an opportunity to collect new information and research topics about downy mildews, and to publish this spin-off volume. It contains 15 invited reviews and original research papers. This book gives the most recent and comprehensive update of knowledge about downy mildews.

We hope that it will be of use to students, academics and researchers in plant pathology, mycology, crop protection, taxonomy, plant biology and ecology, as well as to others involved in the agricultural and horticultural advisory services and industry.

A. Lebeda (✉)
Faculty of Science, Department of Botany,
Palacký University in Olomouc,
Šlechtitelů 11,
783 71 Olomouc-Holice, Czech Republic
e-mail: ales.lebeda@upol.cz

G. J. Holmes
Valent USA Corporation,
202 Davis Grove Circle, Suite 103,
Cary, NC 27519, USA
e-mail: gerald.holmes@valent.com

B. Mauch-Mani
Laboratory of Cellular and Molecular Biology,
University of Neuchâtel,
Rue Emile-Argand 11,
CH-2009 Neuchâtel, Switzerland
e-mail: brigitte.mauch@unine.ch

M. J. Jeger
Division of Biology, Imperial College London,
Silwood Park Campus,
Ascot SL5 7PY, UK
e-mail: m.jeger@ic.ac.uk

Eur J Plant Pathol (2011) 129:3–14
DOI 10.1007/s10658-010-9714-x

Phylogenetic investigations in the genus *Pseudoperonospora* reveal overlooked species and cryptic diversity in the *P. cubensis* species cluster

Fabian Runge · Young-Joon Choi · Marco Thines

Accepted: 20 October 2010 / Published online: 17 December 2010

Abstract *Pseudoperonospora cubensis* is one of the most devastating diseases of cucurbitaceous crops. The pathogen has a worldwide distribution and occurs in all major cucurbit growing areas. It had been noticed for the first time at the end of the 19th century, but it became a globally severe disease as recently as 1984 in Europe and 2004 in North America. Despite its economic importance, species concepts in *Pseudoperonospora* are debated. Here, we report that the genus *Pseudoperonospora* contains cryptic species distinct from the currently accepted ones. *Pseudoperonospora* on *Celtis* is split into two phylogenetic lineages and *Pseudoperonospora humuli* is confirmed as a species distinct from the Cucurbitaceae-infecting lineages. A cryptic species occupying a basal position within the *Pseudoperonospora cubensis* complex is revealed to be present on *Humulus japonicus*, thus providing evidence that the host jump that gave rise to *Pseudoperonospora cubensis* likely occurred from hops. Notably, Cucurbitaceae infecting pathogens are present in two cryptic sister species or subspecies. Clade 1 contains primarily specimens from North America and likely represents *Pseudoperonospora cubensis* s.str.. Pre-epidemic isolates in clade 2 originate from Japan and Korea, suggesting this cryptic species or subspecies is indigenous to East Asia. Recent samples of this lineage from epidemics in Europe and the United States cluster together with clade 2. It thus seems possible that this lineage is associated with the recent severe epidemics of cucurbit downy mildew and is now naturalised in North America and Europe.

Keywords *cox2* · Cucurbit downy mildew · Cryptic species · nrITS · Peronosporaceae · *Pseudoperonospora humuli* · Species delimitation · *ypt1*

F. Runge
Institute of Botany 210, University of Hohenheim,
70593 Stuttgart, Germany

Y.-J. Choi
Department of Organism & Evolutionary Biology,
Harvard University,
22 Divinity Ave,
Cambridge, MA 02138, USA

M. Thines
Biodiversity and Climate Research Centre (BiK-F),
Senckenberganlage 25,
60325 Frankfurt (Main), Germany

M. Thines (✉)
Department of Biological Sciences, Institute of Ecology,
Evolution and Diversity,
Johann Wolfgang Goethe University,
Siesmayerstr. 70,
60325 Frankfurt (Main), Germany
e-mail: marco.thines@senckenberg.de

Introduction

The genus *Pseudoperonospora* includes six widely recognised species, of which *P. humuli* and *P. cubensis* are economically most important. *Pseudoperonospora humuli* occurs in all of the hop-growing countries of the northern hemisphere and in Argentina (Royle and Kremheller 1981). The pathogen was first reported in Japan in 1905 (Miyabe and Takahashi 1906), and has caused severe losses in the major hop growing areas in the 20th century. Johnson et al. (1994) reported that Washington State had the highest disease incidences in 1988, 1990, and 1991. In 2002 and 2003 the disease re-emerged in Argentina and the reduction in hop cone yield of infested fields varied between 20% and 34% (Pérez et al. 2009). But due to advances in forecasting and disease management, hop downy mildew control seems feasible (Mozny et al. 1993; Dolinar and Žolnir 1994; Gent et al. 2009).

Downy mildew of cucurbits caused by *P. cubensis* occurs on wild and cultivated Cucurbitaceae in all major cucurbit growing countries (Cohen 1981; Holmes et al. 2004), and more than 60 host species have been reported (Lebeda 1990, 1999), including all cucurbits of economic interest. It is most widespread on *Cucumis sativus* (about 80 countries) and *Cucumis melo* (more than 50 countries) (Lebeda 1990). In 2006 the European countries Spain, the Netherlands, and Germany together produced about 2 million tons of cucumbers. In East Asia, in Japan and in the Republic of Korea, altogether more than 1 million tons of cucumbers were grown and Canada, Mexico, and the United States produced more than 1 million tons of cucumbers (www.ers.usda.gov) in total. The US cucumber production had a value of more than 180 million US$ in 2009 (Battaglia 2010), highlighting the value of this crop.

In the 1930s and 1940s, in the USA most of the efforts against cucurbit downy mildew were centred on chemical control and breeding for resistance (Palti and Cohen 1980). Until the 1940s, downy mildew was the most important disease of cucumber (Selby 1899; Holmes et al. 2006). This changed with the release of cultivar Palmetto in 1948, which had moderate resistance to downy mildew. In 1966, cultivar Poinsett was released with high resistance to downy mildew. This resistance has been bred into most of the popular cucumber cultivars (Holmes et al. 2006). Also in *Cucumis melo* some effective sources of resistance are available (Pitrat et al. 1998; Lebeda 1999) and the cantaloupe cultivar Smith Perfect is in the pedigree of most cantaloupe material in use in the USA today (Palti and Cohen 1980). In addition Lebeda and Widrlechner (2004) demonstrated broad genetic variation in resistance to *P. cubensis* among wild and weedy *Cucurbita* species. Because of the efficient breeding for resistance, US growers have hardly used fungicides to control downy mildew since the late 1960s (Holmes et al. 2006).

This situation changed in the mid 1980s in Europe and around 2004 in the USA. From 1984 onwards, downy mildew has become the economically most important pathogen of cucurbit vegetables in Central Europe (Lebeda 1990; Lebeda and Urban 2004), and disease control through growing of previously resistant cultivars has not been effective (Lebeda and Schwinn 1994; Lebeda and Widrlechner 2003; Lebeda and Urban 2004). Subsequently, in the Czech Republic the yield of field cucumbers decreased by ~80% in 1985 and 83% in 1989 (Lebeda 1991) and a considerable shift in pathogenicity in the Czech pathogen population was evident during 1986–2000 (Lebeda and Gadasová 2002). Cappelli et al. (2003) reported that in 2002 the disease spread rapidly in Italy and because the control measures used were not effective, growers suffered from severe yield losses. Since 2004 *P. cubensis* has re-emerged as a destructive pathogen in the USA. Outbreaks were found on cucumber about 2 months earlier than previously and it had been decades since growers had faced a similar level of destruction from downy mildew. In 2004 North Carolina, Delaware, Maryland, and Virginia lost 40% of their crop with a value of approximately 20 million US$ (Holmes et al. 2006). Downy mildew occured in several regions in Michigan in 2005, including major cucumber production regions, where the disease was so severe that many of the state's largest growers considered no longer growing the crop (Hausbeck 2007).

As the commercially available resistant cucumber cultivars were no longer effective in controlling downy mildew, fungicides were necessary to manage the disease (Urban and Lebeda 2006, 2007; Colucci et al. 2008a,b). In 1996 the global fungicide market was estimated at approximately 6.25 billion US$, of which 16.7% were chemicals to control downy mildews (Gisi 2002). It can be expected that the recent epidemics will lead to a significant increase in these figures. Unfortunately, not only the deployment of resistant cultivars, but also the effectiveness of systemic fungicides has been compromised by the high adaptability of *P. cubensis* (Ishii et al. 2001; Shetty et al. 2002; Urban and Lebeda 2004, 2006, 2007; Keinath et al. 2007; Colucci et al. 2008a, b; Gugino et al. 2009).

Overcoming host defence or developing fungicide resistance may be due to high intraspecific genetic variation in *P. cubensis*, and strong evidence supports the presence of races of *P. cubensis* differing in virulence (Bains and Jhooty 1976; Palti and Cohen 1980). Thomas et al. (1987) distinguished five different pathotypes. Horejsi et al. (2000) and Shetty et al. (2002) revealed the presence of different races in Asia and the USA. On the basis of field evaluations of cucumbers Horejsi et al. (2000) hypothesised that Europe and the United States may harbour different races of *P. cubensis*. But no evidence for race differences in the pathogen was observed between North America and Europe. However, European locations tended to have higher downy mildew scores compared to the North American locations and resistance of cucumber to *P. cubensis* held up against isolates from the US, but not against European isolates (Lebeda 1992). Lebeda and Urban (2004) determined 34 different pathotypes of *P. cubensis* and concluded that there were no substantial geographical patterns in the occurrence of various pathotypes among 106 locations in the Czech Republic. Holmes et al. (2006) mentioned that the post 2004 downy mildew incidences might indicate the presence of a more virulent type of pathogen than in previous years, corresponding to a new race or pathotype (Holmes et al. 2006). But the paucity of characters suitable for species delineation in *Pseudoperonospora* (Shin and Choi 2003) renders it possible that the new epidemics were caused by a cryptic species of *Pseudoperonospora*.

Of the six widely accepted species, *P. humuli* and *P. cubensis* are the economically most important and are known since the end of the 19th and the beginning of the 20th century from the northern hemisphere. However, all species of *Pseudoperonospora* are morphologically similar, leading to a species classification that is primarily based on the host matrix. Earlier investigations have revealed that *P. cubensis* is closely related to *P. humuli*, despite parasitising the only distantly related Cucurbitaceae (Fraymouth 1956; Constantinescu 2000; Choi et al. 2005). Based on morphological characteristics, Shin and Choi (2003, 2006) and Choi et al. (2005) indicated that *P. cubensis* and *P. humuli* might be conspecific. Molecular phylogenetic analysis of Choi et al. (2005) revealed that evolutionary recent host shifts between Cannabinaceae and Cucurbitaceae have taken place in the genus *Pseudoperonospora*. The directionality of the host jumps could not be inferred, as based on nrITS, the sequences of *P. cubensis* from cucurbit specimens are poorly differentiated from *P. humuli* from hops. But a high degree of sequence similarity does not rule out the existence of closely related, yet distinct species, and Voglmayr (2008) comments that for detailed insights into the evolutionary processes in closely related downy mildew species, the ITS-based phylogenies offer too little resolution, necessitating multilocus analyses, including additional molecular markers with high phylogenetic resolution.

It was the aim of this study using a combined analysis of *cox*2, *ypt*1, and nrITS sequence data for obtaining a higher resolution of the *P. cubensis* species complex for inferring, whether *P. cubensis* and *P. humuli* are distinct species, whether additional cryptic species might be present in this group, and whether the new *Pseudoperonospora* epidemics in Europe and North America might be caused by a genetically distinct lineage which should be hindered from further spread by appropriate quarantine regulations.

Materials and methods

Oomycete material

The oomycete material used in this study is listed in Table 1. Herbarium samples originated from the herbarium of the New York Botanical Garden (NY), the Herbarium of the Plant Diseases Division of New Zealand (PDD), the Farlow Herbarium of the Harvard University (FH), the Mycological Herbarium of the Korea University (KUS-F), and the Herbarium of the University of Hohenheim (HOH). The laboratory strains originated from North America, the Czech Republic, and Germany.

DNA-extraction, PCR and sequencing

DNA was extracted using the method best suited for herbarium samples as described in Telle and Thines (2008). For amplification of the mitochondrial *cox*2 gene, primers designed by Hudspeth et al. (2000)

Table 1 Oomycete material used in this study

Pathogen	Host	DNA accession number	Year	Country	Source or herbarium number	GenBank accession number *cox2*	GenBank accession number *ypt1*	GenBank accession number nrITS
P. cannabina	*Cannabis sativa*	NY409	1958	Poland	NY, *ex* Kochman, Mycotheca Polonica, Fasc. I, No. 8	HM635956	HM636004	HM636052
P. cannabina	*Cannabis sativa*	U181	1936	Latvia	FH, *ex* Herbario J. Smarods, No. 316	HM635955	HM636003	HM636051
P. celtidis	*Celtis occidentalis*	NY411	1981	USA, PA	NY, *ex* Carnegie Museum (CM) Herbarium, No. 14551	HM635949	HM635997	HM636045
P. cubensis	*Citrullus vulgaris*	NY419	1887	USA, LA	NY, *ex* Flora Ludiviciana, No. 1/22	HM635918	HM635966	HM636014
P. cubensis	*Citrullus vulgaris*	NY435	1935	USA, FL	NY, *ex* University of Florida Herbarium, No. 10591	HM635917	HM635965	HM636013
P. cubensis	*Cucumis anguria*	NY416	1923	USA, St. Croix	NY, No. 881	HM635924	HM635972	HM636020
P. cubensis	*Cucumis melo*	D320	2006	Korea	KUS-F, No. 22001	HM635912	HM635960	HM636008
P. cubensis	*Cucumis melo*	NY420	1903	USA, SC	NY, *ex* United States Department of Agriculture, No. 1038	HM635910	HM635958	HM636006
P. cubensis	*Cucumis melo*	NY421	1905	USA, SC	NY, *ex* United States Department of Agriculture, No. 1040	HM635919	HM635967	HM636015
P. cubensis	*Cucumis melo*	NY442	1902	USA, CT	NY, *ex* Reliquiae Seymourianae	HM635915	HM635963	HM636011
P. cubensis	*Cucumis melo*	U187	1899	USA, ME	FH, *ex* Herbarium of W.G. Farlow	HM635914	HM635962	HM636010
P. cubensis	*Cucumis melo*	U189	1910	USA, MA	FH, *ex* United States Department of Agriculture, Mycological Collections	HM635909	HM635957	HM636005
P. cubensis	*Cucumis sativus*	NY423	1892	USA, MA	NY, *ex* Herbarium of the Massachusetts Agricultural College, Dept. Bot., No. 949	HM635922	HM635970	HM636018
P. cubensis	*Cucumis sativus*	NY429	1889	USA, NJ	NY, *ex* Economic Fungi. A.B. Seymour and F.S. Earle, No. 41	HM635930	HM635978	HM636026
P. cubensis	*Cucumis sativus*	NY446	1909	Germany	NY, *ex* Sydow, Phycomyces et Protomycten, No. 266	HM635911	HM635959	HM636007
P. cubensis	*Cucumis sativus*	U190	1889	USA, NJ	FH, *ex* Herbarium of W.G. Farlow	HM635921	HM635969	HM636017
P. cubensis	*Cucumis sativus*	U191	1921	USA, WI	FH, *ex* Herbarium of the University of Wisconsin	HM635923	HM635971	HM636019
P. cubensis	*Cucumis sativus*	U195	1912	Czech Republic	FH, *ex* F. Petrak, Flora Bohemiae et Moraviae exsiccata, Lfg. 9 No. 446	HM635913	HM635961	HM636009
P. cubensis	*Cucumis sativus*	U197	1918	Philippines	FH, *ex* Flora of the Philippines Herbarium, No. 30627	HM635942	HM635990	HM636038
P. cubensis	*Cucurbita maxima*	NY430	1905	USA, NJ	NY, *ex* Herbarium of George Massee	HM635926	HM635974	HM636022
P. cubensis	*Cucurbita maxima*	NY452	1907	USA, AL	NY, *ex* Wellesley College Herbarium (WELC), No. 67485	HM635929	HM635977	HM636025
P. cubensis	*Cucurbita pepo*	NY427	1943	USA, MA	NY, *ex* Univ. Massachusetts Fungus Herbarium (MASS), No. 3431	HM635928	HM635976	HM636024

P. cubensis	*Cucurbita pepo*	NY434	1935	USA, FL	NY, *ex* University of Florida Herbarium, No. 10172	HM635933	HM635981	HM636029
P. cubensis	*Cucurbita pepo*	NY447	1935	USA, FL	NY, *ex* Herbarium of Erdman West	HM635931	HM635979	HM636027
P. cubensis	*Lagenaria leucantha*	NY448	1925	USA, FL	NY, *ex* University of Florida Herbarium	HM635920	HM635968	HM636016
P. cubensis	*Melothria crassifolia*	NY431	1935	USA, FL	NY, *ex* University of Florida Herbarium, No. 11222	HM635927	HM635975	HM636023
P. cubensis	*Sicyos angulatus*	U192	1920	USA, WI	FH, *ex* Herbarium of the University of Wisconsin	HM635916	HM635964	HM636012
P. humuli	*Humulus lupulus*	10124	2005	Germany	HOH, No. HUH1134	HM635946	HM635994	HM636042
P. humuli	*Humulus lupulus*	NY454	1940	Czech Republic	NY, *ex* Cryptogamae exsiccatae a Museo Hist. Natur. Vindobonensi, No. 3404	HM635944	HM635992	HM636040
P. humuli	*Humulus lupulus*	NY455	1946	Germany	NY, *ex* Herbarium K. Starcs Flora Bavarica, No. 2073	HM635945	HM635993	HM636041
P. humuli	*Humulus lupulus*	NY458	1957	Poland	NY, *ex* Kochman, Mycotheca Polonica, Fasc. I, No. 7	HM635943	HM635991	HM636039
P. humuli	*Humulus lupulus*	NY459	1980	Argentina	NY, *ex* Universidad Nacional de la Plata, No. 42094	HM635948	HM635996	HM636044
P. humuli	*Humulus lupulus*	NY460	1925	Germany	NY, *ex* F. Petrak, Mycotheca Carpatica, No. 421	HM635947	HM635995	HM636043
P. urticae	*Urtica dioica*	HUH1111	2009	Italy	HOH, No. HUH1111	HM635952	HM636000	HM636048
P. urticae	*Urtica dioica*	PDD1	1975	Austria	PDD, No. 64564	HM635953	HM636001	HM636049
Pseudoperonospora sp.	*Celtis sinensis*	D232	2004	Korea	KUS-F, No. 20975	HM635954	HM636002	HM636050
Pseudoperonospora sp.	*Cucumis sativus*	21226	2007	Czech Republic	HOH, No. HUH1130	HM635937	HM635985	HM636033
Pseudoperonospora sp.	*Cucumis sativus*	22238	2007	Czech Republic	HOH, No. HUH1131	HM635939	HM635987	HM636035
Pseudoperonospora sp.	*Cucumis sativus*	23231	2007	USA, MI	HOH, No. HUH1132	HM635940	HM635988	HM636036
Pseudoperonospora sp.	*Cucumis sativus*	24230	2007	Germany	HOH, No. HUH1133	HM635938	HM635986	HM636034
Pseudoperonospora sp.	*Cucumis sativus*	NY425	1924	Japan	NY, *ex* Herb. Agr. Dept. Imp. Univ. Kyoto, No. 98	HM635934	HM635982	HM636030
Pseudoperonospora sp.	*Cucumis sativus*	U196	1904	Japan	FH, *ex* Herbarium of Marioka Imperial College of Agriculture and Forestry	HM635932	HM635980	HM636028
Pseudoperonospora sp.	*Cucumis sativus*	U201	1934	Japan	FH, *ex* Herbarium of Morioka Imperial College of Agriculture and Forestry, No. 14	HM635941	HM635989	HM636037
Pseudoperonospora sp.	*Cucurbita moschata*	D190	2002	Korea	KUS-F, No. 19205	HM635925	HM635973	HM636021
Pseudoperonospora sp.	*Cucurbita pepo*	NY440	1890	Japan	NY, *ex* Herbarium of Kingo Miyabe, No. 107	HM635935	HM635983	HM636031
Pseudoperonospora sp.	*Humulus japonicus*	D149	2003	Korea	KUS-F, No. 19582	HM635936	HM635984	HM636032
Pseudoperonospora sp.	*Humulus lupulus*	NY461	1909	USA, WI	NY, *ex* Herbarium of the University of Wisconsin, No. 8	HM635950	HM635998	HM636046
Pseudoperonospora sp.	*Lagenaria siceraria*	D265	2005	Korea	KUS-F, No. 21327	HM635951	HM635999	HM636047

were employed as described previously (Choi et al. 2007), but using the MangoTaq Polymerase (Bioline, Luckenwalde, Germany), adding BSA to a final concentration of 8 μg/ml in the PCR reaction mix. For amplification of the Ras-related protein (*ypt1*) gene the primers Ypt1F described in Chen and Roxby (1996) and Ypt4R described in Moorman et al. (2002) were used. Amplification was carried out with MangoTaq Polymerase and the addition of BSA to a final concentration of 8 μg/ml. Amplification of the nrITS region was carried out as described by Thines (2007) using the primers LR-0 (reverse complement to LR-0R, Moncalvo et al. 1995) and DC-6 (Cooke et al. 2000). Sequencing was carried out by a commercial sequencing provider (GATC Biotech, Konstanz, Germany) using the primers employed for PCR.

Phylogenetic inference

Minimum Evolution analysis was done with MEGA 4.0 (Tamura et al. 2007), under the Tamura-Nei substitution model. All other parameters were set to default values. For testing tree robustness, 1000 bootstrap replicates (Felsenstein 1985) were carried out. For Maximum Likelihood analysis RAxML (Stamatakis 2006) on the webserver at CIPRES was used (http://8ball.sdsc.edu:8889/cipres-web/Home.do), with likelihood search and estimation of invariable sites. All other parameters were set to default values. The analysis was repeated five times with 100 bootstrap replicates each. The bootstrap values obtained were averaged, because the rapid bootstrapping algorithm (Stamatakis et al. 2008) can lead to deviation in the range of several percent. For Bayesian analysis MrBAYES (Huelsenbeck and Ronquist 2001), version 3.1.2 was used. Four incrementally heated chains were run for 10,000,000 generations, with a sampling frequency of 1000, under the general time reversible (GTR) model and assuming gamma-distributed substitution rates. The first 2500 trees sampled this way were discarded, and the remaining 7500 trees were used to compute a majority based consensus tree and for inferring posterior probabilities. According to Choi et al. (2005) *P. cannabina* is considered to be the most basal species of the genus *Pseudoperonospora* and therefore the phylogenetic trees were rooted with sequences from this species.

Results

All phylogenetic analyses, including Minimum Evolution (ME), Maximum Likelihood (ML), and Bayesian Analysis (BA), yielded highly similar topologies, and no strongly supported inconsistencies were observed. The best scoring tree from the ME analysis, with the support values from all phylogenetic analyses is given in Fig. 1. Specimens of *Pseudoperonospora cannabina* formed a monophyletic group with high to maximum support in all analyses. *Pseudoperonospora urticae* clustered with *P. celtidis* with moderate support. *Pseudoperonospora celtidis* was split into two distinct lineages; one representing *P. celtidis* originating from *Celtis occidentalis* in the USA and the other representing an undescribed species of *Pseudoperonospora* on *Celtis sinensis* in Korea. The third lineage clustering with the two lineages from *Celtis* originated from *Humulus lupulus*, representing a rare, undescribed species.

The closely related species *P. humuli* and *P. cubensis* formed a monophyletic group together with two cryptic lineages with high to maximum support in all analyses. *Pseudoperonospora* sp. from *Humulus japonicus* in Korea is placed basal to the other lineages in the *P. humuli-P. cubensis* cluster. Clade 3 contains all *P. humuli* strains from *Humulus lupulus* and was supported with high to maximum support in all analyses. This lineage was placed sister to the two *Pseudoperonospora* clades 1 and 2 pathogenic to the Cucurbitaceae with moderate support. The sister group relationship of the clades 1 and 2 received low bootstrap support in ML analysis (72%) but was unsupported in the other analyses. Clade 2, which monophyly was highly supported in all analyses, contains only *Pseudoperonospora* strains from East Asia, i.e. Japan and Korea, except for the post-epidemic lineages represented by recently sampled laboratory strains originating from the USA, the Czech Republic, and Germany. The two Korean strains in clade 2 were separated from the other

Fig. 1 Phylogenetic tree for the downy mildew genus ▶ *Pseudoperonospora* based on Minimum Evolution analysis. Numbers on branches denote statistical support in Minimum Evolution, Maximum Likelihood, and Bayesian Analysis, in the respective order. Numbers on vertical lines are clade numbers referred to in the text. DNA accession numbers are given behind host taxon names, followed by country of origin and year of collection

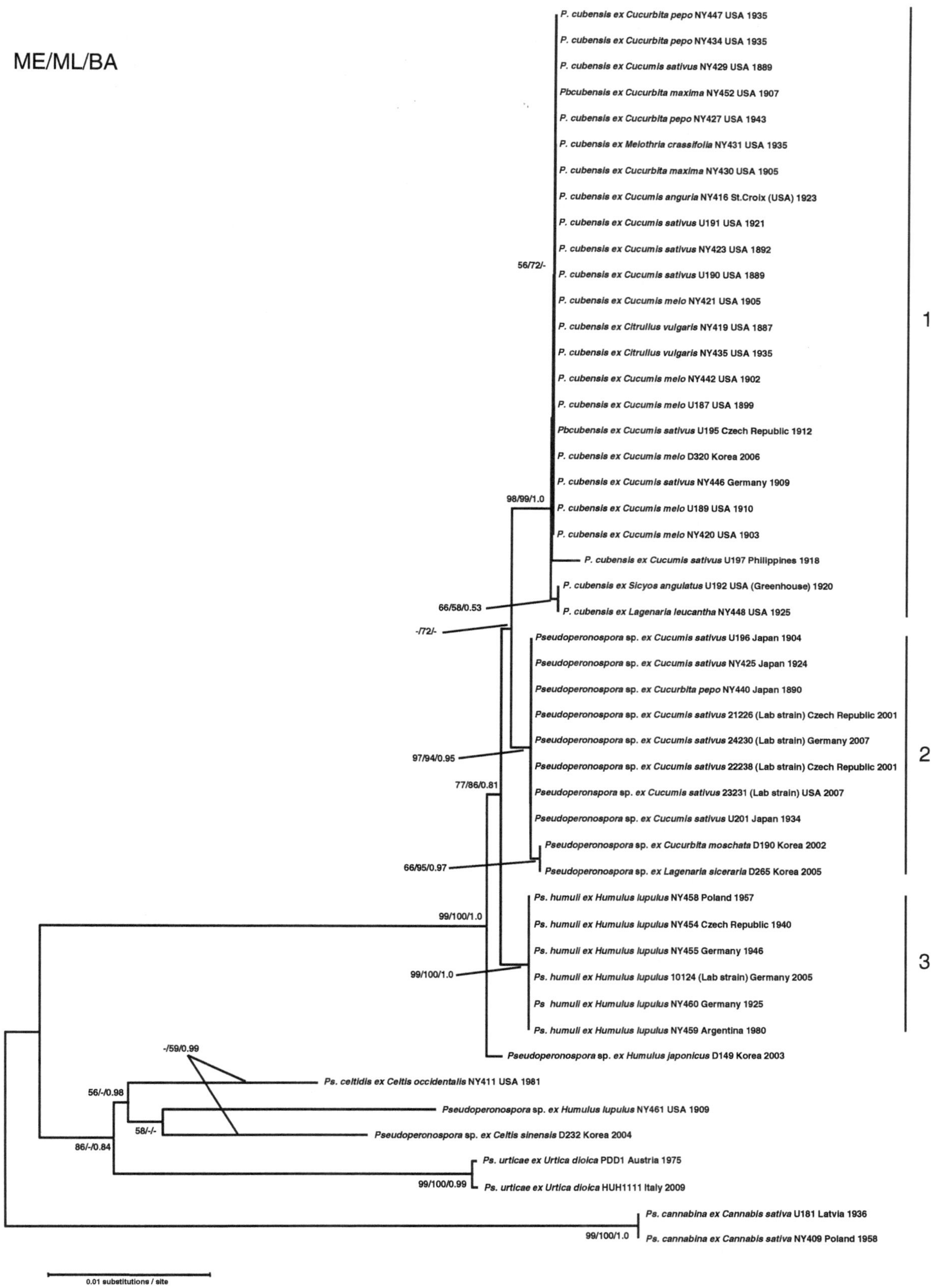
ME/ML/BA
P. cubensis ex Cucurbita pepo NY447 USA 1935
P. cubensis ex Cucurbita pepo NY434 USA 1935
P. cubensis ex Cucumis sativus NY429 USA 1889
Pbcubensis ex Cucurbita maxima NY452 USA 1907
P. cubensis ex Cucurbita pepo NY427 USA 1943
P. cubensis ex Melothria crassifolia NY431 USA 1935
P. cubensis ex Cucurbita maxima NY430 USA 1905
P. cubensis ex Cucumis anguria NY416 St.Croix (USA) 1923
P. cubensis ex Cucumis sativus U191 USA 1921
P. cubensis ex Cucumis sativus NY423 USA 1892
56/72/-
P. cubensis ex Cucumis sativus U190 USA 1889
P. cubensis ex Cucumis melo NY421 USA 1905
P. cubensis ex Citrullus vulgaris NY419 USA 1887
P. cubensis ex Citrullus vulgaris NY435 USA 1935
P. cubensis ex Cucumis melo NY442 USA 1902
P. cubensis ex Cucumis melo U187 USA 1899
Pbcubensis ex Cucumis sativus U195 Czech Republic 1912
P. cubensis ex Cucumis melo D320 Korea 2006
P. cubensis ex Cucumis sativus NY446 Germany 1909
98/99/1.0
P. cubensis ex Cucumis melo U189 USA 1910
P. cubensis ex Cucumis melo NY420 USA 1903
P. cubensis ex Cucumis sativus U197 Philippines 1918
P. cubensis ex Sicyos angulatus U192 USA (Greenhouse) 1920
66/58/0.53
P. cubensis ex Lagenaria leucantha NY448 USA 1925
1
-/72/-
Pseudoperonospora sp. ex Cucumis sativus U196 Japan 1904
Pseudoperonospora sp. ex Cucumis sativus NY425 Japan 1924
Pseudoperonospora sp. ex Cucurbita pepo NY440 Japan 1890
Pseudoperonospora sp. ex Cucumis sativus 21226 (Lab strain) Czech Republic 2001
Pseudoperonospora sp. ex Cucumis sativus 24230 (Lab strain) Germany 2007
97/94/0.95
Pseudoperonospora sp. ex Cucumis sativus 22238 (Lab strain) Czech Republic 2001
2
77/86/0.81
Pseudoperonspora sp. ex Cucumis sativus 23231 (Lab strain) USA 2007
Pseudoperonospora sp. ex Cucumis sativus U201 Japan 1934
Pseudoperonospora sp. ex Cucurbita moschata D190 Korea 2002
66/95/0.97
Pseudoperonospora sp. ex Lagenaria siceraria D265 Korea 2005
Ps. humuli ex Humulus lupulus NY458 Poland 1957
99/100/1.0
Ps. humuli ex Humulus lupulus NY454 Czech Republic 1940
Ps. humuli ex Humulus lupulus NY455 Germany 1946
99/100/1.0
Ps. humuli ex Humulus lupulus 10124 (Lab strain) Germany 2005
3
Ps humuli ex Humulus lupulus NY460 Germany 1925
Ps. humuli ex Humulus lupulus NY459 Argentina 1980
Pseudoperonospora sp. ex Humulus japonicus D149 Korea 2003
-/59/0.99
Ps. celtidis ex Celtis occidentalis NY411 USA 1981
56/-/0.98
Pseudoperonospora sp. ex Humulus lupulus NY461 USA 1909
58/-/-
Pseudoperonospora sp. ex Celtis sinensis D232 Korea 2004
86/-/0.84
Ps. urticae ex Urtica dioica PDD1 Austria 1975
99/100/0.99
Ps. urticae ex Urtica dioica HUH1111 Italy 2009
Ps. cannabina ex Cannabis sativa U181 Latvia 1936
99/100/1.0
Ps. cannabina ex Cannabis sativa NY409 Poland 1958
0.01 substitutions / site

specimens and clustered together with low support in ME and high support in ML and BA. The monophyly of *P. cubensis* s.str. (clade 1) received high to maximum support in all analyses and contains pre-epidemic strains, in particular from the USA, but also a few specimens from Europe and Asia. Within clade 1 most specimens formed a uniform group. Specimens from *Sicyos angulatus* and *Lagenaria leucantha* clustered together with low support in all analyses. No clear-cut host preferences for clade 1 and clade 2 could be observed, both clades include hosts in the genera *Cucumis*, *Cucurbita*, and *Lagenaria*.

Discussion

Judging from the results of Choi et al. (2005) *P. cannabina* is the most basal species of the genus *Pseudoperonospora* and therefore the phylogeny was rooted with this species. In the present study, the species *P. cannabina* and *P. urticae* were monophyletic, while *P. celtidis* s.l. has to be divided into two different species. One of the phylogenetic lineages occurring on *Celtis* represents *P. celtidis* parasitic to the North American species *Celtis occidentalis*, the type host of *P. celtidis* (Waite 1892; Wilson 1907b). The second species is currently undescribed and is parasitic to the East Asian species *Celtis sinensis*, but more specimens should be investigated for clarifying the possibility of morphological delineation between the two species and also for clarifying their geographic range. We thus refrain from introducing the new species in this manuscript. Interestingly, a third phylogenetic lineage is clustering with the two isolates from *Celtis* species. The host for this lineage is *Humulus lupulus* and possibly represents *Pseudoperonospora celtidis* var. *humuli* described by Davis (1910) based on morphological similarities to *Pseudoperonospora celtidis*. Based on the assumption that only one downy mildew species occurs on a specific host plant (e.g. Constantinescu 1989), this variety has been treated as conspecific with *P. humuli*. Whether hop is the native host of the pathogen or an accidental host of a pathogen of another host (e.g. another species of *Celtis*) needs to be determined by a broader sampling of North American specimens of downy mildew from *Humulus lupulus* and species of *Celtis*.

Because of the unsatisfactory resolution of ITS sequences alone for the delineation of closely related species (Voglmayr 2008), we used a multilocus approach in this study, using two additional markers with high phylogenetic resolution. Previous studies (Choi et al. 2007, 2008, 2009; Thines et al. 2009) have proven that in addition to nrITS the mitochondrial *cox2* region offers good phylogenetic resolution in closely related species. For obtaining an even higher resolution, we also included the highly variable Ras-related protein (*ypt1*) gene for additional phylogenetic resolution.

Our results provide evidence that the degradation of *P. humuli* to a synonym of *P. cubensis* as suggested by Choi et al. (2005) should not be upheld. Due to the higher resolution of the present phylogenetic reconstruction, *P. humuli* was found to be distinct from *P. cubensis* with high support. Interestingly, a distinct phylogenetic lineage of *Pseudoperonospora* was observed from *Humulus japonicus* from South Korea, this clade was represented by two specimens from this host. For one of these only the *ypt1* and the *cox2* locus could be obtained and is thus not included in Fig. 1. This cryptic species is basal to all other lineages in the *P. cubensis* complex, suggesting a host jump from hop to Cucurbitaceae gave rise to *P. cubensis*. All other specimens of *Pseudoperonospora* parasitising *Humulus lupulus* were placed in a monophyletic clade, which received high to maximum support in all analyses. This clade contained specimens from several regions of the world, probably reflecting anthropogenic spread. A sample from Japan, the type locality of *P. humuli*, was also placed within this clade based on *ypt1* sequence, but had to be excluded from the multigene phylogeny as other loci did not amplify from this historic specimen. Further investigations encompassing a larger set of samples and the type specimen of *P. humuli* will be necessary to ascertain that this group (clade 3) represents *P. humuli* and to determine the origin of this species.

After the re-emergence of *P. cubensis* as the most destructive pathogen in cucurbit production since the 1980s in Europe (Lebeda 1990; Lebeda and Urban 2004) and since 2004 in the USA (Holmes et al. 2006), uncertainty remained about the reason for the breach of the resistance of the cultivars previously used. Although the occurrence of new pathotypes in *P. cubensis* is known (Lebeda et al. 2006), no molecular phylogenetic basis to distinguish among them is available to date (Sarris et al. 2009). Choi et al. (2005) have shown a generally low diversity in nrITS of the *P. cubensis* species cluster. However, the present investigations using multi-locus sequences reveal a significant degree of genetic divergence within this species cluster, offering an unexpected, yet plausible explanation for the recently occurring epidemics of cucurbit downy mildew. It is noteworthy that two distinct phylogenetic lineages are parasitic to Cucurbitaceae, which are now widely distributed, but previously might have been geographically restricted.

Clade 1 contains specimens from all around the world but exhibits some genetic variation in Northern America and is considered to represent *P. cubensis*. The type species of *Pseudoperonospora* is *Peronospora cubensis* Berk. & Curt., and since Rostovzev (1903) based the description of *P. cubensis* on the original specimen from Cuba and ranked the Russian specimen only as a variety of this species (Wilson 1907a), no nomenclatorial complications with the previous description of Berkeley & Curtis (Berkeley 1869) arise. The higher degree of genetic variation within the North American specimens compared to samples from other parts of the world renders it likely that *Pseudoperonospora cubensis* originated from this region and was later distributed by international trade.

Clade 2 contains primarily East Asian specimens. As these mostly date from about half a century before the recent epidemics in Europe and the USA, it is probably indigenous to Asia. But it also contains the recent Czech and German isolates, as well as the US American laboratory strain of a genotype commonly found on cucumbers in the USA (Lina María Quesada, Hausbeck lab, personal communication). In spring of 1889 *Pseudoperonospora cubensis* was only known in Japan and the Americas (Halsted 1891). Given the geographic isolation, this suggests that the pathogen in Japan was not *P. cubensis* s.str., but a member of clade 2. This is in line with one specimen from 1890 and others from the turn of the century from Japan being placed in clade 2 in this study. In the light of the present results it thus seems likely that the resistance-overcoming lineage might have originated from East Asia. As specimens from Korea showed some genetic divergence, while Japanese samples were identical in sequence compared to the laboratory strains representing isolates from recent outbreaks, it is possible that the causal agent was introduced from Japan. The cryptic species or subspecies became invasive most likely through anthropogenic introduction into Europe and later into Northern America. The paths and invasion routes by which the pathogen entered into Europe and the United States should be clarified by population genetic studies to improve future risk assessment.

In earlier studies a variety of pathotypes of *Pseudoperonospora* on cucurbits was found (Thomas et al. 1987; Cohen et al. 2003; Lebeda and Urban 2004) and the question arose, if the recent epidemics are caused by a pathogen with a spontaneous mutation giving rise to higher virulence of the prevalent population or whether it represents a newly introduced pathogen with a high degree of virulence plasticity (Holmes et al. 2006). Although the differentiation in pathotypes is well known, hitherto no molecular data to distinguish these pathotypes has been available (Sarris et al. 2009). A set of twenty different pathotypes currently present in Europe was tested on differences in their nrITS and *cox2* sequences (data not shown). All pathotypes were identical in these loci and are also identical to recent strains from North America, providing further evidence that the current epidemics are not caused by a single race that has evolved to higher virulence, but by a new phylogenetic lineage with a high degree of adaptability and virulence plasticity, which likely represents a new subspecies or possibly even a new species of *Pseudoperonospora*. In Italy observations of downy mildew from the last millennium suggested that squash was not susceptible (Ciccarese et al. 1990). A more recent study of Cappelli et al. (2003) however obtained isolates on squash and linked it due to its high pathogenicity to pathotype 5 (the most virulent pathotype designated in Thomas et al. 1987). It seems possible that the observations were not merely a result of pathotype differentiation, but that the observations of Ciccarese et al. (1990) were referring to the 'old' pathogen *P. cubensis* s.str., and that the pathogen observed by Cappelli et al. (2003) was representing the 'new' and more virulent pathogen of the cryptic lineage uncovered here. Whether these previously geographically separated lineages are still able to exchange genetic material will be an important question for future research.

Both *P. cubensis* s.str. and the cryptic lineage have a very broad host range with several hosts in common. Thus these remain the downy mildew pathogens with the broadest known host ranges. We refrain from the formal introduction of the cryptic lineage as a new taxon in this manuscript, as morphology of the two species or subspecies is highly similar, and thus the morphological variation caused by ecological conditions (Iwata 1942; Cohen 1981; Dudka et al. 2007) made it impossible to distinguish between them. Measurements of sporangiophores from the same host species harvested under the same controlled conditions might reveal morphological differences suitable for delineation (Runge and Thines 2010) and should be done on a broad set of isolates in future studies. In addition a multitude of isolates should be phylogenetically investigated using several loci, in order to determine if the phylogenetic lineage should be considered a distinct subspecies or species.

The recommendation of Palti (1974) that precautions must be taken not to transfer cucurbits infected with downy mildew from one region to another and his suggestion that varieties bred for resistance in one region have to be retested in any region into which they may be introduced was a remarkable foresight in the light of the present study. Although both cucurbit-affecting lineages are now widespread in cucurbit growing areas, precautions should to be taken for not distributing unknown cryptic pathogen species that might still reside in the *P. cubensis* species cluster, like the cryptic species uncovered on hop. Recent reports suggest an ongoing spread of *P. cubensis* in Korea (Choi and Shin 2008) and Taiwan (Ko et al. 2008) and recently, Voglmayr et al. (2009) reported a member of the *P. cubensis* species complex colonizing the Balsaminaceae. This highlights both the high adaptability and current radiation in this species complex and underscores the necessity for multilocus genotyping of these morphologically cryptic lineages. In addition it should be clarified by extensive population genetic studies, if the cryptic subspecies or species related to *P. cubensis* are still capable of hybridising with each other, for assessing the risk of the emergence of hyper-virulent hybrids.

Acknowledgements The curators of the herbaria BP, BRIP, DAR, FH, and NY are gratefully acknowledged for sending specimens for investigation. Aleš Lebeda, Catherine Erhard, and Peggy Marx are gratefully acknowledged for providing some of the *Pseudoperonospora cubensis* strains that were used in this study. The present study was financially supported by the research funding programme "LOEWE—Landes-Offensive zur Entwicklung Wissenschaftlich-ökonomischer Exzellenz" of Hesse's Ministry of Higher Education, Research, and the Arts and by a grant from the German Science Foundation (DFG) awarded to MT.

References

Bains, S. S., & Jhooty, J. S. (1976). Overwintering of *Pseudoperonospora cubensis* causing downy mildew of muskmelon. *Indian Phytopathology, 29*, 213–214.

Battaglia, R. J. (2010). *Wisconsin vegetables—2009*. Retrieved from http://www.nass.usda.gov/Statistics_by_State/Wisconsin/Publications/Vegetables/vegannual.pdf, January.

Berkeley, M. J. (1869). On a collection of fungi from Cuba. *The Journal of the Linnean Society, 10*, 363.

Cappelli, C., Buonaurio, R., & Stravato, V. M. (2003). Occurrence of *Pseudoperonospora cubensis* pathotype 5 on squash in Italy. *Plant Disease, 87*, 449.

Chen, Y., & Roxby, R. (1996). Characterization of *Phytophthora infestans* gene involved in the vesicle transport. *Gene, 181*, 89–94.

Choi, Y.-J., & Shin, H.-D. (2008). First record of downy mildew caused by *Pseudoperonospora cubensis* on bottle gourd in Korea. *Plant Pathology, 57*, 371.

Choi, Y.-J., Hong, S.-B., & Shin, H.-D. (2005). A re-consideration of *Pseudoperonospora cubensis* and *P. humuli* based on molecular and morphological data. *Mycological Research, 109*, 841–848.

Choi, Y.-J., Shin, H.-D., Hong, S.-B., & Thines, M. (2007). Morphological and molecular discrimination among *Albugo candida* materials infecting *Capsella bursa-pastoris* world-wide. *Fungal Diversity, 27*, 11–34.

Choi, Y.-J., Shin, H.-D., Ploch, S., & Thines, M. (2008). Evidence for uncharted biodiversity in the *Albugo candida* complex, with the description of a new species. *Mycological Research, 112*, 1327–1334.

Choi, Y.-J., Shin, H.-D., & Thines, M. (2009). Two novel *Peronospora* species are associated with recent reports of downy mildew on sages. *Mycological Research, 113*, 1340–1350.

Ciccarese, F., Amenduni, M., & Cirulli, M. (1990). Field reaction to powdery mildew and downy mildew of some cultivars of Cucurbitaceae species in Southern Italy. *Phytopathologia Mediterranea, 29*, 14–18.

Cohen, Y. (1981). Downy mildew of cucurbits. In D. M. Spencer (Ed.), *The downy mildews* (pp. 341–354). London: Academic.

Cohen, Y., Meron, I., Mor, N., & Zuriel, S. (2003). A new pathotype of *Pseudoperonospora cubensis* causing downy mildew in cucurbits in Israel. *Phytoparasitica, 31*, 458–466.

Colucci, S. J., Thornton, A. C., Adams, M. L., & Holmes, G. J. (2008a). *Evaluation of fungicides for control of downy mildew of cucumber I, 2007. Plant Disease Management Reports (online). Report 1:V043*. St. Paul: The American Phytopathological Society. doi:10.1094/PDMR02.

Colucci, S. J., Thornton, A. C., Adams, M. L., & Holmes, G. J. (2008b). *Evaluation of fungicides for control of downy mildew of cucumber II, 2007. Plant Disease Management Reports (online). Report 1:V045*. St. Paul: The American Phytopathological Society. doi:10.1094/PDMR02.

Constantinescu, O. (1989). *Peronospora* complex on Compositae. *Sydowia, 41*, 79–107.

Constantinescu, O. (2000). The fine structure of the sporangium in *Pseudoperonospora humuli* (Chromista, Oomycota, Peronosporales). *Cryptogamie, Mycologie, 21*, 93–101.

Cooke, D. E. L., Drenth, A., Duncan, J. M., Wagels, G., & Brasier, C. M. (2000). A molecular phylogeny of phytophthora and related oomycetes. *Fungal Genetics and Biology, 30*, 17–32.

Davis, J. J. (1910). A new hop mildew. *Science, N.S., 31*, 752.

Dolinar, M., & Žolnir, M. (1994). Schwellenorientiertes Entscheidungsschema für epidemiebezogene Bekämpfung der Hopfenperonospora (*Pseudoperonospora humuli* Miy. *et* Tak.). *Die Bodenkultur, 45*, 49–56.

Dudka, I. O., Anishchenko, I. M., & Terent'eva, N. G. (2007). The variability of *Peronospora alta* Fuckel conidia in dependence on the ecological conditions. In A. Lebeda & P. T. N. Spencer-Phillips (Eds.), *Advances in downy mildew research* ((pp, Vol. 3, pp. 39–46). Kostelec na Hané, Czech Republic: Palacký University in Olomouc and JOLA.

Felsenstein, J. (1985). Confidence limits on phylogenies: an approach using the bootstrap. *Evolution, 39*, 783–791.

Fraymouth, J. (1956). Haustoria of the Peronosporales. *Transactions of the British Mycological Society, 39*, 79–107.

Gent, D. H., Nelson, M. E., Farnsworth, J. L., & Grove, G. G. (2009). PCR detection of *Pseudoperonospora humuli* in air samples from hop yards. *Plant Pathology, 58*, 1081–1091.

Gisi, U. (2002). Chemical control of downy mildews. In P. T. N. Spencer-Phillips, U. Gisi, & A. Lebeda (Eds.), *Advances in downy mildew research* (pp. 119–159). Dordrecht: Kluwer.

Gugino, B. K., Wyenandt, A., MacGrath, M. T., & Ojiambo, P. S. (2009). Fighting downy mildew [Electronic version]. *American Vegetable Grower*, February.

Halsted, B. D. (1891). Notes upon Peronosporaceae for 1891. *Botanical Gazette, 16*, 338–340.

Hausbeck, M. (2007). Monitoring downy mildew on cucurbits in 2006. Retrieved June 12, 2010, from http://www.veggies.msu.edu/Research/GR06-099DMonitoringDM_2007.pdf.

Holmes, G. J., Main, C. E., & Keever, Z. T., III. (2004). Cucurbit downy mildew: A unique pathosystem for disease forecasting. In P. T. N. Spencer-Phillips & M. Jeger (Eds.), *Advances in downy mildew research, vol. 2* (pp. 69–80). Dordrecht: Kluwer.

Holmes, G., Wehner, T., & Thornton, A. (2006). An old enemy re-emerges [Electronic version]. *American Vegetable Grower*, pp 14-15, February.

Horejsi, T., Staub, J. E., & Thomas, C. (2000). Linkage of random amplified polymorphic DNA markers to downy mildew resistance in cucumber (*Cucumis sativus* L.). *Euphytica, 115*, 105–113.

Hudspeth, D. S. S., Nadler, S. A., & Hudspeth, M. E. S. (2000). A COX2 molecular phylogeny of the Peronosporomcetes. *Mycologia, 92*, 674–684.

Huelsenbeck, J. P., & Ronquist, F. (2001). MRBAYES: Bayesian inference of phylogenetic trees. *Bioinformatics, 17*, 754–755.

Ishii, H., Fraaije, B. A., Sugiyama, T., Noguchi, K., Nishimura, K., Takeda, T., et al. (2001). Occurrence and molecular characterization of strobilurin resistance in cucumber powdery mildew and downy mildew. *Phytopathology, 91*, 1166–1171.

Iwata, Y. (1942). Specialization in *Pseudoperonospora cubensis* (Berk. *et* Curt.) Rostov. II. Comparative studies of the morphologies of the fungi from *Cucumis sativus* L. and *Cucurbita moschata* Duchesne. *Annals of the Phytopathological Society of Japan, 11*, 172–185.

Johnson, D. A., Alldredge, J. R., & Allen, J. R. (1994). Weather and downy mildew epidemics of hop in Washington State. *Phytopathology, 84*, 524–527.

Keinath, A. P., Holmes, G. J., Everts, K. L., Egel, D. S., & Langston, D. B., Jr. (2007). Evaluation of combinations of chlorothalonil with azoxystrobin, harpin, and disease forecasting for control of downy mildew and gummy stem blight on melon. *Crop Protection, 26*, 83–88.

Ko, Y., Chen, C. Y., Liu, C. W., Chen, S. S., Maruthasalam, S., & Lin, C. H. (2008). First report of downy mildew caused by *Pseudoperonospora cubensis* on Chayote (*Sechium edule*) in Taiwan. *Plant Disease, 92*, 1706.

Lebeda, A. (1990). Biology and ecology of cucurbit downy mildew. In A. Lebeda (Ed.), *Cucurbit downy mildew* (pp. 13–46). Prag: Czechoslovak Scientific Society for Mycology by Czechoslovak Academy of Sciences.

Lebeda, A. (1991). Resistance in muskmelons to Czechoslovak isolates of *Pseudoperonospora cubensis* from cucumbers. *Scientia Horticulturae, 45*, 255–260.

Lebeda, A. (1992). Screening of wild *Cucumis* species against downy mildew (*Pseudoperonospora cubensis*) isolates from cucumbers. *Phytoparasitica, 20*, 203–210.

Lebeda, A. (1999). *Pseudoperonospora cubensis* on *Cucumis* spp. and *Cucurbita* spp.—resistance breeding aspects. *Acta Horticulturae, 492*, 363–370.

Lebeda, A., & Gadasová, V. (2002). Pathogenic variation of *Pseudoperonospora cubensis* in the Czech Republic and some other European countries. *Acta Horticulturae, 588*, 137–141.

Lebeda, A., & Schwinn, F. J. (1994). The downy mildews—an overview of recent research progress. *Journal of Plant Diseases and Protection, 101*, 225–254.

Lebeda, A., & Urban, J. (2004). Disease impact and pathogenicity variation in Czech populations of *Pseudoperonospora cubensis*. In A. Lebeda & H. S. Paris (Eds.), *Progress in Cucurbit genetics and breeding research. Proc. Cucurbitaceae 2004. 8th EUCARPIA Meeting on Cucurbit Genetics and Breeding* (pp. 267–273). Olomouc, Czech Republic: Palacký University in Olomouc.

Lebeda, A., & Widrlechner, M. P. (2003). A set of Cucurbitaceae taxa for differentiation of *Pseudoperonospora cubensis* pathotypes. *Journal of Plant Diseases and Protection, 110*, 337–349.

Lebeda, A., & Widrlechner, M. P. (2004). Response of wild and weedy *Cucurbita* L. to pathotypes of *Pseudoperonospora cubensis* (Berk. & Curt.) Rostov. (cucurbit downy mildew). In P. T. N. Spencer-Phillips & M. Jeger (Eds.), *Advances in downy mildew research, vol. 2* (pp. 203–210). Dordrecht: Kluwer.

Lebeda, A., Widrlechner, M. P., & Urban, J. (2006). Individual and population aspects of interactions between cucurbits and *Pseudoperonospora cubensis*: Pathotypes and races. In G. J. Holmes (Ed.), *Proceedings of Cucurbitaceae 2006* (pp. 453–467). North Carolina: Universal Press, Raleigh.

Miyabe, K., & Takahashi, Y. (1906). A new disease of hop-vine caused by *Peronoplasmopara humuli* n. sp. *Transactions of the Sapporo Natural History Society, 1*, 149–157.

Moncalvo, J. M., Wang, H. H., & Hseu, R. S. (1995). Phylogenetic relationships in *Ganoderma* inferred from the internal transcribed spacers and 25S ribosomal DNA sequences. *Mycologia, 87*, 223–223.

Moorman, G. W., Kang, S., Geiser, D. M., & Kim, S. H. (2002). Identification and characterization of *Pythium* species associated with greenhouse floral crops in Pennsylvania. *Plant Disease, 86*, 1227–1231.

Mozny, M., Krejci, J., & Kott, I. (1993). CORAC, Hops protection management systems. *Computers and Electronics in Agriculture, 9*, 103–110.

Palti, J. (1974). The significance of pronounced divergences in the distribution of *Pseudoperonospora cubensis* on its crop hosts. *Phytoparasitica, 2*, 109–115.

Palti, J., & Cohen, Y. (1980). Downy mildew of cucurbits (*Pseudoperonospora cubensis*): the fungus and its hosts, distribution, epidemiology and control. *Phytoparasitica, 8*, 109–147.

Pérez, B. A., Martínez, E., Noetinger, F., & Wright, E. R. (2009). Hop downy mildew caused by *Pseudoperonospora humuli* in Argentina. *Plant Disease, 93*, 839.

Pitrat, M., Dogimont, C., & Bardin, M. (1998). Resistance to fungal diseases foliage in melon. In J. D. McCreight (Ed.), *Cucurbitaceae '98. Evaluation and enhancement of cucurbit germplasm* (pp. 167–173). Alexandria: ASHS.

Rostovzev, S. J. (1903). Beiträge zur Kenntnis der Peronosporeen. *Flora, 92*, 405–430.

Royle, D. J., & Kremheller, H. T. H. (1981). Downy mildew of the hop. In D. M. Spencer (Ed.), *The downy mildews* (pp. 395–419). London: Academic.

Runge, F., & Thines, M. (2010). Host matrix has major impact on the morphology of *Pseudoperonospora cubensis*. *European Journal of Plant Pathology*, in this issue.

Sarris, P. F., Abdelhalim, M., Kitner, M., Skandalis, N., Panopoulos, N. J., Doulis, A. G., et al. (2009). Molecular polymorphisms between populations of *Pseudoperonospora cubensis* from Greece and the Czech Republic and the phytopathological and phylogenetic implications. *Plant Pathology, 58*, 933–943.

Selby, A. D. (1899). Additional host plants of *Plasmopara cubensis*. *Botanical Gazette, 27*, 67–68.

Shetty, N. V., Wehner, T. C., Thomas, C. E., Doruchowski, R. W., & Shetty, K. P. V. (2002). Evidence for downy mildew races in cucumber tested in Asia, Europe, and North America. *Scientia Horticulturae, 94*, 231–239.

Shin, H.-D., & Choi, Y.-J. (2003). A first check-list of Peronosporaceae from Korea. *Mycotaxon, 86*, 249–267.

Shin, H.-D., & Choi, Y.-J. (2006). *Peronosporaceae of Korea*. Suwon: National Institute of Agricultural Science and Technology.

Stamatakis, A. (2006). RAxML-VI-HPC: maximum likelihood-based phylogenetic analyses with thousands of taxa and mixed models. *Bioinformatics, 22*, 2688–2690.

Stamatakis, A., Hoover, P., & Rougemont, J. (2008). A rapid bootstrap algorithm for the RAxML web-servers. *Systematic Biology, 57*, 758–771.

Tamura, K., Dudley, J., Nei, M., & Kumar, S. (2007). MEGA4: Molecular Evolutionary Genetics Analysis (MEGA) software version 4.0. *Molecular Biology and Evolution, 24*, 1596–1599.

Telle, S., & Thines, M. (2008). Amplification of cox2 (620 bp) from 2 mg of up to 129 years old herbarium specimens, comparing 19 extraction methods and 15 polymerases. *Plos ONE, 3*, e3584.

Thines, M. (2007). Characterisation and phylogeny of repeated elements giving rise to exceptional length of ITS2 in several downy mildew genera (Peronosporaceae). *Fungal Genetics and Biology, 44*, 199–207.

Thines, M., Telle, S., Ploch, S., & Runge, F. (2009). Identity of the downy mildew pathogens of basil, coleus, and sage with implications for quarantine measures. *Mycological Research, 113*, 532–540.

Thomas, C. E., Inaba, T., & Cohen, Y. (1987). Physiological specialization in *Pseudoperonospora cubensis*. *Phytopathology, 77*, 1621–1624.

Urban, J., & Lebeda, A. (2004). Resistance to fungicides in population of cucurbit downy mildew in the Czech Republic. *Acta fytotechnica et zootechnica, 7*, 327–329.

Urban, J., & Lebeda, A. (2006). Fungicide resistance in cucurbit downy mildew—methodological, biological and population aspects. *Annals of Applied Biology, 149*, 63–75.

Urban, J., & Lebeda, A. (2007). Variation for fungicide resistance in Czech populations of *Pseudoperonospora cubensis*. *Journal of Phytopathology, 155*, 143–151.

Voglmayr, H. (2008). Progress and challenges in systematics of downy mildews and white blister rusts: new insights from genes and morphology. *European Journal of Plant Pathology, 122*, 3–18.

Voglmayr, H., Piatek, M., & Mossebo, D. C. (2009). *Pseudoperonospora cubensis* causing downy mildew disease on *Impatiens irvingii* in Cameroon: a new host for the pathogen. *Plant Pathology, 58*, 394.

Waite, M. B. (1892). Description of two new species of *Peronospora*. *The Journal of Mycology, 7*, 105–109.

Wilson, G. W. (1907a). An historical review of the proposed genera of phycomycetes: I. Peronosporales. *The Journal of Mycology, 13*, 205–209.

Wilson, G. W. (1907b). Studies in North American Peronosporales—II. Phytophthoreae and Rhysotheceae. *Bulletin of the Torrey Botanical Club, 34*, 415.

Eur J Plant Pathol (2011) 129:15–24
DOI 10.1007/s10658-010-9650-9

Host matrix has major impact on the morphology of *Pseudoperonospora cubensis*

Fabian Runge · Marco Thines

Accepted: 11 June 2010 / Published online: 4 July 2010

Abstract Oomycetes contain some of the economically most important pathogens of flowering plants. Most have a rather narrow host range, often being restricted to single host species. In downy mildews and other obligate biotrophic plant parasites, like powdery mildews and rusts, delimitating species on grounds of morphological characteristics is often hardly possible and thus often based on only subtle differences. This has led to the widespread application of a broad species concept for these organisms. Consequently, despite the fact that morphological differences were reported for *Pseudoperonospora cubensis* from different host species, the corresponding new pathogen species were not accepted as being independent, and the host range of *Pseudoperonospora cubensis* is reported to encompass more than 50 host species in the Cucurbitaceae in temperate to tropical climates. However, recent studies have reported narrow host ranges for other downy mildew genera and advocated a narrow species concept. Here, we report successful colonisation of five different tribes of the Cucurbitaceae by a strain of *Pseudoperonospora cubensis* and demonstrate that the host matrix has a major impact on the morphology of the pathogen. On the basis of five morphological criteria significant differences could be found for all hosts. These differences were more pronounced in phylogenetically unrelated than in related hosts. Our results provide evidence for a broad host range of *Pseudoperonospora cubensis* and demonstrate that species delimitation based on morphological characters is not feasible in *Pseudoperonospora* on Cucurbitaceae. Also in other biotrophic plant pathogens, the situation could be similar, thus necessitating thorough morphological, molecular phylogenetic and cross inoculation experiments for species recognition.

Keywords Cucurbit downy mildew · Obligate biotrophic plant pathogens · Host-pathogen interaction · Cucurbitaceae · Peronosporaceae · Species concepts

F. Runge
Institute of Botany 210, University of Hohenheim,
70593 Stuttgart, Germany

M. Thines (✉)
Biodiversity and Climate Research Centre (BiK-F),
Georg-Voigt-Str. 14-16,
60325 Frankfurt (Main), Germany
e-mail: marco.thines@senckenberg.de

M. Thines
Johann Wolfgang Goethe University,
Department for Biological Sciences, Insitute of Ecology, Evolution and Diversity,
Siesmayerstr. 70,
D-60323 Frankfurt (Main), Germany

Introduction

Oomycetes are fungal-like organisms that are unrelated to the Mycota, but belong to the kingdom Straminipila (Dick 2001), which also includes seaweeds and diatoms. The most species-rich order of the oomycetes is the order Peronosporales, which contains mainly plant parasitic species, but also some fungus and animal pathogens. The largest family are the Peronosporaceae, which include the hemibiotrophic genus *Phytophthora* and the obligate biotrophic downy mildews. According to Voglmayr (2008) and Thines et al. (2009a) 18 genera are included in the Peronosporaceae with more than 800 described species. Among these, several cause economically important diseases of crops, including *Bremia lactucae* (lettuce downy mildew), *Plasmopara halstedii* (sunflower downy mildew), *Plasmopara viticola* (grape downy mildew), and *Pseudoperonospora cubensis* (cucurbit downy mildew). Earlier reports, stating that downy mildews have narrow host ranges, often limited to a single host species (Gäumann 1918, 1923; Crute 1981; Brandenburger 1985), have recently been confirmed by molecular phylogenetic studies (Cunnington 2006; Choi et al. 2007a, 2009a; García-Blázquez et al. 2008; Thines et al. 2009b). However, there is some evidence that a few downy mildew species are able to parasitise a variety of distantly related host species (Choi et al. 2005; Göker et al. 2009), similar to the situation described for the obligate biotrophic oomycete *Albugo candida* (Choi et al. 2007b, 2008, 2009b; Thines et al. 2009c), which causes white blister disease on a broad range of Brassicaceae, Capparaceae and Cleomaceae. As with rusts and in powdery mildews, in downy mildews it is often almost impossible to distinguish species on the basis of morphological characters. Therefore, before molecular phylogenetic investigations revealed the evolutionary diversity of these organisms, a broad species concept was usually applied, and some specialised forms were assumed to be present in some host families (Yerkes and Shaw 1959; Skidmore and Ingram 1985). One of these broad species is *Pseudoperonospora cubensis*, which is one of the most important and devastating diseases in cucurbitaceous crops. However, contrary to *Hyaloperonospora parasitica* (Göker et al. 2009) and *Peronospora farinosa* (Choi et al. 2007a), its integrity has not been questioned, but rather been confirmed by molecular phylogenetic studies (Choi et al. 2005). *Pseudoperonospora cubensis* has a worldwide distribution and is present in all cucurbit growing areas (Holmes et al. 2004). The host range of *P. cubensis* is reported to include over 50 cucurbits, among which are the following crop species—*Benincasa hispida, Citrullus lanatus*, *Cucumis sativus, Cucumis melo, Cucurbita pepo, Cucurbita maxima, Lagenaria siceraria*, and *Luffa cylindrica* (Lebeda and Widrlechner 2003; Thomas 1986). Although Sawada (1931) introduced some new segregate species of *Peronoplasmopara* (a synonym of *Pseudoperonospora*) parasitic to Cucurbitaceae on the basis of morphological differences, these were not considered different species by subsequent authors (Iwata 1942, 1953; Palti and Cohen 1980; Waterhouse and Brothers 1981), also because it was unclear if the variation observed might have been due to environmental factors. Dudka et al. (2007) examined the dependence of the sporangial dimensions of *Peronospora alta* on climatic conditions and assumed that humidity has a major impact on sporangial dimensions, and Iwata (1942) showed that the morphology of sporangiophores of *P. cubensis* varies with different temperatures. Waterhouse and Brothers (1981) focused their investigations on measurements of the sporangia of *P. cubensis* and found a significant influence of the host plant in this character. Delanoe (1972) reported that there are differences in morphology of sporangiophores and sporangia of *Plasmopara halstedii* isolated from different parts of the host plant. The investigations of Waterhouse and Brothers (1981) that demonstrated an impact of the host matrix on the morphology of sporangia were not followed up by investigations using a more comprehensive set of hosts and additional characters. Also, it is unclear, if climatic factors such as light, humidity and temperature were tightly controlled in their experiments.

The aim of this study was to investigate the potential dependence of the morphology of plant pathogens on the host matrix, using the example of *P. cubensis*. These investigations might shed light on the question of whether it is justified to erect new species or confirm previously described species solely on the basis of morphological characters and host matrix, which has been common practice for many plant pathogens. *Pseudoperonospora cubensis* was chosen for these investigations, because of its reported broad host range. For evaluating these reports, and for investigating the impact of the host matrix on the morphology of the pathogen, six species belonging to five different tribes of the Cucurbitaceae (Kocyan et al. 2007), were used in this study—*Cucurbita maxima*, *Citrullus lanatus* and *Cucumis sativus*, *Sicyos angulatus*, *Luffa cylindrica*, and *Bryonia dioica* belonging to the tribes Cucurbiteae, Benincaseae, Sicyeae, Luffeae, and Bryonieae respectively.

Material and methods

Inoculation experiments were carried out using the *P. cubensis* strain P.C. 26/01 that originated from the Olomouc Region of the Czech Republic. The strain was originally isolated from *Cucumis sativus* and is maintained on *C. sativus* at the Institute of Botany of the University of Hohenheim in climate chambers (16°C, 14 h light, 10 h darkness). Inoculations for continuous cultivation of P.C. 26/01 and cross-inoculations were done using a dab-off technique as described previously (Runge and Thines 2009). For inoculations, the lower surface of uninfected leaves was moistened with deionised water. Subsequently, sporulating leaf-parts were gently dabbed onto the moist leaf surface of the uninfected leaf several times, touching previously unaffected areas of the leaf for inoculation. The dab-off technique leads to different spore concentrations over the leaf surface. Thus, the spore concentration ranges from areas without sporangia to areas with very high amounts of inoculum. This technique has the advantage that optimal inoculum concentrations, which will differ from species to species and even from cultivar to cultivar, need not be determined, because these will be met through the differences in inoculum load throughout the leaves. Fully mature leaves of *Cucurbita maxima*, cultivar "Kaempe Melon", *Citrullus lanatus*, cultivar "Sugar Baby", *Cucumis sativus*, cultivar "Chinese Slangen", *Sicyos angulatus*, *Luffa cylindrica*, and *Bryonia dioica* were taken from plants in 5 to 10-leaf stage grown in greenhouses at the University of Hohenheim, except for the leaves of *S. angulatus*, which were cut from outdoor plants (Botanical Garden of the University of Hohenheim), while care was taken to select leaves of comparable age. After inoculation the leaves were transferred to transparent boxes (approximately 30 cm long, 20 cm wide, 5 cm high) on water-soaked paper towels to ensure that 100% relative humidity (RH) was maintained within the boxes. Cross-inoculations were done in three technical replicates. Cross inoculations using *Bryonia dioica* and *Sicyos angulatus* were repeated for testing reproducibility over time. As these tests were successful, infection trials for the other cucurbitaceous hosts were done only once, in three replicates carried out at the same time. Two days after first sporulation was observed sporangiophores were picked from the leaf surface with precision tweezers, transferred to a drop of water on a microscopic slide and covered with a coverslip. The morphology of sporangia and sporangiophores were investigated using a Biomed (Leitz, Wetzlar, Germany) light microscope. Pictures of sporangia and sporangiophores were taken using a Canon PowerShot A640 photocamera (Canon, Tokyo, Japan). Before each picture series a picture of a stage micrometer was taken to calibrate the measurements, which were conducted using the AxioVision LE software (Carl Zeiss Imaging Solutions, München, Germany). The characters examined were the length of the sporangiophores ($n=25$), the length of the ultimate branchlets ($n=100$), and the length and the width of the sporangia ($n=100$). The fifth character, the ratio of length to width of the sporangia was calculated for each individual sporangium. Subsequently, the data were statistically analysed using STATISTICA '99 software (StatSoft, Tulsa, OK, USA). The Mann-Whitney-*U*-Test (Mann and Whitney 1947) was applied for investigating the significance of the morphological differences of the pathogen on the host species investigated.

Results

Four to eight days post inoculation, first sporulation could be observed (*Sicyos angulatus*: 4 d; *Cucurbita maxima*, *Cucumis sativus*: 5 d; *Luffa cylindrica*: 6 d; *Bryonia dioica*: 7 d; *Citrullus lanatus*: 8 d). The parasitised leaves with the respective symptoms are shown in Fig. 1. In the case of *Cucurbita maxima* (Fig. 1a) some areas became necrotic. In these necrotic areas, sporangiophores occurred in sparse groups. Also on the leaves of *Citrullus lanatus* (Fig. 1b) and *Luffa cylindrica* (Fig. 1e) there was only sparse sporulation, with sporangiophores dispersed throughout almost round, chlorotic to necrotic spots. In case of *B. dioica* (Fig. 1f) round chlorotic to necrotic spots could also be observed. In contrast to the former two hosts the sporulation was dense on leaves of *Bryonia dioica*, with the result that the infection spots were exhibiting a blackish colour. In case of *Sicyos angulatus* (Fig. 1d) the sporulation was restricted to few areas; however, in these areas the sporulation was very profuse and therefore appeared floccose and dark grey in colour. On *Cucumis sativus* (Fig. 1c), the native host of the isolate, sporulation occurred on almost the whole lower leaf surface and

Fig. 1 Symptoms of *P. cubensis* (black arrows) on the parasitised leaves of the host plants *Cucurbita maxima* (**a**), *Citrullus lanatus* (**b**), *Cucumis sativus* (**c**), *Sicyos angulatus* (**d**), *Luffa cylindrica* (**e**) and *Bryonia dioica* (**f**)

some parts became densely covered with sporangiophores, resulting in a downy and greyish appearance. In all replicates the symptoms on the respective hosts were similar. No differences in morphology and shape of the fully mature sporangiophores and sporangia from the different replicates for a specific host could be observed.

The morphological characterisation of *P. cubensis* from the different hosts is given in Table 1, and the results of the statistical analyses of the morphology on different hosts are summarised in Fig. 2. *Pseudoperonospora cubensis* differed in almost all characters in at least one of the alternative hosts markedly in morphology compared to its original host, as well as in cross comparisons. In the analysis of the length of the sporangiophores two groups were apparent. One included *Cucurbita maxima*, *Citrullus lanatus*, and *Bryonia dioica* with mean sporangiophore lengths of 289 μm, 301 μm, and 321 μm, respectively, and no significant differences compared to each other. The second group included *Cucumis sativus*, *Sicyos angulatus*, and *Luffa cylindrica* with mean lengths of 387 μm, 377 μm, and 393 μm, respectively. When comparing members of the first group with the ones of the second group, significant differences were observed in each possible combination. In the length of the ultimate branchlets, only *Bryonia dioica* with a mean of 9.45 μm showed significant differences to all other hosts (*Cucurbita maxima*: 8.33 μm, *Cucumis sativus*: 8.35 μm, *Sicyos angulatus*: 8.74 μm, *Luffa cylindrica*: 8.18 μm), except for *Citrullus lanatus*, which was exhibiting an intermediate mean length of 9.19 μm and therefore showed significant differences to none of the others. On the basis of the length of the sporangia *Citrullus lanatus*, *Cucumis sativus*, and *Luffa cylindrica* formed a homogeneous group (mean lengths of 21.6 μm, 21.1 μm, and 21.7 μm, respectively). Sporangia taken from *Cucurbita maxima* and *Bryonia dioica* were longer (23.4 μm and 24.1 μm, respectively) and showed no significant differences to each other. Shortest sporangia were found in *Sicyos angulatus* with a mean of 19.5 μm. Differences between the three groups were highly significant ($p<0.001$). In the width of sporangia *Cucurbita maxima*, *Citrullus lanatus*, and *Bryonia dioica* (means of 16.0 μm, 15.5 μm, and 15.7 μm, respectively) formed a group with no internal significant differences. Also *Cucumis sativus* (mean 15.4 μm) was not significantly different from *Citrullus lanatus*. But all other combinations (*Sicyos angulatus*: mean of 13.1 μm, *Luffa cylindrica*: mean of 16.5 μm) showed significant deviation. The shape of the sporangia, expressed by the ratio of length to width, revealed that *Sicyos angulatus* did not differ significantly in shape from *Cucurbita maxima* and *Bryonia dioica* (means of 1.49, 1.47, and 1.54, respectively), but taking into account their length and width it is apparent that sporangia on *Sicyos angulatus* were significantly

Table 1 Morphological characteristics of *P. cubensis* on different Cucurbitaceae

Host species			
	Cucurbita maxima	*Citrullus lanatus*	*Cucumis sativus*
Sporangiophore length	(205-)244–289–345(-425)	(153-)226–301–369(-397)	(186-)280–387–494(-606)
Length of ultimate branchlets	(3.08-)6.26–8.33–10.8(-16.2)	(3.63-)6.11–9.19–13.8(-18.4)	(2.93-)6.03–8.35–11.3(-15.2)
Sporangial length	(13.6-)21.2–23.4–25.9(-32.9)	(14.3-)19.6–21.6–23.4(-27.1)	(14.2-)19.0–21.1–23.6(-30.1)
Sporangial width	(11.7-)14.5–16.0–17.4(-23.4)	(10.0-)13.9–15.5–17.1(-19.7)	(10.3-)14.3–15.4–18.0(-21.5)
Ratio length to width of sporangia	(1.01-)1.30–1.47–1.62(-1.84)	(1.01-)1.31–1.41–1.52(-1.88)	(1.00-)1.22–1.38–1.47(-1.77)
	Sicyos angulatus	*Luffa cylindrica*	*Bryonia dioica*
Sporangiophore length	(214-)296–377–483(-570)	(253-)314–393–439(-500)	(198-)258–321–429(-465)
Length of ultimate branchlets	(2.45-)6.29–8.74–12.9(-18.7)	(2.11-)5.04–8.18–9.74(-14.7)	(4.91-)7.42–9.45–12.9(-18.0)
Sporangial length	(12.0-)16.3–19.5–21.2(-27.1)	(16.3-)19.4–21.7–23.5(-30.1)	(18.8-)21.8–24.1–27.1(-32.4)
Sporangial width	(9.78-)12.1–13.1–14.2(-17.5)	(12.2-)14.8–16.5–18.1(-21.4)	(11.6-)14.8–15.7–16.9(-20.1)
Ratio length to width of sporangia	(1.01-)1.18–1.49–1.59(-1.80)	(1.03-)1.23–1.32–1.42(-1.68)	(1.25-)1.44–1.54–1.64(-1.95)

All measurements given in the form (minimum-) border of 30%—*mean*—border of 30% (-maximum). All values rounded to three counting digits

smaller. *Cucurbita maxima* and *Bryonia dioica* showed weakly significant ($p<0.05$) differences in the shape of sporangia. The shape of the sporangia of *Cucumis sativus* and *Citrullus lanatus* (1.38 and 1.41, respectively) was not significantly different rendering it impossible to distinguish sporangia from these hosts. All of the other combinations showed significant differences. Sporangia from *Luffa cylindrica* were, with a ratio of 1.32, more globose than sporangia from all other hosts.

A graphical display of all significant differences of the respective host-combinations at different p-levels (Fig. 3) illustrates that only two combinations, *Cucurbita maxima-Bryonia dioica* and *Citrullus lanatus-Cucumis sativus*, had no highly significant differences. The combination *Cucurbita maxima-Citrullus lanatus* showed only one highly significant difference. On the other hand the *Luffa cylindrica-Bryonia dioica*-combination showed highly significant differences in four of the five characters investigated. The *Citrullus lanatus–Cucumis sativus*-combination was the sole one with a single significant difference only in all characters investigated. Except for this combination, all other combinations had at least two significant differences at a p-value level below 0.05. *Pseudoperonospora cubensis* from *Bryonia dioica* had overall the most statistically supported differences (18 of 25 possible), especially due to the longest ultimate branchlets and the longest and most elongated sporangia on this host. Also *P. cubensis* from *Luffa cylindrica* revealed many supported differences to other hosts in several characteristics (17), in particular due to the shortest ultimate branchlets, the broadest and roundest sporangia, and the longest sporangiophores on this host compared to the other hosts. Also *P. cubensis* from *Sicyos angulatus* proved to be equally divergent in morphology. *Pseudoperonospora cubensis* from *Citrullus lanatus* was the host combination with the fewest differences to others, because most characters investigated were close to the mean of the other isolates.

Discussion

First sporulation was observed after 4–8 days post inoculation. It is supposed that the first sporulation reflects the sporulation time under optimal inoculum load, as inoculations were done using the dab-off technique, which ensures a heterogeneous inoculum

a - Length of sporangiophores	B. dioica	L. cylindrica	S. angulatus	C. sativus	C. lanatus	C. maxima
Cucurbita maxima						X
Citrullus lanatus					X	0,25826
Cucumis sativus				X	0,00394	0,00076
Sicyos angulatus			X	0,63760	0,00714	0,00070
Luffa cylindrica		X	0,47539	0,79194	0,00006	0,00000
Bryonia dioica	X	0,00099	0,01698	0,01386	0,40223	0,11612

b - Length of ultimate branchlets	B. dioica	L. cylindrica	S. angulatus	C. sativus	C. lanatus	C. maxima
Cucurbita maxima						X
Citrullus lanatus					X	0,22008
Cucumis sativus				X	0,18718	0,99480
Sicyos angulatus			X	0,68205	0,41974	0,77943
Luffa cylindrica		X	0,57014	0,90112	0,13090	0,97497
Bryonia dioica	X	0,00826	0,04512	0,00996	0,29016	0,00714

c - Length of conidia	B. dioica	L. cylindrica	S. angulatus	C. sativus	C. lanatus	C. maxima
Cucurbita maxima						X
Citrullus lanatus					X	0,00004
Cucumis sativus				X	0,05220	0,00000
Sicyos angulatus			X	0,00012	0,00000	0,00000
Luffa cylindrica		X	0,00000	0,05572	0,97667	0,00003
Bryonia dioica	X	0,00000	0,00000	0,00000	0,00000	0,26125

d - Width of conidia	B. dioica	L. cylindrica	S. angulatus	C. sativus	C. lanatus	C. maxima
Cucurbita maxima						X
Citrullus lanatus					X	0,13863
Cucumis sativus				X	0,30914	0,00490
Sicyos angulatus			X	0,00000	0,00000	0,00000
Luffa cylindrica		X	0,00000	0,00000	0,00039	0,01433
Bryonia dioica	X	0,00027	0,00000	0,02697	0,60638	0,25377

e - Ratio length to width of conidia	B. dioica	L. cylindrica	S. angulatus	C. sativus	C. lanatus	C. maxima
Cucurbita maxima						X
Citrullus lanatus					X	0,01122
Cucumis sativus				X	0,15830	0,00010
Sicyos angulatus			X	0,00000	0,00001	0,20117
Luffa cylindrica		X	0,00000	0,00066	0,00001	0,00000
Bryonia dioica	X	0,00000	0,59028	0,00000	0,00000	0,01440

Legend
p<0,001
p<0,01
p<0,05
p=0,05

Fig. 2 Statistical analysis of the measurements of morphological characters of *P. cubensis* isolated from different hosts. The numbers in the lower right matrices are p-values for the significance of morphological differences of *P. cubensis* on the respective hosts. The p-levels are visualised by different shadings as given in the legend in the lower right corner

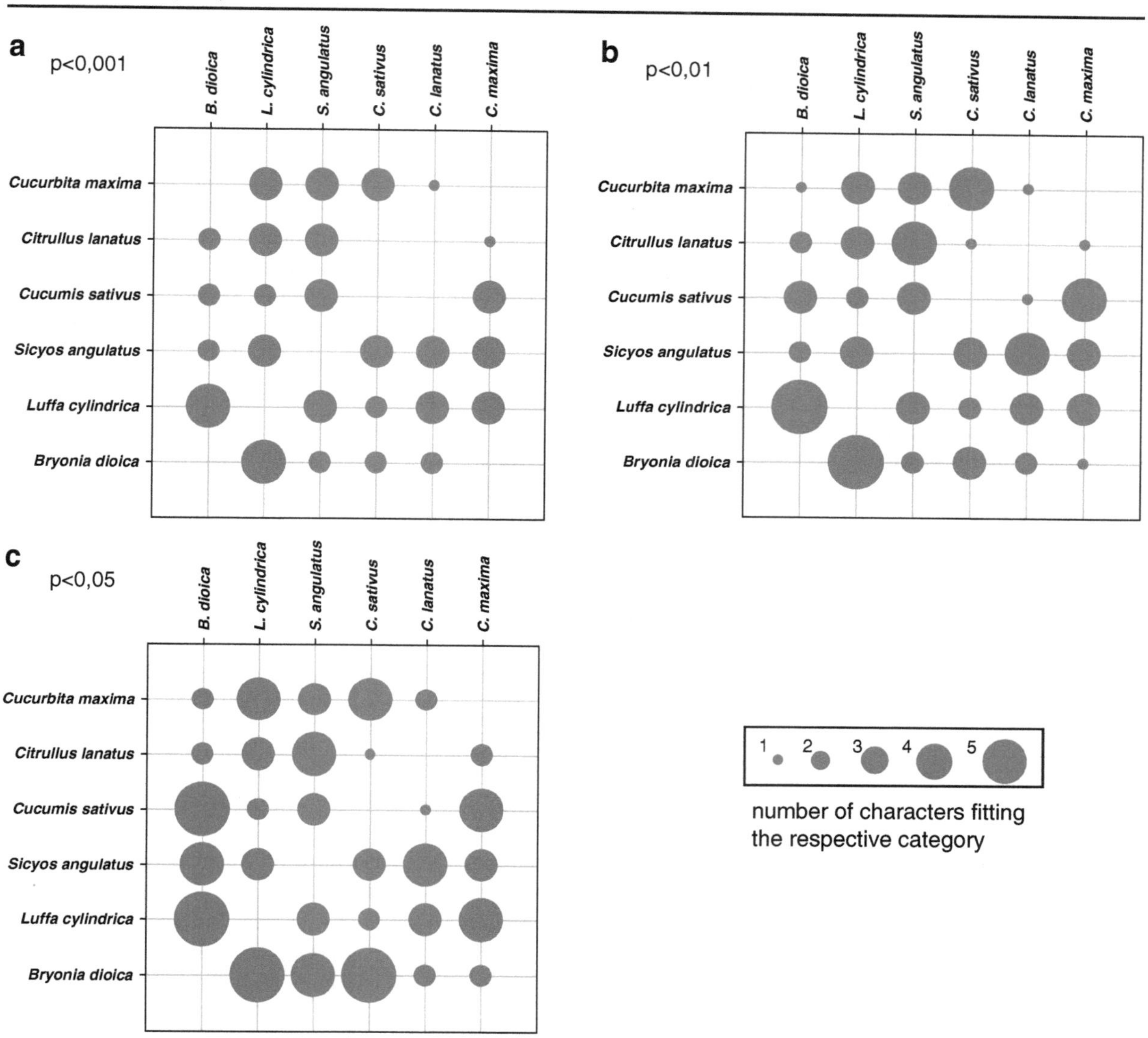

Fig. 3 Graphic display of the distinctiveness of the morphology of *P. cubensis* on various hosts. Number of differences in all characters investigated for the respective host-pathogen combinations with the respective p-levels are visualised by the size of the grey circles

load over different areas of the leaves. Usually *P. cubensis* produces a new generation of sporangiophores and spores 7–8 days past inoculation (Urban and Lebeda 2007). The fact that in our investigations sporulation occurred already after 5 days on *C. sativus* supports previous results of the presence of distinct physiological forms of *P. cubensis* (Shetty et al. 2002; Lebeda and Widrlechner 2003). Our finding that symptoms and onset of the disease differed widely on the different hosts investigated provide further evidence for the existence of a broad genetic variation in resistance to *P. cubensis* among cucurbits as discussed by Lebeda (1991) and Lebeda and Widrlechner (2004).

Recent molecular phylogenetic studies have shown that the broad species concept advocated by Yerkes and Shaw (1959) is not tenable for most downy mildews (Riethmüller et al. 2002; Voglmayr 2003; Voglmayr et al. 2004; Cunnington 2006; Choi et al. 2007b, 2009b; Göker et al. 2009; Thines et al. 2009b). For most of the species of the downy mildews, it is hardly possible to distinguish between phylogenetically related species based on morphological characters, but thorough morphological investigations in *Peronospora* species from Lamiaceae have revealed differences that enable species delimitation in the *Peronospora lamii* species complex (Choi et al. 2009b; Thines et al. 2009b).

However, contrary to the situation in other downy mildew genera affecting multiple species within a specific host family, genetic distances within *P. cubensis* are very short (Choi et al. 2005), which hinders easy molecular phylogenetic delimitation of species that might be present in *Pseudoperonospora* parasitic to Cucurbitaceae. But as short genetic distances do not necessarily rule out the presence of independent species, multigene analyses are often necessary for distinguishing closely related species (Blair et al. 2008). In addition, morphological investigations and cross-inoculations are necessary for clarification. However, the broad host spectrum of the *P. cubensis* strains previously characterised (Lebeda 1991; Shetty et al. 2002; Lebeda and Widrlechner 2003, 2004) and confirmed in this study using the strain P.C. 26/01, casts doubts on the credibility of the species segregated from *P. cubensis* on the basis of host matrix and morphological differences by Sawada (1931). However, in previous studies characterising the infectivity of *P. cubensis* strains on various hosts, potential differences in the morphology of the strains used were not investigated. Therefore, the morphological species introduced by Sawada (1931), which were not considered independent species in comprehensive treatises (Iwata 1942, 1953; Palti and Cohen 1980; Waterhouse and Brothers 1981), could still be credible, even when considering possible variations caused by differing temperatures or humidity (Iwata 1942; Dudka et al. 2007). However, this study demonstrates that the morphology of the pathogen is largely dependent on the host matrix. This is in line with the results of Waterhouse and Brothers (1981), who focused on sporangial dimensions on selected hosts and found significant differences between these hosts. To exclude both temperature and humidity effects, we used controlled conditions for both parameters and also applied identical light conditions after inoculation. Cohen (1981) suggested that the sporangial yield per unit area of affected leaf tissue in a moist chamber depends upon many factors, the most important of which were plant species and variety, age and size of lesions, nutritional status of the host, temperature, and illumination. However, investigations using *Cucmis melo* as a host showed that the variation in nutrient solutions did not affect the size of sporangia (Bains and Jhooty 1978). An influence of the organ type infected is likely; as Delanoe (1972) pointed out that there are significant differences in morphology of sporangiophores and sporangia of *Plasmopara halstedii* isolated from different parts of sunflowers. Therefore, only fully mature leaves in comparable stages of development were used in this study. It is most likely due to highly homogenous sporulation conditions applied in this study that the influence of the host matrix on the morphology of the pathogen could be demonstrated clearly, and it should be taken into consideration that for field material the variation in the morphological characters is expected to be much higher. The finding of Choi et al. (2005) that the morphology of sporangia and sporangiophores of *P. cubensis* were indistinguishable for all specimens from *Cucumis*, *Cucurbita*, and *Citrullus* investigated, could well be the result of a high variation caused by different sporulation conditions (like temperature and humidity), leaf ages or organ affected. Unfortunately, no statistical analyses were presented to underscore their results.

When comparing the morphological results of this study in the light of the molecular phylogeny of the Cucurbitaceae presented by Kocyan et al. (2007), some patterns were becoming apparent. *Bryonia dioica* belonging to the Bryonieae is the most basal member of the Cucurbitaceae investigated here, and its phylogenetic distance to the other hosts used in this study is large. Interestingly, this is also the host on which *P. cubensis* shows the most divergent morphology when compared to the other hosts. *Luffa cylindrica* and *Sicyos angulatus*, belonging to the Luffeae and Sicyeae respectively, the hosts with the second most significant differences, also occupy rather basal phylogenetic positions. *Citrullus lanatus* and *Cucumis sativus*, the hosts on which the morphology of the pathogen was found to be almost identical, belong to the same tribe, the Benincaseae. *Cucurbita maxima* of the sister tribe Cucurbiteae occupied an intermediate position, both in the phylogenetic distance from the native host (*Cucumis sativus*) of the isolate used in this study and in morphological differences induced in the pathogen compared to other hosts. Thus, it could be suggested that the phylogenetic distance of the hosts correlates with the morphological differences induced in a pathogen, with greater distances between hosts being more likely to induce significant differences in pathogen morphology. However, at present this can only be regarded as a tendency and further investigations with a broader set of Cucurbitaceae will be necessary to clarify if the observed pattern follows a general rule. Considering the high variation of the sporangial dimensions depending on the host matrix it seems highly unlikely that the species segregated from *P. cubensis* by Sawada (1931) are tenable. In addition to the sporangial dimensions, which were found to be influenced the most by the host matrix in this study, in line with results of Waterhouse and Brothers (1981), our results provide evidence that the morphology of sporangiophores can also be influenced by the host matrix. Further investigations are necessary to clarify which morphological characters are significantly influenced by the host matrix, and which are not. This will help clarifying, if variation observed in newly reported hosts could rather be the result of morphological plasticity of the corresponding character or due to the observation of an unknown species with deviating morphology.

Judging from our results, species of downy mildews described on the basis of morphological differences with no phylogenetic support, as with some *Plasmopara* species occurring on Apiaceae (Voglmayr et al. 2004), should be treated with caution and be verified by multigene analyses and cross-infection experiments in order to rule out a possible influence of the host matrix on the morphological differences observed. However, the possibility that two potential pathogen species may infect the same host species cannot be the sole criterion for species rejection, as for both *Hyaloperonospora* (Göker et al. 2009) and *Plasmopara* (Komjáti et al. 2007) it is known that phylogenetically clearly distinct species are able to infect the same host species.

Acknowledgements Aleš Lebeda is gratefully acknowledged for providing the *Pseudoperonospora cubensis* strain that since 2007 is used as a reference strain in our laboratory and was also used in this study. The present study was financially supported by the research funding programme "LOEWE—Landes-Offensive zur Entwicklung Wissenschaftlich-ökonomischer Exzellenz" of Hesse's Ministry of Higher Education, Research, and the Arts and by a grant from the German Science Foundation (DFG) awarded to MT. GR, BL and JL are acknowledged for support throughout the compilation of this manuscript.

References

Bains, S. S., & Jhooty, J. S. (1978). Relationship between mineral nutrition of muskmelon and development of downy mildew caused by *Pseudoperonospora cubensis*. *Plant and Soil, 49*, 85–90.

Blair, J. E., Coffey, M. D., Park S.-Y., Geiser D. M., & Kang S. (2008). A multi-locus phylogeny for *Phytophthora* utilizing markers derived from complete genome sequences. *Fungal Genetics and Biology, 45*, 266–277.

Brandenburger, W. (1985). *Parasitische Pilze an Gefäßpflanzen in Europa*. Stuttgart: Gustav-Fischer.

Choi, Y.-J., Hong, S.-B., & Shin, H.-D. (2005). A reconsideration of *Pseudoperonospora cubensis* and *P. humuli* based on molecular and morphological data. *Mycological Research, 109*, 841–848.

Choi, Y.-J., Hong, S.-B., & Shin, H.-D. (2007a). Reconsideration of *Peronospora farinosa* infecting *Spinacia oleracea* as distinct species, *Peronospora effusa*. *Mycological Research, 110*, 381–391.

Choi, Y.-J., Shin, H.-D., Hong, S.-B., & Thines, M. (2007b). Morphological and molecular discrimination among *Albugo candida* materials infecting *Capsella bursa-pastoris* world-wide. *Fungal Diversity, 27*, 11–34.

Choi, Y.-J., Shin, H.-D., Ploch, S., & Thines, M. (2008). Evidence for uncharted biodiversity in the *Albugo candida* complex, with the description of a new species. *Mycological Research, 112*, 1327–1334.

Choi, Y.-J., Shin, H.-D., & Thines, M. (2009a). Two novel *Peronospora* species are associated with recent reports of downy mildew on sages. *Mycological Research, 113*, 1340–1350.

Choi, Y.-J., Shin, H.-D., & Thines, M. (2009b). The host range of *Albugo candida* extends from Brassicaceae over Cleomaceae to Capparaceae. *Mycological Progress, 8*, 329–335.

Cohen, Y. (1981). Downy mildew of cucurbits. In D. M. Spencer (Ed.), *The downy mildews* (pp. 341–354). London: Academic.

Crute, I. R. (1981). The host specifity of peronosporaceous fungi and the genetics of the relationship between host and parasite. In D. M. Spencer (Ed.), *The downy mildews* (pp. 237–253). London: Academic.

Cunnington, J. H. (2006). DNA sequence variation supports multiple host-specialised taxa in the *Peronospora viciae* complex (Chromista: Peronosporales). *Nova Hedwigia, 82*, 23–29.

Delanoe, D. (1972). Biologie et Epidemiologie du mildiou du tournesol. *CETIOM Informations Techniques, 26*, 1–61.

Dick, M. W. (2001). *Straminipilous fungi: Systematics of the peronosporomycetes including accounts of the marine straminipilous protists, the plasmodiophorids and similar organisms*. Dordrecht: Kluwer Academic Publishers.

Dudka, I. O., Anishchenko, I. M., & Terent'eva, N. G. (2007). The variability of *Peronospora alta* Fuckel conidia in dependence on the ecological conditions. In A. Lebeda & P. T. N. Spencer-Phillips (Eds.), *Advances in downy mildew research, vol. 3* (pp. 39–46). Kostelec na Hané (Czech Republic): Palacký University in Olomouc and JOLA.

García-Blázquez, G., Göker, M., Voglmayr, H., Martín, M. P., Tellería, M. T., & Oberwinkler, F. (2008). Phylogeny of Peronospora, parasitic on Fabaceae, based on ITS sequences. *Mycological Research, 112*, 502–512.

Gäumann, E. (1918). Über die Formen der *Peronospora parasitica* (Pers.) Fries. Ein Beitrag zur Speziesfrage bei den parasitischen Pilzen. *Botanisches Centralblatt, Beihefte, 35*, 395–533.

Gäumann, E. (1923). Beiträge zu einer Monographie der Gattung Peronospora Corda. *Beiträge zur Kryptogamenflora der Schweiz, 5*, 1–360.

Göker, M., Voglmayr, H., García-Blázquez, G., & Oberwinkler, F. (2009). Species delimitation in downy mildews: the case of *Hyaloperonospora* in the light of nuclear ribosomal ITS and LSU sequences. *Mycological Research, 113*, 308–325.

Holmes, G. J., Main, C. E., & Keever, Z. T., III. (2004). Cucurbit downy mildew: A unique pathsystem for disease forecasting. In P. T. N. Spencer-Phillips & M. Jeger (Eds.), *Advances in downey mildew research, vol. 2* (pp. 69–80). Dordrecht: Kluwer.

Iwata, Y. (1942). Specialization in *Pseudoperonospora cubensis* (Berk. *et* Curt.) Rostov. II. Comparative studies of the morphologies of the fungi from *Cucumis sativus* L. and *Cucurbita moschata* Duchesne. *Annals of the Phytopathological Society of Japan, 11*, 172–185.

Iwata, Y. (1953). Specialization in *Pseudoperonospora cubensis* (Berk. *et* Curt.) Rostov. IV. Studies on the fungus from Oriental pickling melon (*Cucumis melo* var. *conomon* Makino). *Bulletin of the Faculty of Agriculture, Mie University, 1*, 30–35.

Kocyan, A., Zhang, L. B., Schaefer, H., & Renner, S. S. (2007). A multi-locus chloroplast phylogeny for the Cucurbitaceae and its implications for character evolution and classifcation. *Molecular Phylogenetics and Evolution, 44*, 553–577.

Komjáti, H., Walcz, I., Virányi, F., Zipper, R., Thines, M., & Spring, O. (2007). Characteristics of a *Plasmopara angustiterminalis* isolate from *Xanthium strumarium*. *European Journal of Plant Pathology, 119*, 421–428.

Lebeda, A. (1991). Resistance in muskmelons to Czechoslovak isolates of *Pseudoperonospora cubensis* from cucumbers. *Scientia Horticulturae, 45*, 255–260.

Lebeda, A., & Widrlechner, M. P. (2003). A set of Cucurbitaceae taxa for differentiation of *Pseudoperonospora cubensis* pathotypes. *Journal of Plant Disieases and Protection, 110*, 337–349.

Lebeda, A., & Widrlechner, M. P. (2004). Response of wild and weedy *Cucurbita* L. to pathotypes of *Pseudoperonospora cubensis* (BERK. & CURT.) ROSTOV. (cucurbit downy mildew). In P. T. N. Spencer-Phillips & M. Jeger (Eds.), *Advances in downey mildew research, vol. 2* (pp. 203–210). Dordrecht: Kluwer.

Mann, H., & Whitney, D. (1947). On a test of whether one of two random variables is stochastically larger than the other. *Annals of Mathematical Statistics, 18*, 50–60.

Palti, J., & Cohen, Y. (1980). Downy mildew of cucurbits (*Pseudoperonospora cubensis*): the fungus and its hosts, distribution, epidemiology and control. *Phytoparasitica, 8*, 109–147.

Riethmüller, A., Voglmayr, H., Göker, M., Weiß, M., & Oberwinkler, F. (2002). Phylogenetic relationships of the downy mildews (Peronosporales) and related groups based on nuclear large subunit ribosomal DNA sequences. *Mycologia, 94*, 834–849.

Runge, F., & Thines, M. (2009). A potential perennial host for *Pseudoperonospora cubensis* in temperate regions. *European Journal of Plant Pathology, 123*, 483–486.

Sawada, K. (1931). Descriptive catalogue of the Formosan Fungi. *Report of the Government Research Institute, Department of Agriculture, Formosa, 51*, 76–78.

Shetty, N. V., Wehner, T. C., Thomas, C. E., Doruchowski, R. W., & Shetty, K. P. V. (2002). Evidence for downy mildew races in cucmber tested in Asia, Europe, and North America. *Scientia Horticulturae, 94*, 231–239.

Skidmore, D. I., & Ingram, D. S. (1985). Conidial morphology and the specialization of *Bremia lactucae* Regel (Peronosporaceae) on hosts in the family Compositae. *Botanical Journal of the Linnean Society, 91*, 503–522.

Thines, M., Voglmayr, H., & Göker, M. (2009a). Taxonomy and phylogeny of the downy mildews (Peronosporaceae). In K. Lamour & S. Kamoun (Eds.), *Oomycete genetics and genomics: Diversity, interactions, and research tools* (pp. 47–75). Weinheim: Wiley-VCH.

Thines, M., Telle, S., Ploch, S., & Runge, F. (2009b). Identity of the downy mildew pathogens of basil, coleus, and sage with implications for quarantine measures. *Mycological Research, 113*, 532–540.

Thines, M., Choi, Y.-J., Kemen, E., Ploch, S., Holub, E. B., Shin, H.-D., et al. (2009c). A new species of *Albugo* parasitic to *Arabidopsis thaliana* reveals new evolutionary patterns in white blister rusts (Albuginaceae). *Persoonia, 22*, 123–128.

Thomas, C. E. (1986). Downy and powdery mildew resistant muskmelon breeding line MR-1. *Horticultural Science, 21*, 329.

Urban, J., & Lebeda, A. (2007). Variation of fungicide resistance in czech populations of *Pseudoperonospora cubensis*. *Journal of Phytopathology, 155*, 143–151.

Voglmayr, H. (2003). Phylogenetic study of *Peronospora* and related genera based on nuclear ribosomal ITS sequences. *Mycological Research, 107*, 1132–1142.

Voglmayr, H. (2008). Progress and challenges in systematics of downy mildews and white blister rusts: new insights from genes and morphology. *European Journal of Plant Pathology, 122*, 3–18.

Voglmayr, H., Riethmüller, A., Göker, M., Weiß, M., & Oberwinkler, F. (2004). Phylogenetic relationships of *Plasmopara*, *Bremia* and other genera of downy mildews with pyriform haustoria based on Bayesian analysis of partial LSU rDNA sequence data. *Mycological Research, 108*, 1011–1024.

Waterhouse, G. M., & Brothers, M. P. (1981). The taxonomy of *Pseudoperonospora*. *Mycological Papers, 148*, 1–28.

Yerkes, W. D., & Shaw, C. G. (1959). Taxonomy of *Peronospora* species on Cruciferae and Chenopodiaceae. *Phytopathology, 49*, 499–507.

Eur J Plant Pathol (2011) 129:25–60
DOI 10.1007/s10658-010-9658-1

Cucurbit downy mildew (*Pseudoperonospora cubensis*)—biology, ecology, epidemiology, host-pathogen interaction and control

Aleš Lebeda · Yigal Cohen

Accepted: 5 July 2010 / Published online: 25 July 2010

Abstract Cucurbit downy mildew, caused by the oomycete *Pseudoperonospora cubensis*, is a devastating, worldwide-distributed disease of cucurbit crops in the open field and under cover. This review provides recent data on the taxonomy, biology, ecology, host range, geographic distribution and epidemiology of *P. cubensis*. Special attention is given to host-pathogen interactions between *P. cubensis* and its economically-important cucurbit hosts (*Cucumis sativus, C. melo, Cucurbita pepo, C. maxima,* and *Citrullus lanatus*); pathogenic variability in *P. cubensis* at the species, genus, and population levels; and, differentiation of pathotypes and races. Genetics and variability of host resistance and cellular and molecular aspects of such resistance are considered. A focus is given to methods of crop protection, including prevention and agro-technical aspects, breeding for resistance—classical and transgenic approaches, chemical control and fungicide resistance. Novel technologies in biological and integrated control are also discussed. This review also summarizes the most important topics for future research and international collaboration.

A. Lebeda (✉)
Department of Botany, Faculty of Science,
Palacký University in Olomouc,
Šlechtitelů 11,
CZ-783 71 Olomouc, Czech Republic
e-mail: ales.lebeda@upol.cz

Y. Cohen
Bar Ilan University, Faculty of Life Sciences,
52100 Ramat Gan, Israel
e-mail: ycohen@mail.biu.ac.il

Keywords Biological control · Breeding for resistance · Classification · Chemical control · Cisgenic plants · Epidemiology · Fungicide resistance · Genetics of resistance · Geographic distribution · Host range · Disease cycle · Integrated pest management · Life cycle · Migration · Overwintering · Pathotypes · Races · Symptomatology · Systematics · Transgenic plants

Introduction

Pseudoperonospora cubensis (causal agent of cucurbit downy mildew) is one of the most economically important and widespread plant pathogens. It is a biotrophic plant parasite belonging to the kingdom Chromista, class Peronosporomycetes (Thomas 1996; Dick 2001a; Göker et al. 2007; Voglmayr 2008). Although it has been studied by mycologists and plant pathologists for more than 100 years, there is still a lack of information about this pathogen in many specific areas (Lebeda 1990, 1999; Lebeda and Schwinn 1994). Together with *Bremia lactucae* (Lebeda et al. 2002, 2008a, b; Michelmore and Wong 2008) and *Hyaloperonospora parasitica* (Slusarenko and Schlaich 2003; Holub 2008) it is one of the most studied Peronosporomycete biotrophic parasites of plants. The last comprehensive and internationally-available reviews on this pathogen were published thirty years ago (Palti and Cohen 1980; Cohen 1981). In the present review we discuss a wide range of specific issues related to *P. cubensis* including the biology, diversity, ecology, distribution, host range, epidemiology, host-pathogen interactions, genetics of resistance, breeding and control. *P. cubensis* is used as a "case study" to demonstrate the diverse, complex interactions of downy mildews with their host plants.

Parts of this review were published before in Czech (Lebeda 1990; Lebeda et al. 2006a; Lebeda and Urban 2005). The current review was broadened and supplemented with the literature published in the last twenty years. This literature mainly considers the topics of variability of host-pathogen interactions, variation of pathogenicity, resistance of host plants, breeding for resistance and pathogen control, focusing mainly on efficacy of and resistance to fungicides. The aims of this paper are to summarize the current information on *P. cubensis*, critically discuss the areas in which we still lack basic data, and introduce them into the wider context of biotrophic parasitism within the downy mildews.

Taxonomy

History of nomenclature

Pseudoperonospora cubensis (Berk. et Curt.) Rostov. was first described by Berkeley in 1868 in herbarium plant material originated from Cuba, and hence its species name *cubensis* (Skalický 1961). It was also identified as a new genus, *Pseudoperonospora* Berkeley, together with the description of the species *Pseudoperonospora cubensis*. *P. cubensis*, therefore, served as the type species of the *Pseudoperonospora* genus (Dick 2001b, 2002b). The pathogen was first observed and described on live plants by Rostovzev in The Botanical Gardens of Moscow (Russia) in 1903 (Skalický 1961).

During the last century some misleading synonyms were used to describe *P. cubensis*, such as *Peronospora cubensis* Berk. et Curt., *Peronospora cubensis* Berk. et Curt. var. *atra* Zimmerm., *Plasmopara cubensis* (Berk. et Curt.) Humphrey, and *Peronoplasmopara cubensis* (Berk. et Curt.) Clinton (Dick 2001b).

Recent systematic classification

According to recent taxonomic classification, *Pseudoperonospora cubensis* belongs to kingdom Chromista, subdivision Peronosporomycotina, class Peronosporomycetes (originally described as Oomycetes), order Peronosporales (downy mildews), family Peronosporaceae (Göker et al. 2007; Voglmayr 2008). Peronosporomycetes is comprised of ca 900 (perhaps up to 1500) species, 75 genera and 19 families (Dick 2001a, c), and new genera were recently described (Göker et al. 2007; Voglmayr 2008).

The genus *Pseudoperonospora* belongs to a taxonomic group on the border between genera that regularly produce zoospores (*Pythium* spp.) and genera that never produce zoospores (*Peronospora* spp., *Albugo* spp.) (Göker et al. 2007). *P. cubensis* does produce zoospores but its sporophores resemble those of *Peronospora* spp. It is a species with highly distinct host specificity attacking above-ground parts, mainly leaves, of only the Cucurbitaceae. The formation of zoospores in this species depends on environmental conditions. It occurs in water only and is inhibited at high temperatures. The whole zoosporangium content is cleaved and biflagellate zoospores, 10–13 μm in diameter, are released (Thomas 1996). However, there are also reports on direct germination by germ tubes (Lange et al. 1989a, c). Sporangia easily dislodge from the sporangiophore and are distributed by air or rain splash (similar to other *Pseudoperonospora* spp.) (Lange et al. 1989c). *Pseudoperonospora* spp. has dichotomously-branched sporangiophores with terminal growth; the sporangia of similar age are present at the ends of sterigma (Choi et al. 2005; Voglmayr 2003). The family Peronosporaceae includes at least 17 genera (Göker et al. 2007; Voglmayr 2008). Probably the most widespread genera in Europe and North America are *Bremia, Peronospora, Hyaloperonospora* and *Plasmopara* of which representatives cause serious diseases of cultivated plants (Lebeda and Schwinn 1994; Voglmayr 2008). The individual genera are mainly characterized by the shape and branching of the sporophore/sporangiophores and the ability to discharge zoospores. From this point of view, *Pseudoperonospora* represents a transitional type between *Plasmopara* and *Peronospora* (Dick 2001b, 2002a, b; Voglmayr 2003; Choi et al. 2005; Göker et al. 2007). According to Riethmüller et al. (2002) the genus *Pseudoperonospora* is monophyletic, however, it has close relationship with *Peronospora*, as they share similar haustoria, conidiosporangiophore morphology, and conidiosporangium colour. Recently, several haustorial types (e.g. clavate-branched, ellipsoid-pyriform, hyphal) were recognized for downy mildews, including *Pseudoperonospora* (clavate-branched), which are considered an important diagnostic feature (Voglmayr et al. 2004).

Based on the variability of the genome size, Voglmayr and Greilhuber (1998) stated that the phylogenetic position, and the definition of individual genera, has not yet been clearly resolved. Recent molecular-phylogenetic studies done with Peronosporales and Peronosporaceae (Göker et al. 2003, 2007; Riethmüller et al. 2002; Voglmayr 2008) demonstrated that the genus *Pseudoperonospora* is a unique monophyletic group. This supported the previous opinion of taxonomists (Constantinescu 2000; Waterhouse and Brothers 1981), and disagreed with Skalický's (1966) concept on the division of the genus. Based on sporangial ultrastructure and the phenetic characters shared by the species of the genus, Constantinescu (2000) suggested that a distinct genus for *Pseudoperonospora* is justified (Riethmüller et al. 2002). Indeed, comparative morphological and molecular (ITS rDNA) studies of *P. cubensis* and *P. humuli* showed that the genus *Pseudoperonospora* is a distinct taxonomical unit. However, recent studies also demonstrated that both species are very similar (Choi et al. 2005; Gent et al. 2009; Sarris et al. 2009), and *P. humuli* was suggested as a synonym of *P. cubensis* (Choi et al. 2005). Previous (Choi et al. 2005) and recent (Sarris et al. 2009) studies of ITS rDNA showed only a limited intraspecific variability of *P. cubensis*. Nevertheless, the limited ITS rDNA intraspecific variability stands in contrast with the very broad pathogenic variability of *P. cubensis* (Lebeda and Widrlechner 2003; Lebeda et al. 2006b). Recently, the mitochondrial cytochrome oxidase (COX) gene cluster, and two nuclear loci, ITS, NADH gene regions and ß-tubulin, were sequenced. Conserved single nucleotide polymorphisms (SNPs) were found that consistently differentiate *P. cubensis* and *P. humuli* (Mitchell et al. 2009). Host range and pathogenicity studies demonstrated that these species have distinct pathogenic capabilities (Gent et al. 2009), *P. cubensis* may infect hop, but caused very little disease, whereas *P. humuli* never infected cucurbits (Mitchell et al. 2009). Current studies are focused on identifying a suite of effector proteins from *P. cubensis*, and characterizing them based on *in planta* localization, putative function, and in part, contribution to overall pathogen virulence (Day and Hausbeck 2009).

All these results unambiguously showed that a detailed understanding of inter and intraspecific variability of *Pseudoperonospora* spp. must be based on more complex experimental approach (e.g. Gent et al. 2009). From a purely phytopathological viewpoint, pathogenicity is a much more significant taxonomic criterion as it enables specification of intraspecific pathogenic variants (pathotypes and races) of *P. cubensis* (Lebeda et al. 2006b).

Biology

Symptoms of infection

P. cubensis is a leaf pathogen, attacking exclusively the leaves of cucurbitaceous plants (Cohen 1981; Thomas 1996). However, the formation of sporangiophores was observed also on stems, leaf petioles, tendrils and peduncles of heavily infected melons. On cucumber, fruit watery spots were recorded followed by pathogen sporulation (Palti and Cohen 1980). Host plants may be infected at all developmental stages (seedlings, young and adult plants) but symptoms on young, newly developing leaves are rather rare (Lebeda 1990). However, cotyledons are actually more susceptible than true leaves. Symptoms differ markedly among cucurbit species (Fig. 1). In some species (cucumber, *Luffa*) *P. cubensis* causes irregular, localized, yellow lesions, restricted by leaf veins whereas in cantaloupe and watermelon, lesions are not restricted by leaf veins and are more circular and irregular. Unlike some other members of the Peronosporaceae (Lebeda and Schwinn 1994), *P. cubensis* does not produce systemic infection of the whole plant (Cohen 1981).

The incubation period, from penetration until visible external symptoms, is 4–12 days under field conditions, depending on the environmental conditions and inoculum load (Cohen 1977), and resistance/susceptibility of the host plant (Lebeda and Widrlechner 2003). The first symptoms of *P. cubensis* infection in Cucurbitaceae are pale-yellow, oily lesions on the upper side of the leaf, sometimes restricted by leaf veins which are described as angular leaf spots (Lebeda 1986b). The size of primary lesions

Fig. 1 Symptoms of *Pseudoperonospora cubesis* in leaves of cucurbits taken from the field. Note that the dark sporulating areas in squash are smaller (5–8 mm) than those in cucumber (5–20 mm)

varies from 3–10 mm. During the development of the disease, lesions coalesce and form larger lesions, and may eventually cover the entire leaf. Artificially-inoculated plants or leaf discs may show irregular, chlorotic or even necrotic lesions, especially when inoculum concentration is high and/or the host is highly susceptible (Lebeda 1986b, 1990).

During the reproductive phase of disease development, a thin layer of dark brown, grey or violet-black sporangiophores bearing sporangia appear on the lower (abaxial) surface of the leaves. Under extremely heavy infection, leaves become necrotic followed by death of the whole plant (within 4 to 10 days from first symptoms, depending on weather conditions, inoculum concentration, and host genotype) (Lebeda, 1990). Heavy infection may significantly reduce yield quantity and quality (Lebeda and Urban 2004b).

The symptoms of the disease can be quite variable not only among different species of cucurbits, but also genotypes (cultivars) of the same host species. They could be also influenced by weather conditions, e.g., atypical water-soaked lesions may be seen under extremely humid conditions on some host species or genotypes (Lebeda 1986b, 1990). Genetically-resistant cultivars of melon inoculated in growth chambers produce water-soaked lesions of 1-2 mm in diameter (Thomas et al. 1987; Thomas 1996). Such cultivars produce no symptoms under epiphytothic field conditions.

Parasitism

P. cubensis (like all other members of the family Peronosporaceae (Göker et al. 2007)) is an obligate biotrophic parasite, absolutely dependent on its host plant for growth and survival. It cannot survive outside its host except as oospores. It attacks living host tissues in order to survive and reproduce (Palti and Cohen 1980). Necrosis of the affected plant tissue will lead to death of *P. cubensis*. Like other plant pathogenic Peronosporaceae, *P. cubensis* does not produce toxins (Švábová and Lebeda 2005); it releases, to a limited extent, the enzymes necessary for the primary penetration of the cell wall (Lebeda et al. 2001a). At the initial stage of development, *P. cubensis* can temporarily support the growth of host cells and increase their number of organelles (Lebeda and Schwinn 1994).

An early necrotic reaction of the infected plant tissue (hypersensitive reaction) is typical for resistant hosts (Cohen et al. 1989) whereas in susceptible hosts necrosis appears as a late reaction of the infected tissue. The obligate biotrophic nature of *P. cubensis* is also characterized by the fact that it cannot be cultivated on artificial nutritive media (Cohen and Eyal 1977; Lebeda 1986b).

Life and disease cycle

Asexual reproduction

The primary and the main infective unit is the asexual spore (conidiosporangium, zoosporangium; Fig. 2). Sporangia are ovoid or elliptic in shape, and measure 15 to 25×20 to 35 μm (Skalický 1961). At maturity sporangia are light-grey to deep-purple in colour (Thomas 1996). They easily dislodge from the sporangiophores and are distributed by wind or water splash. After deposition on the leaf surface of a host plant, sporangia require contact with water (e.g., rain or dew) in order to germinate. Germination is indirect: the multinucleate protoplast differentiates into 5 to 15 biflagellate zoospores measuring 8–12 μm, that

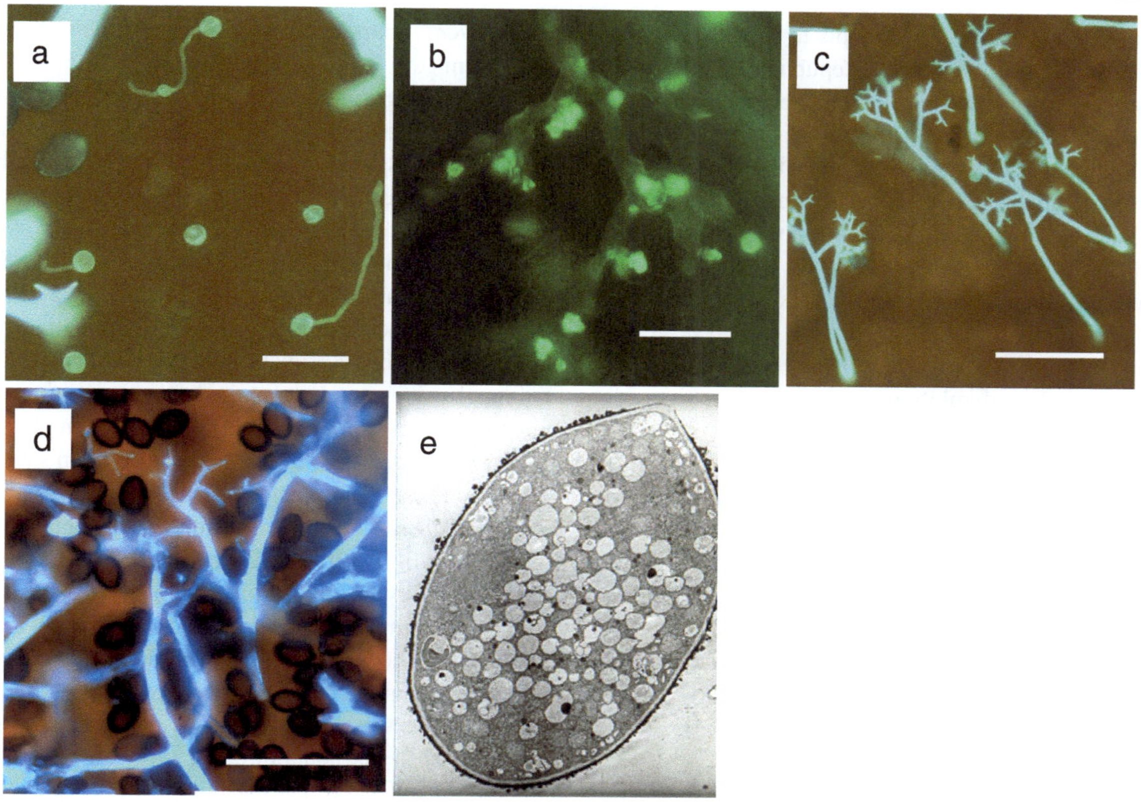

Fig. 2 Developmental stages in the pathogenesis of *Pseudoperonospora cubensis* in cucumber. **a** germinating cystospores on leaf surface. Bar=25 μm. **b** intercellular mycelium with haustoria. The lighter spots are the callose collars surrounding the haustorial neck. Bar=25 μm. **c** sporangiophores emerging from stomata. Bar=100 μm. **d** sporangiophores (blue) and sporangia (dark brown) on leaf surface. Bar=100 μm **e** ultrasructure of a sporangium. ×2000

emerge through a papilum (Palti and Cohen 1980). The zoospores actively swim in the direction of stomatal apertures where they settle, lose their flagella and encyst (Cohen 1981). A germ tube subsequently grows from the cyst (Fig. 2), produces an appressorium from which a penetration hypha develops and penetrates into the stomatal aperture to the substomatal cavity of the leaf tissue. The penetration via stomata is the most frequent mechanism of penetration of *P. cubensis* (Cohen 1981). Rarely, a direct (epidermal) penetration occurs (Lebeda 1990). In *Bremia lactucae*, 95% of penetrations occur directly *via* the epidermis (Lebeda and Reinink 1991). In *P. humuli*, the zoospores swim towards the stomatal apertures by thigmotropism, settle, and produce a germ-tube that penetrates into the stoma (Royle and Kremheller 1981).

Under suitable environmental conditions and in a susceptible host, the colonization of the parasite in tissue proceeds relatively quickly and sporangiophores emerge from stomata within 5 to 7 days (Fig. 2), mainly on the lower side of the leaves where stomata are more frequent (Cohen 1981). On susceptible hosts, a new infection cycle takes place once in 7 to 14 days, depending on the environmental conditions. *P. cubensis* is polycyclic with regards to its disease cycle (Kranz 2003).

Sexual reproduction

Sexual reproduction is rare, and so far has not been proven in most countries where *P. cubensis* prevails. Sexual reproduction of *P. cubensis,* as in other species of the *Peronosporaceae*, proceeds via the production of oospores (Michelmore 1981). It occurs at the end of the season when the infected tissues become necrotic (Bedlan 1989; Lebeda 1990). In Europe, the only unambiguously observed occurrence of oospores came from Austria (Bedlan 1989). The other records of oospore occurrence came from Israel (Cohen et al. 2003), India (Mahrisi and Siradhana 1984; Singh and Sokhi 1989), Iran (Zaker and Ommati 1991) and China

(Zhang et al. 2006). The attempts to find oospores in cucurbit plants in the Czech Republic, including the possibility to produce oospores experimentally, were unsuccessful (Lebeda and Urban 2004a; Lebeda unpubl. data). Similar attempts were made in the USA without success (Kanetis and Holmes, unpubl. data). It is, therefore, unclear whether this pathogen survives in Central Europe or USA by oospores (Lebeda 1986a; Lebeda and Schwinn 1994; Lebeda and Urban 2004a).

Ecology and epidemiology

Environmental and ecological conditions have great impact on progress of the disease cycle, pathogenic processes, symptom expression and epidemiology (Cohen 1981). Current advances on comparative ecology and epidemiology of zoosporic plant pathogens were summarized by Jeger and Pautasso (2008). We focus here on the most important ecological factors that influence the epidemiology of *P. cubensis*.

Sources of inoculum

In some oomycetes e.g., *Bremia lactucae, Sclerospora sorghi* and *Phytophthora infestans* (in some regions) the pathogen emerges in the new season from oospores harboring in plant debris in soil. Others might emerge from oospores carried by seeds, e.g., *P. infestans* in tomato seeds and *Plasmopara halstedii* in sunflower seeds. In *P. infestans*, inoculum often emerges from mycelia harboring in potato tubers or tomato seeds. In *P. cubensis* none of these mechanisms is known. The pathogen produces oospores very rarely; mycelia, sporangia or zoospores do not occur in seeds or fruits, nor survive in soil or in plant debris. Therefore, an overwintering mechanism is unknown. The survival of *P. cubensis* on wild cucurbits is discussed below. In Israel, downy mildew occurs every year since records have been taken (about 70 years), but the source of initial inoculum was never recorded. Nowadays, the fact that cucurbits are grown year around in the open field, net-houses or greenhouses, assures that inoculum will occur year around.

Factors influencing early stages of infection

The lifespan of sporangia is very short. Normally, it does not exceed 48 h and in many cases not more than several hours after the dislodge of the sporangium from the sporagiophore (Cohen and Rotem 1971a). During this short period, the sporangia must land on the leaf surface of a susceptible host and germinate. Dispersed sporangia laying on cucumber plants lose infectivity as temperature rises. However, infectivity was better retained when plants were incubated at low RH (5–28%) than at high RH (84–90%) (Cohen and Rotem 1971a). The presense of free water on the leaves is essential for germination and for the formation of primary infectious structures. The minimal wetting period required for germination and penetration is approximately 2 h (Cohen 1981). Germination occurs in different frequencies on susceptible and resistant host genotypes, as well as on nonhost plants (Cohen 1976). The optimum temperature for germination is 10–20°C (Cohen 1977).

A short drying period (ca 10–15 min) applied to sporangia during germination leads to disruption of integrity of the internal mitochondrial membranes of sporangia, thus, preventing the formation of zoospores and hence cessation of the entire infection process (Cohen 1977). Six hours of dew period are considered optimal for infection. During this period the pathogen completes its penetration into the stomata and becomes independent of the presence of free water on the leaf surface (Cohen 1981). *P. cubensis* can also use guttation droplets for its germination. This event appears later as symptoms developing on the periphery of leaf lamina blades, especially in greenhouses. Production of sporangia can occur at temperatures from 5 to 30°C, with optimum production at 15–20°C (Cohen et al. 1971).

The release of zoospores from the zoosporangium does not occur under anaerobic conditions or in the presence of respiration inhibitors. This process is temperature-dependent within the range of 9 to 30°C. Zoospores may persist in a water environment for 18 h at low temperatures, but they immediately encyst at higher temperatures. Optimal temperature for cyst germination is 25°C (Cohen 1981). In *P. humuli* on hop, zoospores swim towards the stomatal openings, land on stomatal apertures, encyst, and produce a germ-tube which grows into the meshophyll (Royle and Kremheller 1981). The first penetrations are observable approximately 5 h after the adhesion of the sporangia to the leaf surface (Lebeda 1990). Light is a factor that may support the development of infection even in a short dew period (Cohen et al.

1971). The initial stages of the disease cycle of *P. cubensis* (from the release of zoospores to the formation of the first hyphae) take place in both susceptible and resistant hosts (Cohen 1981). In resistant hosts, the growth stops after the formation of the first haustorium. The formation of haustoria was not observed in non-hosts (Cohen 1981), neither in a resistant melon Cohen et al. 1989). This fact does not correspond to the observations made with other downy mildews (e.g., *Lactuca–Bremia lactucae*) in which haustoria were seen also in nonhost plants (Lebeda et al. 2001b, 2002, 2008b). Ultrastructure of *P. cubensis* in leaves of *C. melo* is summarized in Fig. 3.

Factors influencing colonisation and symptom development

During the incubation period, while the colonization of host tissue is in progress, mycelia of *P. cubensis* grow in intercellular spaces and haustoria develop inside the mesophyll cells (Lange et al. 1989b). Under natural (field) conditions, the incubation period lasts for 4–12 days, depending on climatic conditions and host genotype (Lebeda 1986b). At the initial phase of the infectious process a temperature regime of 25–30°C during the day and 10–15°C during the night, is favourable (Palti and Cohen 1980). Better illumination in the incubation period supports the development of hypha and haustoria in the tissue leading to the formation of larger lesions on leaves. Low light intensity leads to the reduction in number and size of lesions due to the weak development of hypha and haustoria (Cohen 1981). The incubation period (the period lapsed between inoculation and the appearance of first symptoms) is significantly influenced by inoculum concentration. The first symptoms may appear 3 to 4 days after inoculation at high inoculum concentration (ca 1000 sporangia/cm^2 leaf), as compared to seven or more days at low doses (ca 10 sporangia/cm^2 per leaf) (Cohen and Eyal 1977).

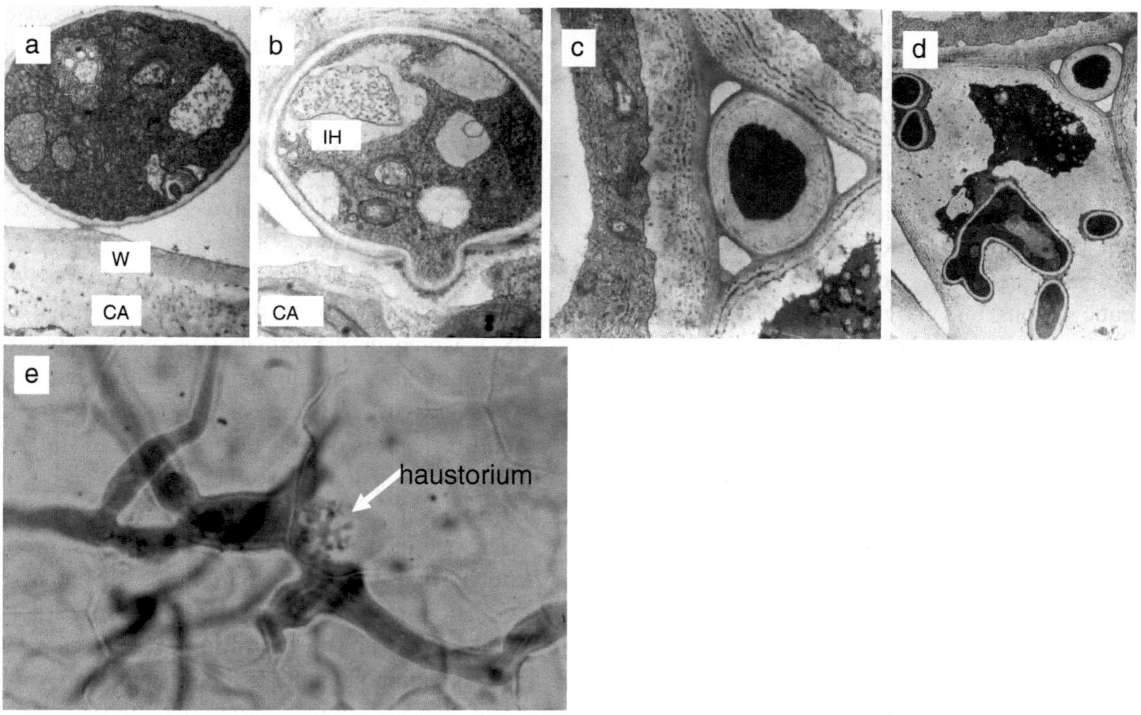

Fig. 3 Ultrastructure of *Pseudoperonosapora cubensis* in leaves of *Cucumis melo* PI124111F (**a**–**d**) and bright field microscopy in Ananas Yokneam /AY/ (**e**) (modified from Cohen et al. 1989). **a** Interface between hypha and host cell at 20 h post inoculation (hpi). Note heavy callose in the host cell (×13,500); W—host cell wall, CA—callose like material. **b** Attempted penetration, 96 hpi (×8250); IH—intercellular hypha. **c** Intercellular hypha surrounded by three host cells, 96 hpi. Note—heavy callose in both the host and pathogen (×20,250); **d** Necrotic lobbed haustorium in a necrotic host cell heavily embedded with callose, 96 hpi (×6175); **e** Hyphae and haustorium in the susceptible AY, 96 hpi, aniline blue staining (×160)

The rate of tissue colonization and symptom development are significantly influenced by temperature. Lower temperatures, although allowing colonization, delay symptom development while higher temperatures enhance symptom development, speed up the progress of lesions from chlorotic to necrotic and terminate the development of the pathogen earlier (Cohen 1977). The sporulation potential of necrotic lesions is very low and the vitality of the formed spores decreases quickly (Lebeda 1990). Hot and dry spells in the field enhance the necrotization of the lesions and terminate the survival of *P. cubensis* in the leaves, thus bringing to an end the development of the disease and its spread (Cohen 1981).

Factors influencing sporulation

Asexual sporulation is a process in which sporangiophores emerge from stomatal openings, ramify dichotomously, and produce sporangia on sterigmata. In a conducive environment, the whole process lasts ca 6 h (Cohen 1977). Sporangiophores emerge from stomatal openings only when relative humidity is 90% or above, regardless of whether the leaves are incubated in light or darkness. In contrast, differentiation of sporangia on sterigmata takes place in darkness only. Infected cotyledon leaves placed on wet filter paper in Petri dishes under continuous light conditions produce abundant white sporangiophores (Cohen 1977). Sporulation occurs on lesions of certain physiological age only. Normally it will happen on chlorotic lesions but not necrotic ones. The lesions should contain enough photosynthetic carbohydrates to support the process, as sporangiophores and sporangia are rich in glucans (Perl et al. 1972). Photosynthate accumulation occurs when plants are exposed to a continous strong light, preferably blue or red, which enhances photosynthesis. The glucan polymers accumulated in the leaf degrade during darkness to hexoses (a monosaccharide with six carbon atoms) which are consumed by the mycelia of *P. cubensis* for the production of sporangia (Ibid). Hexose formation occurs faster at higher temperature and is not dependent on relative humidity. Therefore, the first part of the dew period required for sporulation may be replaced by a dry dark period without affecting sporulation (Cohen 1977). Under optimal environmental conditions during incubation (temperature and humidity as mentioned above), the sporulation takes place as early as 4 to 5 days after inoculation (Palti and Cohen 1980; Lebeda 1986b). The quality and intensity of light significantly influence the formation of sporangia. Maximal production of sporangia occurs at night and is greatly inhibited by light. It was found experimentally that blue light is most inhibitory to sporulation whereas green and red light had a smaller influence. The inhibitory effect of light on the formation of spores is strongly dependent on temperature (increased inhibition with higher temperature) (Cohen and Eyal 1977). Light is desirable for *P. cubensis* before spore formation starts; then practically every factor stimulating photosynthesis of the host enhances spore production in the following dark period (e.g. long light period, more intensive illumination, adequete light spectrum, higher temperature) (Cohen 1981). Cohen et al. (1971) suggested that at least 6 h of darkness period is required for the formation of sporangia because this is the period needed for conversion of accumulated assimilates into compounds required by the pathogen for sporangial formation.

Experiments with *C. melo* showed that plant nutrition also plays an important role in disease development (Bains and Jhooty 1978; Mahrisi and Siradhana 1988). In general, it was shown that unbalanced treatments by N, P and K reduced *P. cubensis* infection. However, balanced treatments with the most suitable level of nutrients for *C. melo* development and fruit yield were also the best for disease development.

A continual layer of free water on the leaves (water film) allows the growth of sporangiophores (Cohen 1981). Moreover, the formation of sporangiophores is induced by high humidity and is often related to dew on the leaves. An increase of temperature and humidity causes rapid maturation and intensive liberation of sporangia (Cohen et al. 1971; Cohen and Eyal 1977).

The age of the infected leaf and the host (variety, genotype) also affect symptom development and sporulation intensity. Symptoms emerge earlier on young leaves (first to third true leaf), develop faster and sporulation potential is higher compared to older leaves. Under optimal conditions, the number of sporangia in susceptible varieties of *Citrullus lanatus* reaches $4 \times 10^3/\text{cm}^2$, in *Cucumis sativus* $7 \times 10^4/\text{cm}^2$ and in *Cucumis melo* up to $1 \times 10^5/\text{cm}^2$ of leaf area (Cohen 1981). The density of sporangiophores is usually higher on small-area lesions then on large-area lesions. This fact can be explained by better supply of nutrients from the surrounding healthy tissue. Sporulation is much more intense on chlorotic lesions than in necrotic lesions. Physical and biotic factors causing necrosis simultaneously limit sporulation. For example, at lower temperatures, sporulation occurs later but lasts longer (e.g. lesions can be fertile for at least 16 days) (Cohen 1977; Cohen and Eyal 1977).

Factors influencing dispersion

Low humidity and dry leaf surface are optimal for the dispersion of sporangia. Temperature and light have very low influence on dispersal (Cohen 1981). The most significant mechanisms of *P. cubensis* spores disperse autonomously by wind and water. Dispersal by wind (anemochory) is considered as the primary and most effective way of dispersion by which the spores are transferred by wind to distances of several hundred kilometers (Lebeda 1990). Wind dispersal caused intensive infections of cucumbers with *P. cubensis* in southern Sweden and Finland (Forsberg 1986; Tahvonen 1985) in the second half of the 80 s, due to migration from Central Europe where only one year earlier strong epidemics of *P. cubensis* were first recorded (Lebeda 1986a, 1990; Lebeda and Schwinn 1994). Similar long-distance distribution from the south to the north occurs in the eastern USA (Holmes et al. 2004, Ojiambo et al. 2009). Due to such long-distance transport, infection can progress in the areas where *P. cubensis* can not overwinter (Lebeda 1990). Dispersal by water (hydrochory), is a secondary mechanism of spore distribution over short distances (from leaf to leaf and plant to plant) within cucurbit fields (Lebeda 1990).

Factors influencing sporangial viability and overwintering of *P. cubensis*

Asexual spores (conidiosporangia) do not survive for a long time under common environmental conditions. When detached from sporangiophores, or when positioned on non-living or necrotic leaves, they lose viability and infection ability rather quickly (i.e., 24–72 h) (Cohen and Rotem 1971a). Their long-lasting maintenance or preservation is possible only under low (−18°C) or ultralow (−80°C) temperatures (Lebeda 1986b). Because of the obligate biotrophic nature of *P. cubensis* survival of mycelium in dead leaves is not possible (Lebeda 1990).

Detached sporangia were shown to survive better in cloudy than in sunny days and to withstand up to 23.5 MJ/m^2 and 1.2 MJ/m^2 of solar and UV irradiance, respectively (Kanetis et al. 2010; Ojiambo et al. 2009).

The main way of survival under inconvenient conditions (overwintering) is the formation of thick-walled resting oospores. Their occurrence was recorded in Japan, China, India, and Asian part of former USSR, Israel and Italy (Bedlan 1989; Cohen et al. 2003; Lebeda 1990; Palti and Cohen 1980; Cohen et al. 2003; Zhang et al. 2006). In Central Europe, the occurrence of oospores has not been verified so far, suggesting that *P. cubensis* may not overwinter in this region (Bedlan 1989; Lebeda 1990). Oospores were detected in Austria only, in older leaves of greenhouse cucumbers (Bedlan 1989). In the Czech Republic, oospores have not been unambiguously detected so far despite the fact that their occurrence is probable (Lebeda and Urban 2004a). In Central Europe, the main source of inoculum emigrates every year by air streams from Southeast Europe, where the pathogen can overwinter on living plants (Lebeda 1990; Lebeda and Schwinn 1994).

P. cubensis can also survive the winter *via* so-called green bridge, on protected cultures of cucurbit plants (e.g., in greenhouses). In Michigan USA, the early attacks in field-grown cucumber are thought to originate from greenhouses in Ontario, Canada (Day and Hausbeck 2009). In areas with a suitable climate, the perennial mycelia can overwinter on some host species (e.g. *Citrullus* spp., *Cucumis* spp.) even under field conditions as proved e.g. in India and southern USA (Palti and Cohen 1980; Holmes et al. 2004).

Long distance migration of sporangia

Spore trapping studies conducted in Israel (Cohen and Rotem 1971a, b) showed that in a regular summer day the peak of sporangial dispersal occurred at 8am. It starts at sunrise (6am), when the temperature rises, and the RH decreases. Dispersal (as indicated by the spore trap) continues until 4pm at very low rates. These late-dispersed sporangia have a better chance, compared to the early-dispersed sporangia, to remain viable and infect when the sun sets and dew accumulates. Sporangia of *P. cubensis* dislodge from sporophores by a twisting mechanism of the sporophore. This twisting occurs when RH decreases, and is responsible for dislodging the sporangium from its sporophore. Sporangia can dislodge by water splash also, which ensures their rapid germination and infection. However, if splash dispersal happens with the aid of sprinkling irrigation, sporangial viability may be hampered if penetration was not completed before the end of the irrigation. The dispersed sporangia were shown to withstand high temperatures, provided that air relative humidity is low (Cohen and Rotem 1971a).

The best data on sporangial migration came from the USA where Nusbaum (1944, 1948) showed that infected cucurbits grown in the southeast in the winter time served as a source of air-borne sporangia to cucurbits crops grown along the East coast of the USA and Canada. He showed, on the basis of visual observations of disease outbreaks, that sporangia migrate in air trajectories, progressively reaching northern states in the spring and summer, and finally reaching Canada in late summer.

Recently, a very detailed migration, forecasting and epidemiological study showed the spatial and temporal movement of *P. cubensis* sporangia from southern Florida or Mexico to the north East coast of the USA. For this purpose a network of ca 40 representatives from the USA and Mexico was established (Holmes et al. 2004; Ojiambo et al. 2009). In comparison with Nusbaum (1944, 1948), the current system is aided by prognosis, it is rapid and more precise and availability of the forecasts to farmers is via the Internet. The precision resulted from two improvements: 1) use of meteorological models to actually track spore movement; and 2) a large network of collaborators who report disease outbreaks. A unique feature of this forecasting system is that growers can sign up on the forecasting web site to receive alerts of new disease outbreaks or risk of disease outbreak, via email and text messages on their cell phones (Holmes et al. 2004; Ojiambo et al. 2009). Recently, molecular and bioinformatic approaches are also introduced to the epidemiological studies of *P. cubensis* (Day and Hausbeck 2009).

Another significant example of long-distance travel of *P. cubensis* occured in 1985 when inoculum was distributed via air streams from Central Europe and Poland to Finland and Sweden causing extensive damage to cucurbits (Forsberg 1986; Tahvonen 1985). In Czechoslovakia, the pathogen spreads from Hungary via South Slovakia to South and Central Moravia, and later to East and Central Bohemia, i.e., the pathogen movement was from south-east to the western parts of Czechoslovakia (Lebeda 1990).

In Israel, *Cucurbita* spp. did not serve as hosts of *P. cubensis* until 2002 (Cohen et al. 2003). Thus, infection of *Cucurbita* spp. since 2002 could have been due to sporangia belonging to a pathotype capable of infecting *Cucurbita* spp. that may have emigrated from Southern or Central Europe (possibly Poland). Polish genotypes of *P. infestans* were identified in Israel on potato crops (Cohen 2002b).

Disease in protected crops (plastic and glasshouses)

Plastic houses, glass houses and net houses are widely used for growing cucurbits. In Israel, cucumbers are grown under plastic during the winter season assuring fruit supply year around. Watermelon and melon are grown under cover in the Arava valley during the winter allowing export of fruits to Europe during Christmas. Farmers face severe problems of downy mildew attacks during the winter, especially in melon and cucumber, and are forced to use frequent fungicide applications. The higher temperatures and relative humidity occurring under such covers in winter normally favour the development of downy mildew. The formation of guttation droplets, along the leaf boundries due to the reduced water potential in the xylem vessels at night, further increases the infection frequency. The closed environment enables the sporangia produced to stay inside the house so as to cause more infection. However, a closed plastic house may be devastating to the disease. This might happen under sunny skies when no ventilation takes place. The temperature may than elevate to 45°C or more with RH of >90%, a combination lethal to the pathogen.

Severe infection of *P. cubensis* on oriental melon (*C. melo* var. *makuwa*), grown in plastic greenhouses, was reported in Korea (Yeon et al. 2002). Downy mildew prevailed until the 2nd harvest and caused 16–34% yield reduction. Melon yield and downy mildew incidence were negatively correlated, and appropriate control methods had be taken at initial stage of the disease (Yeon 2007). As a part of appropriate control measures, an early warning model for occurrence of cucurbit downy mildew in non-heated greenhouses was developed based on disease records and microclimatic parameter analysis (Yang et al. 2007).

The microclimate manipulation by the polyethylene mulch could be one of the solutions for reducing *P. cubensis* infection in greenhouses. The ability of the mulch to suppress *Phytophthora infestans* in tomato and *P. cubensis* in cucumber was studied successfully (Shtienberg et al. 2010). The disease-suppressing effect of mulch appeared to come from a reduction in both the frequency of nights when dew formed and the number of dew hours per night when it formed. Mulching also reduced RH in the canopy, which may have reduced sporulation (Shtienberg et al. 2010).

Host range

P. cubensis affects the Cucurbitaceae family only (Palti and Cohen 1980; Lebeda and Widrlechner 2003). This family is relatively large and very heterogeneous. Currently, it includes more than 118 genera with 825 species (Lebeda et al. 2007b). Cohen (1981) reported that approximately 12 species of Cucurbitaceae are cultivated and nine of them are affected by *P. cubensis* under natural conditions. In fact, there are probably more cultivated hosts with economic significance (Lebeda 1990; Lebeda et al. 2007b). In addition to cultivated species, *P. cubensis* also attacks various semi-cultivated, weedy and wild genera and species of Cucurbitaceae (Cohen 1981; Lebeda and Widrlechner 2003, 2004). Artificial inoculations and observations made in nature, demonstrated that ca 60 species and 20 genera belong to the host list of *P. cubensis* (Lebeda 1992b, 1999).

The longest list of host species belongs to the genus *Cucumis*. It includes over 30 wild species, occurring mainly in arid and semi-arid areas of Africa, and two commonly cultivated species, *C. sativus* L. (cucumber) and *C. melo* L. (muskmelon) that have probably originated from the Indian gene centre (Kirkbride 1993; Lebeda et al. 2007b). Both species and ca eight wild *Cucumis* species are known as the natural hosts of *P. cubensis* (Lebeda and Widrlechner 2003).

Among the frequent hosts of *P. cubensis* are also representatives of genus *Cucurbita*. It includes approximately 14 species native to the area from the USA to Argentina (Lebeda et al. 2007b). Five species (*C. argyrosperma* C. Huber, *C. ficifolia* Bouche, *C. maxima* Duchesne, *C. moschata* Duchesne and *C. pepo* L.), domesticated in these areas before the arrival of Europeans, are currently abundantly-grown in many parts of the world (Lebeda et al. 2007b). All these species are hosts of *P. cubensis* (Lebeda and Widrlechner 2003, 2004). Artificial inoculations under laboratory conditions proved that *P. cubensis* might infect a number of wild and weedy genotypes of *Cucurbita* spp. (Lebeda and Widrlechner 2004).

Citrullus, *Lagenaria*, *Benincasa* and *Luffa* are important host genera (Robinson and Decker-Walters 1997; Rubatzky and Yamaguchi 1997). The genus *Citrullus* includes four species, of which *C. colocynthis* (L.) Schrad. and *C. lanatus* (Thunb.) Matsum. et Nakai (watermelon) were reported as hosts of *P. cubensis* (Lebeda and Widrlechner 2003). The genus *Lagenaria* which originated from Africa includes six species of which only *L. siceraria* is a cultivated crop. Natural hosts of *P. cubensis* are *L. siceraria* (Molina) Standl. and *L. sphaerica* (Sond.) Naud (Lebeda and Widrlechner 2003). *Benincasa* is a monotypic genus represented only by the species *B. hispida* (Thunb.) Cogn., originating from Southeast China. This species is a natural host of *P. cubensis*. Two cultivated species out of seven species of genus *Luffa* originating from tropic Asia, *L. acutangula* (L.) Roxb. and *L. cylindrica* (L.) M. J. Roem. (syn. *L. aegyptiaca* Mill.), are also natural hosts of *P. cubensis* (Lebeda and Widrlechner 2003).

A limited number of cucurbit vegetables are cultivated in Central Europe, mainly *Cucumis sativus*, *C. melo*, *Cucurbita maxima*, *C. pepo* and *Citrullus lanatus*. *Cucurbita foetidissima* and *Lagenaria siceraria* are also marginally grown in Central Europe (Moravec et al. 2004). The major natural host of *P. cubensis* in the Czech Republic is cucumber (*Cucumis sativus*), on which the pathogen causes extensive damage (ca 90% of the available host tissue was infected and necrotic during the second half of growing season in 1986–1990) (Lebeda and Schwinn 1994). In 2003, natural infection was recorded on *C. melo* (Lebeda and Urban 2004a, b). Lebeda (1986a) reported the first infection of *C. melo* in 1984 and the sporadic occurrence of the pathogen on *Citrullus lanatus* in 1985. Wild cucurbits probably do not serve as hosts of *P. cubensis* in the Czech Republic (Lebeda and Urban 2004a). *Cucumis* was the only genus for which specific physiological specialization of the local population of *P. cubensis* could be distinghished. This specialization predominated in Central Europe (Lebeda 1986a). More details regarding this topic are discussed below under "pathogenic variability."

Recently, the occurrence of *P. cubensis* on some weedy and cultivated Cucurbitaceae was published (Choi and Shin 2008; Ko et al. 2008; Salti et al. 2010). Laboratory experiments demonstrated that *P. cubensis* can infect *Bryonia dioica*, the only perennial plant of the Cucurbitaceae in Central and Northern Europe. Based on this finding, it was hypothesised, that *P. cubensis* may overwinter on this host species and serve as a primary source of inoculum (Runge and Thines 2009). However, long-lasting field observations of *B. dioica* in the Czech Republic (Lebeda 1990, unpublished results; Lebeda and Urban 2004a, b), failed to confirm this hypothesis. From the viewpoint of exclusive alignment of *P. cubensis* with Cucurbitaceae it is interesting to refer to Dick (2001a) who reported on the susceptibility of beans to *P. cubensis* when exposed to *Uromyces* sp. (Uredinales).

Geographic distribution

Worldwide distribution

P. cubensis is widely distributed in all continents of the north and south hemispheres where cucurbit plants are cultivated. It mainly occurs in warm, temperate, subtropic and tropic areas on field cultures as well as on protected (glasshouse, plastichouse and shadehouse) crops (Cohen 1981; Lebeda 1990), especially in areas with annual precipitation of >300 mm (Lebeda 1990). Although distributed worldwide, essential differences among geographic areas may be observed in the occurrence of *P. cubensis*, and the damage it causes to various host species (Lebeda 1990; Palti and Cohen 1980; Thakur and Mathur 2002).

The highest frequency of *P. cubensis* is apparent on the genus *Cucumis*. It currently occurs in more then 80 countries on *C. sativus* and in more then 50 countries on *C. melo* (Palti and Cohen 1980; Lebeda 1990; Thakur and Mathur 2002). In the eastern U.S.A. *P. cubensis* has re-surged as a major problem on cucumbers beginning in 2004 (Holmes et al. 2004; Holmes and Ojiambo 2009; Holmes and Thomas 2009). In Europe, *P. cubensis* was originally common in the Mediterranean region. Recently, it has quickly spread to most European countries, reaching Scandinavia (Lebeda 1990). This indicates the high adaptative capability of *P. cubensis* which enabled it to inhabit new geographic regions with very diverse ecological conditions (Lebeda and Schwinn 1994).

The distribution of *P. cubensis* on *Cucurbita* spp. is relatively limited in comparison with *Cucumis*. It was recorded in approximately 40 countries around the world and the main centre of this distribution is Central America and the Caribbean region (Palti and Cohen 1980). In Europe, the documented occurrence of *P. cubensis* comes from Yugoslavia and former USSR (Lebeda 1990).

The distribution of the pathogen on the genus *Citrullus* is even more limited, in about 25 countries, with the main centre of occurrence in Central America. The natural occurrence of *P. cubensis* on *Citrullus lanatus* is common in Florida, USA but not in Europe or the Middle East (Lebeda 1990; Cohen et al. 2003).

Distribution in Europe

P. cubensis is known in Central Europe since the beginning of 20th century (Lebeda 1991; Skalický 1961). During the first half of the 20th century, it was recorded several times, but as sporadic events. In Czechoslovakia, *P. cubensis* was also observed at the beginning of 20th century and then, in 1924 and 1925 (Zacha et al. 1985; Lebeda 1986a; Ackermann 1990).

Broad and epidemic distribution of the pathogen in Central Europe occurred in the second half of the 1980s (Lebeda 1986a; Lebeda and Schwinn 1994). The year 1985 is considered crucial, during which strong epidemics of downy mildew occured in cucumber (*Cucumis sativus* only) all over Central Europe, parts of Western Europe (Czechoslovakia, Poland, Germany, Austria, Switzerland, Hungary), and even in some Eastern and Southern European states (Yugoslavia, Bulgaria, Romania, Moldova, The Ukraine, Byelorussia) (Lebeda 1990). In that period, inoculum distributed via air streams from Poland to Finland and Sweden where it caused, for the first time, extensive damage to cucurbits (Tahvonen 1985; Forsberg 1986). In the same period *P. cubensis* did not occur in Central Europe on *C. melo* or *Cucurbita* spp. in the field or the greenhouse (Lebeda and Gadasová 2002). The repeated occurrence of heavy infections on cucumbers in the following two decades indicated that *P. cubensis* acquired an epidemic character in Central Europe, comparable to many other areas of the world (e.g. Israel, Japan, India) (Lebeda and Schwinn 1994; Cohen et al. 2003; Holmes et al. 2004).

However, in these areas, the epidemics also occur on other crops (e.g. *Cucumis melo*, *Cucurbita* spp., *Citrullus lanatus*).

In the former Czechoslovakia, the first epidemic of cucurbit downy mildew occurred in 1984 on cucumbers (mainly in Southern Slovakia, South, Central and North Moravia and in areas surrounding Prague). The damage in cucumbers was not very high due to the late appearance of the disease (end of July–August) (Zacha et al. 1985; Lebeda 1986a). A year later, 1985, the epidemics were extremely devastating, with an estimated yield loss of 80–90% in cucumbers (Lebeda 1991). The epidemics in the following years 1986–1988 coresponded (by level) to the epidemics in 1985, but due to better crop protection measures losses were lower (Rod 1990). In 1989, losses again went to over 80% (Lebeda 1991). In all these years the disease in Czechoslovakia was a consequence of pathogen spread from Hungary *via* South Slovakia to South and Central Moravia, and later to East and Central Bohemia. The cucumbers in South and West Bohemia (areas of recent Czech Republic near the Austrian and German border) were always affected later (Lebeda 1986a, 1990).

Detailed observations showed that during the last decade, *P. cubensis* occurred every year all over the Czech Republic, not only in the main host growing areas, but also in marginal areas such as foothills and hills, where cucurbit crops are rarely cultivated. Generally, the first infection symptoms on cucumber, not the other Cucurbitaceae, appeared in the second half of July or at the beginning of August. The fluctuation of the distribution and the intensity of infection was significantly influenced by environmental conditions (Lebeda and Urban 2004a,b). Disease index in cucumber crops ranges from strong to very strong severities (DI 3-4, Fig. 4) at harvest time (Lebeda and Urban 2005, 2007).

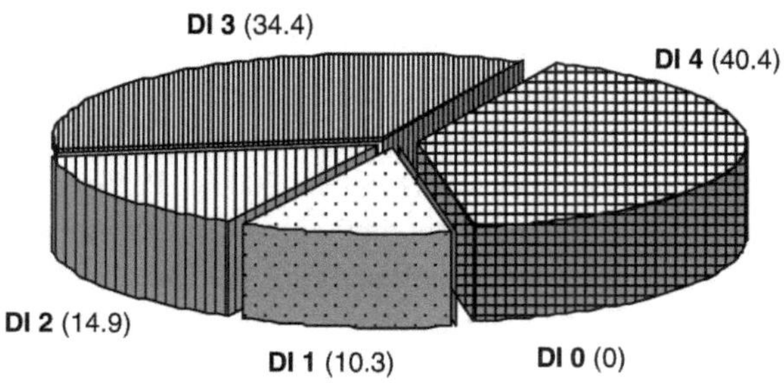

Fig. 4 Frequency (per field) of *P. cubensis* disease intensity / severity/ (DI, %) on cucumber (*Cucumis sativus*) in the Czech Republic in 2003

Host-pathogen interactions Cucurbitaceae–*P. cubensis*

Specificity of relationships between Cucurbitaceae and *P. cubensis*

Species of the Peronosporaceae are characterized by their complicated relationships with their hosts on various levels of biological organization (Lebeda and Schwinn 1994; Göker et al. 2007). Their biotrophic obligate parasitic nature dictates strict host specificity (Crute 1981; Dick 2002a). However, individual species of Peronosporaceae differ in the level of their host specificity, from a single plant species to a relatively large number of species and genera (Lebeda and Schwinn 1994). *P. cubensis* affects a limited number of genera and species within the Cucurbitaceae (Cohen 1981; Lebeda 1999). Specialization in *P. cubensis* is rather diverse and distinct in various pathogen populations (Lebeda et al. 2006b).

P. cubensis is compatible (host-pathogen interaction) with some, but not all, species and genera of Cucurbitaceae (see Host range section). However, certain genotypes (varieties, lines) of host species assign race-specific resistance (Lebeda et al. 2006b). This resistance is a result of highly specific and concurrently relatively simple metabolic and genetic adaptations for the formation of defence mechanisms (Mauch-Mani 2002). It is, therefore, relatively unstable because of the changes that take place in the pathogen from one to another (Lebeda and Schwinn 1994). On the other hand, *P. cubensis* forms physiologically specialized entities (pathotypes and races) characterized by certain types of pathogenicity that enables them to overcome the race-specific resistance of certain host genotypes (Lebeda and Widrlechner 2003; Lebeda et al. 2006b).

The display of compatibility/incompatibility in the interactions between oomycetes and their hosts is well differentiated and has a discontinuous character. For this reason, the classification of pathotypes and physiological races is based on the display of compatible/incompatible reactions on differential host species and genotypes (for details see below) (Lebeda and Schwinn 1994; Lebeda and Widrlechner 2003).

Pathogenic variability of *P. cubensis*

P. cubensis shows an extensive intraspecies variability in pathogenicity (Thomas et al. 1987c; Lebeda et al.

2006b) as with other oomycetes (Lebeda and Schwinn 1994). The first information about the pathogenic variability of *P. cubensis* came from Japan in the 1940s (Iwata 1941). This topic was later elaborated in detail in India, Israel, Japan and USA (Hughes and van Halteren 1952; Cohen 1976; Bains and Sharma 1986; Inaba et al. 1986; Thomas et al. 1987c). Pathogenic variability appears when host genera, species or their lower taxonomic and genetic units interact differentialy with *P. cubensis*. The interactions between the pathogen populations and the host populations can take place at various levels of specifity that are (on the side of pathogen) expressed in the differentiation of different pathogenic groups (Caten 1987). In *P. cubensis*, physiological races and/or phenotypes of virulence, pathotypes and possibly formae speciales were reported (see below).

Formae speciales

A special form (*forma specialis*, f. sp.) is an intraspecific taxonomic unit used only for phytopathogenic fungi. The populations and isolates without morphological differences that are distinguished physiologically are classified as special forms by their parasitic adaptation on various host genera. The main criterion of this specialization is often a certain genus (or other group such as tribes) within the frame of a complete host range (Holliday 2001). The special forms can be distinguished by cross inoculations, where the widest spectrum of host genera known for a certain pathogen is inoculated by the isolates from certain host genera and species.

Such cross inoculations conducted with several representatives of Peronosporaceae showed that their isolates were often compatible only with that host genera or species from which they were isolated. A characteristic example is *Bremia lactucae* where the isolates assigning the specificity to various genera of the family Asteraceae were distinguished as special forms (Skidmore and Ingram 1985). *P. cubensis* isolates from *Cucumis sativus* were compatible with five different genera of the Cucurbitaceae (Lebeda and Widrlechner 2003, 2004), however, they expressed also incomplete compatibility on some other genera (Lebeda and Widrlechner 2003).

Five special forms were originally considered for *P. cubensis*: f.sp. *cucumae*, f.sp. *cucurbitae*, f.sp. *lagenariae*, f.sp. *benincasae* and f.sp. *luffae*, which coresponded to the isolates collected from the genera *Cucumis*, *Cucurbita*, *Lagenaria*, *Benincasa* and *Luffa* (Lebeda 1990). According to the results of compared cross inoculations published by various authors, we found significant differences in pathogenicity among the given isolates (Lebeda et al. 2006b). New experiments with the isolates originated from *Cucumis sativus* and *Cucumis melo* showed that the classification of pathogenic variability of *P. cubensis* on the level of special forms might be misleading (Lebeda and Gadasová 2002), as was also considered by Thomas et al. (1987c). Recent molecular studies showed that isolates of *P. cubensis* from various hosts were almost identical in terms of sequence analysis of ITS rDNA, which was interpreted as *P. cubensis* being a homogeneous taxon (Choi et al. 2005).

To this date, the existence of special forms of *P. cubensis* is neither confirmed nor negated. Additionally classical, genetic and molecular research (focused on host-pathogen specificity and cross-infection ability), based on broad international co-operation (esp. exchange of isolates originating from different Cucurbitaceae), might contribute more knowledge in this area (Lebeda et al. 2006b).

Pathotypes

Differentiation of pathotypes

Pathotypes and physiological races are very common in plant pathogens. They are morphologically identical but differ in their ability to attack different species within a genus, and/or among different cultivars within a single species (Holliday 2001). Pathotypes of *P. cubensis* represent the variability in the pathogen from the viewpoint of host range within the family Cucurbitaceae (Thomas et al. 1987c; Lebeda and Widrlechner 2003; Lebeda et al. 2006b). In principle, pathotypes are physiological forms that differ in host specifity on the level of genera, species or subspecies of various Cucurbitaceae (Lebeda and Gadasová 2002; Lebeda and Widrlechner 2003). Pathotypes may also represent pathogen variability previously considered as formae speciales of *P. cubensis* (see section on formae speciales).

Thomas et al. (1987c) were the first to establish a method to identify pathotypes of *P. cubensis*. They used a set of differential species which included *Cucumis sativus, C. melo reticulatus,*

C. melo conommon, C. melo acidullus, Citrullus lanatus and *Cucurbita pepo*. This set was selected based on the results obtained from inoculation of 26 genotypes of seven genera of Cucurbitaceae from which, for differentiation of pathotypes, they chose the most susceptible genotypes enabling a clear differentiation of compatibility/incompatibility (Thomas et al. 1987c).

Using this set, five pathotypes were distinguished according to the different reaction patterns of eight tested isolates of *P. cubensis* originating from the USA, Israel and Japan. The authors described them as "pathotypes 1 to 5" according to the increasing number of hosts on which a virulent (compatible) reaction occurred (Thomas et al. 1987c). Pathotype 1 was compatible with only one differential genotype, while pathotype 5 was compatible with all six differential genotypes. Based on a similar differential set (including *Luffa cylindrica*) pathotype 6 from Israel was described (Cohen et al. 2003).

The differential set of Thomas et al. (1987c) had, nevertheless, several limitations (Lebeda and Widrlechner 2003): it did not include important host genera (e.g. *Benincasa, Luffa, Lagenaria*); differential genotypes were not precisely taxonomically-defined (on species, subspecies and genotype/accession level); and were not maintained as a complete unit, by any responsible institution. Lebeda and Widrlechner (2003) used an extended set of differential genotypes which included also *Lagenaria siceraria, Benincasa hispida*, and *Cucurbita maxima* (Table 1).

This new differential set was developed for the differentiation of *P. cubensis* pathotypes according to a new denomination system (Lebeda and Widrlechner 2003). It was based on the set of Thomas et al. (1987c), but was expanded to include taxonomically-defined genotypes. The new set consists of 12 genotypes belonging to the six most important host genera of the Cucurbitaceae: *Cucumis, Cucurbita, Citrullus, Benincasa, Luffa* and *Lagenaria* (Table 1). The basic data on specificity and variability of the interactions between *P. cubensis* and these taxons are available. All taxa are well defined on the level of species, sub-species and genotype, and are maintained as accessions in several international gene bank collections (e.g. Plant Introduction Station, USDA, Ames, Iowa, USA). This adjusted differential set enables the characterization of *P. cubensis* pathotypes by distinguishing between 12 "pathogenicity factors" and their combinations. A new system of description and denomination of pathotypes was developed in parallel (Lebeda and Widrlechner 2003). This system is based on numerical tetrade codes (Limpert et al. 1994). Based on a binary evaluation of compatible/incompatible reaction pattern (+ or −) of a certain isolate, a numeric tetrade code was created for this isolate. Numeric composition of the code gives a clear picture

Table 1 Differentials set of cucurbit taxa for determination of *P. cubensis* pathotypes (modified according to Lebeda and Widrlechner (2003))

Number	Differential genotype	Genotype code		Cultivar	Country of origin
		Donor	EVIGEZ		
1	*Cucumis sativus*		H39-0121	Marketer 430	USA
2	*C. melo* subsp. *melo*	PI 292008	H40-1117	Ananas Yoqne´am	Israel
3	*C. melo* subsp. *agrestis* var. *conomon*	CUM 238/1974	H40-0625	Baj-Gua	Japan
4	*C. melo* subsp. *agrestis* var. *acidulus*	PI 200819	H40-0611		Myanmar
5	*Cucurbita pepo* subsp. *pepo*	PI 171622	H42-0117	Dolmalik	Turkey
6	*C. pepo* subsp. *texana*	PI 614687	H42-0130		USA
7	*C. pepo* subsp. *fraterna*	PI 532355	H42-0136		Mexico
8	*Cucurbita maxima*		H42-0137	Goliáš	Czechoslovakia
9	*Citrullus lanatus*		H37-0008	Malali	Israel
10	*Benincasa hispida*	BEN 485	H15-0001		USA
11	*Luffa cylindrica*		H63-0010		?
12	*Lagenaria siceraria*		H63-0009		?

EVIGEZ—the information system on plant genetic resources (Gene Bank VURV Olomouc, Czech Republic)

Taxonomy of genus *Cucurbita* adjusted according to Lebeda et al. (2006b, 2007b)

on the pathogenicity of an isolate and concurrently identifies (numerically describes) its relevant pathotype (Table 2). One can deduce the pathogenicity of an isolate to each differential from its tetrade code.

This set might be extended by the incorporation of new taxa or genotypes of Cucurbitaceae. It will, thus, be possible to make pathotype differentiation a flexible process which continuously develops (Lebeda and Widrlechner 2003; Lebeda et al. 2006b). Current attempts are aimed to also differentiate races of *P. cubensis* (Lebeda et al. 2006b).

Geographic distribution of pathotypes

Thomas et al. (1987c) and other authors (for references see Lebeda et al. (2006b)) studied pathotype variability of *P. cubensis* with a limited number of isolates from Japan, Israel and the USA. Data collected by Lebeda and Widrlechner (2003) and Lebeda et al. (2006b) showed substantial differences in virulence of *P. cubensis* in different geographical regions of the world (Table 3). Pathotypes were not determined for European isolates until the late 1990s (Lebeda and Widrlechner 2003). In 2002, Lebeda and Gadasová (2002) used the above new differential set of cucurbits to identify pathogenic variability among 22 isolates of *P. cubensis* originating from four European countries (mostly from Czech Republic). They distinguished 13 different pathotypes that differed from pathotypes 1 to 5 described by Thomas et al. (1987c). Only one isolate coresponded to pathotype 1. The newly distinguished pathotypes produced between 2 to 9 susceptible reactions on the 12 differential genotypes, suggesting that they each carry 2 to 9 pathogenicity factors (PF, i.e. factors able to overcome resistance of individual differential genotypes). The European population of *P. cubensis* is, therefore, significantly variable in pathotype structure and generally does not resemble the model of Thomas et al. (1987c) which cannot detect such variability (Lebeda and Gadasová 2002) due to its limited range. The pathotype structure of *P. cubensis* population differed significantly across various areas of Europe (Lebeda and Gadasová 2002; Sarris et al. 2009). We assume that similar variabilities might also occur in other countries and continents (Lebeda et al. 2006b; Lebeda and Widrlechner 2003) like it is evident on recent data from USA (Colucci 2008).

Until now, pathogenic variation in *P. cubensis* was studied in detail only in the Czech Republic (Lebeda et al. 2006b), and more recently in USA (Colucci 2008). The pathotype structure of the pathogen in the Czech Republic is quite variable and highly pathogenic, i.e. with isolates having a high number of pathogenicity factors (Fig. 5). Over 40 pathotypes were distinguished among 198 isolates collected during 2001 to 2004 and ca 70% of these isolates

Table 2 Some examples of tetrade numeric codes of pathotypes of *P. cubensis*. Codes were established on the basis of the reaction of selected Czech isolates with cucurbit plants (differential set, see Table 1) (Lebeda and Widrlechner 2003)

Groups of differentials*	1. *Cucumis* spp.				2. *Cucurbita* spp.				3. Other Cucurbitaceae				Code of pathotype
Differential genotype*	1	2	3	4	5	6	7	8	9	10	11	12	
Value	1	2	4	8	1	2	4	8	1	2	4	8	
P. cubensis isolates													
PC 3/00	1	2	0	0	0	0	0	0	0	0	0	0	3.0.0
PC 13/00	1	0	0	8	0	2	0	8	0	0	0	0	9.10.0
PC 1/88	1	2	0	0	0	2	0	0	0	2	0	8	3.2.10
PC 3/98	1	0	0	0	0	2	0	8	0	2	0	8	1.10.10
PC 1/98	1	2	0	0	0	2	0	8	0	2	0	8	3.10.10
PC 4/00	1	0	4	0	0	2	4	8	0	2	0	8	5.14.10
PC 12/00	1	2	0	8	0	2	0	8	0	2	4	8	11.10.14
PC 1/97	1	2	4	8	0	2	0	8	1	2	0	8	15.10.11
PC 24/01	1	2	4	8	1	2	4	8	0	2	4	8	15.15.14
PC 39/01	1	2	4	8	1	2	4	8	1	2	4	8	15.15.15

* see Table 1; 0 = resistant reaction; 1, 2, 4, 8: susceptible reaction

Table 3 Availability of data on pathotypes and races of *P. cubensis* in various countries (modified according to Lebeda et al. (2006b))

Pathogenicity category	Pathotype		Race	
Country	Data available	References	Data available	References
China	?		+?	11
Czech Republic	+	6,7,8,9,10	+	6,7,8,9,10
Bulgaria	+?	1	+?	1
India	+	2	+?	2,11
Israel	+	3,12	+	3,12
Japan	+	12	+?	12
Poland	?		+?	11
USA	+	11	+	4,5,11,12
Others (FR, NL, SP) *	+	7	+	7,9

− = pathotype or race absent; + = pathotype or race present; ? = data not available or not experimentally confirmed; * only one isolate determined

FR France, *NL* the Netherlandsm, *SP* Spain

References (full citations are in References): (1) Angelov et al. (2000); (2) Bains and Sharma (1986); (3) Cohen et al. (2003); (4) Colucci (2008); (5) Horejsi et al. (2000); (6) Lebeda (1999); (7) Lebeda and Gadasová (2002); (8) Lebeda and Urban (2004a,b, 2006); (9) Lebeda and Widrlechner (2003); (10) Lebeda and Widrlechner (2004); (11) Shetty et al. (2002); (12) Thomas et al. (1987c)

carried 9 to 12 pathogenicity factors/PF/) (Lebeda and Urban 2004a,b, 2005, 2007).

A shift towards higher pathogenicity was evident during the evaluation period (Lebeda and Gadasová 2002; Lebeda and Urban 2004a, b, 2007). Pathotypes with low pathogenicity (total number of PF up to 4) were detected only in 2001. Isolates with moderate pathogenicity (total number of PF betwen 5 to 8) and high pathogenicity (total number of PF betwen 9 to 12) predominated in the pathogen populations during 2001–2004. The ratio between the last two pathogenic groups was about 1:1 in 2001 and 2002, but changed to 1:7 in 2003 and 2004 when 87.5–93% of isolates respectively, belonged to the group of highly pathogenic pathotypes (Fig. 5). The data show that PF 5, 9 and 11 were least common. The frequency of PF 5 (squash) and 9 (watermelon) tended to increase during the sampling period, whereas that of PF 11 (*Luffa*) tended to decrease.

Increased pathogenicity of *P. cubensis* was recently observed in Israel and USA. In Israel, pathotype 3 (sensu Thomas et al 1987c, attacking cucumber and melons only) was common since pathotype

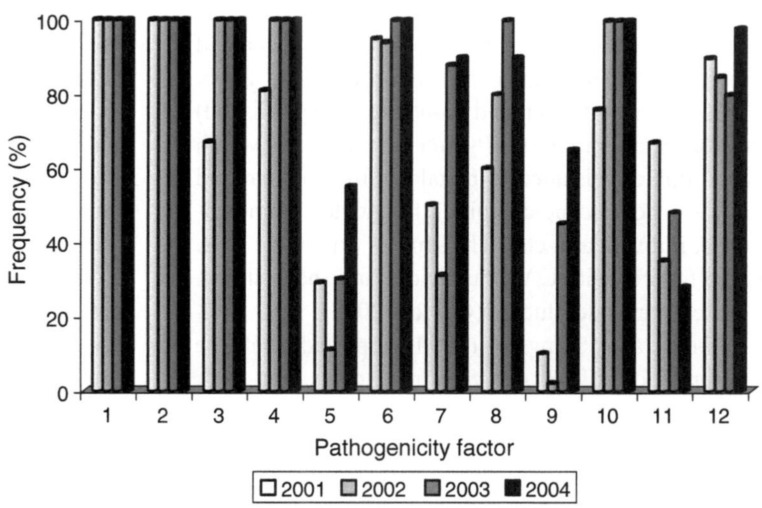

Fig. 5 The frequency of pathogenicity factors (see Tables 1 and 2) in 198 isolates of *P. cubensis* collected in the Czech Republic between the years 2001 and 2004 (Lebeda et al. 2006b)

determination studies were done in 1965 (Y. Cohen, unpublished). In 2003, a new pathotype (number 6) which can also attack squash and watermelon appeared in Israel (Cohen et al. 2003). In the USA, increased virulence to cucumber was observed since 2004 (Holmes et al. 2004; Holmes and Thomas, 2009).

Physiological races

There are indications in the literature (Table 3) that *P. cubensis* might also vary at the species level, suggesting the occurrence of physiological races. Such races (particularly in oomycetes and fungi) are characterized by specialization to different cultivars of one host species (Caten 1987; Holliday 2001). This phenomenon was also described for the interactions of some cucurbits with *P. cubensis* (Lebeda et al. 2006b; Lebeda and Widrlechner 2003) (Table 3).

In 1932, cucumbers that were resistant to downy mildew in Massachusetts, were found to be susceptible in other parts of the USA (Cohen 1981). Another famous example is the sudden breakdown of resistance of the cultivar Palmetto (in South Carolina, 1950) causing severe crop losses due to the selection of a new race of *P. cubensis* (Cohen 1981). A major problem in the last 30 years is the evolution of new field populations of *P. cubensis* resistant to a number of commonly-used fungicides. The first appearance of such a new strain was reported by Reuveni et al. (1980), who detected metalaxyl-resistance in *P. cubensis* from cucumber greenhouses where this fungicide was used repeatedly. Under the influence of repeated applications of different fungicides, a fast selection of resistant strains has occurred (Lebeda and Urban 2004a, b; Urban and Lebeda 2004, 2006, 2007). Recently, Cohen and co-workers (unpublished) have detected strains of *P. cubensis* in Israel which carry double resistance to metalaxyl/mefenoxam and CAA (carboxylic acid amide) fungicides (dimethomorph, iprovalicarb, etc). The same is true in the USA where failures with CAA fungicides have been shown repeatedly in field studies (Colucci 2008).

Lebeda and Schwinn (1994) reported that the differentiation of *P. cubensis* races was not fully unambiguous because the pathogen did not show any significant differences in virulence on *Cucumis sativus* and wild *Cucumis* species (Lebeda 1992a, b). The existence of differential responses (compatibility/incompatibility) of various cucumber cultivars has not been experimentally demonstrated (Lebeda 1992a, 1999; Lebeda and Prášil 1994; Lebeda and Urban 2005), although differences in sporulation intensity are documented (Lebeda and Doležal 1995; Lebeda 1999). Thus, so it is clear that levels of resistance from high to low exist in cucumber, but that no cucumber cultivar has been shown to be completely resistant to *P. cubensis* infection. Nevertheless, the existence of physiological races was proved on *Cucumis melo* (Thomas et al. 1987c; Lebeda 1991; Lebeda et al. 2007a), *Cucurbita pepo* and other *Cucurbita* spp. (Thomas et al. 1987c; Lebeda and Křístková 1992, 1993; Lebeda and Widrlechner 2004). Race-specific interactions also were displayed on *Citrullus* (Thomas et al. 1987c). The recent knowledge about *P. cubensis* races mostly refers to isolates originating from *Cucumis sativus* and *C. melo* (Lebeda 1990; Lebeda et al. 2006b). The problems related to the topic of *P. cubensis* races were discussed in detail by Lebeda et al. (2006b). Our current understanding suggests that *P. cubensis* races do exist. Unfortunately, no suitable differential sets are yet available for the most important host genera, *Cucumis*, *Cucurbita* and *Citrullus* (Lebeda at al. 2006b).

Genetic diversity of P. cubensis

Only limited information is available on the genetic diversity of *P. cubensis* in relationship to geographic distribution, pathogenicity variation of isolates and populations. Amplified Fragment Length Polymorphisms (AFLP) and the nucleotide sequence of the *ITS1-5.8S-ITS2* subunit of ribosomal DNA (*rDNA-ITS*) have been used for studying genetic diversity in *Phytophthora infestans* (Cooke and Lees 2004) and for taxonomic and phylogenetic studies of downy mildew pathogens (e.g. Voglmayr 2008), but not for intra-species population studies of isolates from geographically distant areas (Sarris et al. 2009). In a recent molecular investigation the genetic diversity of *P. cubensis* was compared in populations originating from Crete; Czech Republic and Central Europe; the Western European countries France and the Netherlands (Sarris et al. 2009). All studied *P. cubensis* isolates originated from cucumber (*Cucumis sativus*). AFLP fingerprinting produced ample polymorphisms and isolates were grouped into two separate clusters; one included the Czech (Central Europe) and West

European (the Netherlands, France) isolates, and the other included the isolates of Crete. Significant differences were found between these two populations. Within each group some variations found were attributed to geographic origin, host cultivar, pathogenicity and fungicide resistance. *rDNA ITS* analysis showed no variability among isolates in *ITS1;* however, all *ITS2 rDNA* sequences of Crete and Czech isolates clustered together with isolates from Austria, forming a large cluster together with *P. humuli*, indicating their close taxonomic relationship (see part Taxonomy) (Sarris et al. 2009). These results need to be validated with a larger number of isolates of *P. cubensis* originating from largely distinct areas and well characterized in their phytopathological attributes (pathotype, race and fungicide resistance). This will provide the basis for investigating the sources and shifts in genetic diversity within and between *P. cubensis* populations (Choi et al. 2005; Gent et al. 2009; Sarris et al. 2009), as well as a better background for diseases management. In a most recent study (H. Sierotzki, M. Blum, G. Olaya, M. Waldner-Zulauf, J. Wullschleger, Y. Cohen and U. Gisi, unpublished data) resistance of *P. cubensis* isolates towards CAA fungicides was related to the cellulose synthase A3 (cesA3) gene structure: isolates obtained from US or Israel displayed a different mutation at position 1105 of cesA3: US-Gly ggg to Trp tgg and Israel-Gly ggg to Val gtg.

Host variability in interactions with *P. cubensis*

The extensive intraspecific variability of *P. cubensis* host specificity derives from the large taxonomic and genetic diversity of the Cucurbitaceae. Although host genotypes display clear pathotype or race-specificity (Lebeda et al. 2006b, 2007a), heterogeneous reactions and incomplete resistance/compatibility are also observed (Lebeda and Widrlechner 2003).

Optimal laboratory (growth chamber) conditions for inoculation and disease development enable the interactions between host plants and *P. cubensis* to be precisely described. In contrast, the reaction of the same host genotypes under natural epiphytotic conditions in the field can differ markedly. This is why the phenomenon of field resistance (Lebeda and Jendrulek 1988) to *P. cubensis* was also studied in some cucurbits (Cohen and Rotem 1971b; Lebeda and Doležal 1995; Wehner and Shetty 1997; Lebeda 1999). Field resistance is defined as the interaction of a plant population (cultivar, accession) with a pathogen population during the cultivation period (Lebeda and Jendrulek 1988). Field resistance is therefore a complex epidemiological phenomenon characterized by many different features, such as timing of disease onset, length of latent period, rate of disease progress (low epidemic rate, r), infection frequency, disease incidence (leaf or plant basis), degree of sporulation (Lebeda and Jendrulek 1988; Lebeda and Schwinn 1994). This type of resistance which may provide effective protection of the crop in the field, may not be easily detected or characterized in greenhouse or laboratory tests (Lebeda and Reinink 1991). Field resistance appears with low inoculum levels, is markedly dependent on environmental conditions (Lebeda 1990), and is not directly dependent on the composition of pathogen population(s) (Lebeda 1991b).

The following section of this review provides details on the variation of host-pathogen interactions between the most economically important Cucurbitaceae and *P. cubensis*. They are primarily based on experimental studies done under controlled conditions.

Cucumis spp.

Cucumis sativus (cucumber) is genetically fairly homogenous, and therefore exhibits low variability in its interactions with *P. cubensis* under laboratory conditions (Lebeda and Widrlechner 2003; Lebeda and Urban 2004a). *C. sativus* is highly susceptibile to *P. cubensis*. The available genotypes (including commercial cultivars) do not contain reliable sources of resistance and no unambiguously proved race-specific interactions were detected by Lebeda et al. (Lebeda 1991, 1992a, b; Lebeda and Widrlechner 2003; Lebeda and Urban 2004a, b; Lebeda et al. 2006b). Nevertheless, Shetty et al. (2002) discussed the possibility of race-specificity in cucumbers. *C. sativus* serves mostly as a susceptible control in differential sets for distinguishing pathotypes and races (Lebeda and Widrlechner 2003). Cultivars of *C. sativus* with a high level of field resistance (Table 4) are also known (Lebeda 1999; Lebeda and Doležal 1995; Wehner and Shetty 1997; Bjoern and Kampmann 2000; Doruchowski and Lakowska-Ryk

Table 4 The variability of field resistance in *Cucumis sativus* germplasm against *P. cubensis* (modified according to Lebeda 1999)

Cultivar/accession *C. sativus*	Origin	AUDPC
Commercial cultivars		
Santana F1	The Netherlands	1500
Regina F1	Czech Republic	1297
Niva F1	Israel	1203
Anuschka F1	The Netherlands	1044
Admira F1	Czech Republic	950
Nora F1	Czech Republic	919
Poinsett 76	USA	631
Dalnevostočnyj-6	USSR	528
Germplasm		
PI 169395[a]	Turkey	2162
PI 267741	Japan	1225
PI 263083	China	738
PI 288238	Egypt	441
PI 197085	India	288
C. melo[b]		
PI 124111	India	13

AUDPC area under disease progress curve. Lower values represent higher field resistance

[a] Susceptible control characterized by very low level of field resistance

[b] Resistant control characterized by high level of field resistance

2000; Petrov et al. 2000; Lebeda and Widrlechner 2003). Recently, the development of *Cucumis sativus-hystrix* introgression lines exhibiting resistance to downy mildew was reported (Zhou et al. 2008).

Unlike cucumber, muskmelon (*Cucumis melo*) is a very variable species from morphological, genetic and molecular viewpoints (Lebeda et al. 2007b). Despite this fact, all its forms are easily crossable (Thomas et al. 1987a, b, c). Within the genus *Cucumis*, *C. melo* is the only species with relatively well-investigated race-specificity (Lebeda et al. 2007a) and available effective sources of resistance (Thomas 1982, 1986; Cohen and Eyal 1987; Lebeda and Widrlechner 2003; Lebeda et al. 2006b). Its intraspecific taxonomic units and genotypes display the basic differences in resistance/susceptibility to *P. cubensis* and are therefore used for differentiation of pathotypes (and races) (Lebeda et al. 2006b, 2007a). Thomas et al. (1987a, b) included three subspecies of *C. melo* in their differential set and distinguished three pathotypes of *P. cubensis* on these genotypes.

In previous (Thomas et al. 1987c) and current (Lebeda and Widrlechner 2003) differential sets for identifying pathotypes of *P. cubensis* three genotypes of three taxons of *C. melo* (*C. melo* subsp. *melo*, *C. melo* var. *conomon*, *C. melo* var. *acidulus*) are included. With these three genotypes, Lebeda and Gadasová (2002) discovered significant variability in the interactions with European isolates of *P. cubensis*. In *C. melo*, there are also cultivars with field resistance (Table 4) (Thomas et al. 1987a, b; Lebeda and Schwinn 1994; Lebeda 1999).

The screening of 20 wild *Cucumis* species did not show any significant differences in reaction patterns, most of the accessions exhibited susceptibility, only in some cases was race-specific resistance recorded (Lebeda 1992b; Lebeda and Widrlechner 2003). The elaboration of a differential set for *Cucumis melo* is currently being investigated (Lebeda et al. 2006a, b, 2007a).

Cucurbita spp.

The genus *Cucurbita* is genetically extremely variable (Lebeda et al. 2007b) with the frequent occurrence of race-specific resistance (Lebeda and Widrlechner 2003; Lebeda et al. 2006b) against *P. cubensis*. In a number of genotypes, a high level of resistance or susceptibility with clear expression of race-specificity was observed (Lebeda and Křístková 1992, 1993, 2000; Lebeda and Widrlechner 2004).

C. pepo represents agriculturally the most significant and polymorphic species, and from the viewpoint of interaction with *P. cubensis*, is the most studied species of the genus. It expresses significant race-specificity as seen by the different reaction patterns of subspecies, botanical varieties and cultivars inoculated with different isolates of *P. cubensis* (Lebeda and Widrlechner 2003, 2004). The differences in the resistance/susceptibility to *P. cubensis* were discovered for example among individual morphotypes of *C. pepo* (i.e., the groups of genotypes with a certain characteristic shape of the fruits (Paris 2008)) (Lebeda and Křístková 2000). Certain morphotypes are highly resistant while others are highly susceptible. The variability in the morphotype reactions is influenced by their origin, genetic relationships and cultivation methods (Lebeda and Křístková 2000).

Pathotype specificity was described for *Cucurbita pepo*, *C. maxima* and *C. moschata*. Three accessions of *C. pepo* (*C. pepo* subsp. *pepo*, *C. pepo* subsp.

texana, C. pepo subsp. *fraterna*) and one accession of *C. maxima* are therefore included in the differential set for pathotype determination (Lebeda and Widrlechner 2003).

Wide variability in the reaction with isolates of *P. cubensis* was also shown among wild and weedy *Cucurbita* species ranging from highly resistant to susceptible with the majority of genotypes showing pathotype- or race-specific resistance/susceptibility. The phenomenon of incomplete resistance was also recorded (Lebeda and Widrlechner 2004).

Other important host genera

We have a very limited knowledge of the specificity of interaction between *Citrullus* spp. and *P. cubensis* (Thomas 1970; Thomas et al. 1987c; Cohen et al. 2003; Lebeda and Widrlechner 2003). Within the genus *Citrullus*, significant cultivar variability in the reaction to isolates of *P. cubensis* was observed, with pathotype and race-specificity for *C. lanatus* (Lebeda et al. 2006b). Therefore, this species is included in the original (Thomas et al. 1987c), as well as in the new set of differentials for *P. cubensis* pathotype identification (Lebeda and Widrlechner 2003; Lebeda et al. 2006b).

The interaction of *P. cubensis* with the representatives of the genera *Benincasa*, *Luffa* and *Lagenaria* were studied to a limited extent. In *Benincasa hispida* (the only species of the genus) pathotype-specific resistance to *P. cubensis* was shown (Thomas et al. 1987c; Lebeda and Widrlechner 2003; Lebeda et al. 2006b). A genotype used in the inoculation experiments of Thomas et al. (1987c) was highly resistant to the isolates from Japan, Israel and the USA. In contrast, the genotype included in the new differential set was very susceptible to the European isolates of pathogen (Lebeda and Gadasová 2002; Lebeda et al. 2006b).

Race-specificity is also known in *Luffa cylindrica* and *L. acutangula* (Thomas et al. 1987c; Lebeda et al. 2006b). The genotype of *L. cylindrica* which is part of the differential set for *P. cubensis* pathotype identification (Lebeda and Widrlechner 2003) is resistant to a number of European isolates (Lebeda and Gadasová 2002; Lebeda and Urban 2004a,b; Lebeda et al. 2006b). Thomas et al. (1987c) also reported resistance of this species. However, strong epidemics of downy mildew were recorded in field-grown *Luffa* spp. in China and India (Singh and Singh 1998; Jamadar and Desai 1999; Lebeda and Widrlechner 2003).

Within the genus *Lagenaria*, there is only information about the reactions of *L. siceraria*, for which race-specificity to *P. cubensis* was demonstrated (Lebeda and Widrlechner 2003). The genotype included in the differential set (Lebeda and Widrlechner 2003) showed some susceptibility to European isolates (Lebeda and Gadasová 2002). Thomas et al. (1987c) reported frequent resistance to their isolates.

Genetic aspects in the interaction between Cucurbitaceae and *P. cubensis*

The basic prerequisite for the understanding of the genetic relationships between a host and its pathogen is a parallel genetic study of host resistance and pathogen virulence (Crute 1986). While such detailed studies are available for *Lactuca/Bremia* (Lebeda et al. 2002; Michelmore and Wong 2008), and *Arabidopsis/Hyaloperonospora*, only limited information is avilable for other Peronosporaceae, including cucurbits/*P. cubensis* (Göker et al. 2007; Hardham 2007; Lebeda et al. 2008b).

The resistance of plants against oomycetes follows, in general, the heredity rules of Mendel (Lebeda and Schwinn 1994). The pathotype/race specificity in the interactions between cucurbits and *P. cubensis* follow the "gene-for-gene" theory of Flor (Lebeda and Schwinn 1994; Lebeda 1999; Lebeda and Widrlechner 2004). Pathotype/race specific resistance is usually controlled by one or a few major genes (monogenic up to oligogenic heritability). Field resistance against *P. cubensis* which has also been reported in cucurbits (Lebeda and Doležal 1995) might be of polygenic nature. For example, two recessive genes were reported to control a high level of field resistance against *P. cubensis* an F_1 cucumber cultivar (Doruchowski and Lakowska-Ryk 2000).

Basic information on the genetic control of resistance against *P. cubensis* is available for only a few host species (Lebeda and Widrlechner 2003, 2004; Lebeda et al. 2006b). Monogenic (or oligogenic) resistance occurs in *Cucumis sativus* and *C. melo* (Palti and Cohen 1980; Cohen 1981; Lebeda 1999; etc.). The major genes for specific resistance in *C. sativus* probably have a recessive effect (Cohen 1981). The described sources of resistance of *C. melo* have monogenic or oligogenic character (Cohen 1981; Pitrat 1990; Cohen 1992a; Kenigsbuch and Cohen 1992a, b; Lebeda and Widrlechner 2003). The genetic background of resistance was studied in detail in the accession PI 124111 (*C. melo* var. *reticulatus*), where resistance is probably based on two partially-dominant complimentary genes Pc-1 and Pc-2 (Thomas 1986; Cohen and Eyal 1987). Balass et al. (1993) reported on a high level of resistance in two lines (PI 124111F and 31-10), which were derived from PI 124111. These authors also reported that the genes for resistance were specifically inactivated by low temperature (see below for more details). Kenigsbuch and Cohen (1992a) reported that resistance in another accession PI 124112 (*C. melo* var. *reticulatus*) is controlled by two partially-dominant complementary genes Pc-1 and Pc-3. However, a polygenic control was described for this accession by others (Epinat and Pitrat 1994a,b; Perchepied et al. 2005). Generation mean analysis of resistance to *P. cubensis* in adult *C. melo* plants showed that genetic dominance has a greater importance for expression of resistance. There were observed expression of significant and positive additive gene effects, estimates of heritability were high. Therefore some inbred lines could be used to exploit heterotic effects (Shashikumar et al. 2010). The genetics of host resistance to *P. cubensis* in other cucurbits (e.g. *Cucurbita* spp., *Citrullus* spp.) is still unknown (Lebeda et al. 2007b; Paris 2008).

The genetic basis of virulence variability of *P. cubensis* is known to a very limited extent (Lebeda et al. 2006b). From this aspect, the best known oomycete pathogen is *Bremia lactucae* (Michelmore and Wong 2008). In general, virulence to each host genotype/cultivar is usually determined by one recessive independent allele on one locus (Lebeda and Schwinn 1994). Thomas et al. (1987c) suggested that the significant genetic variability in *P. cubensis* derives from its diploid nature. The double set of genes enables a wider range of responses of *P. cubensis* to selective stresses and therefore increases the ability of adaptation to specific hosts. So far, the contribution of oospores to genetic recombinations of *P. cubensis* is not known because of their rare occurrence.

Currently, a three-fold integrative approach has been undertaken to identify the genetic basis for enhanced virulence observed in recent isolates of *P. cubensis*. This approach involves the following steps: 1) phylogenetic characterization and sequencing of *P. cubensis* isolates, and determination of the genetic basis of virulence; 2) identification of genes which are differentially expressed following infection of *P. cubensis*; 3) isolation of single host cells at various stages of pathogen infection and profiling gene expression during pathogen invasion. It is expected that this approach will provide a framework for determining the basis of pathogenicity and susceptibility in the cucumber/*P. cubensis* interaction (Savory et al. 2008).

Cellular and molecular aspects of interaction

The recognition (compatibility/incompatibility) between a host plant and its oomycete pathogen is determined shortly after infection structures have developed in the plant cell (Lebeda et al. 2008b; Bouwmeester et al. 2009). Recognition occurs when some individual structural constituents of an incompatible pathogen (Pathogen Molecular Patterns, or PAMP) are detected by specific diagnostic molecules of a resistant plant. The detection of "foreign structures (molecules)" induces a sequence of quick biochemical processes that lead to the termination of further growth of the pathogen. If a pathogen is not recognized as "foreign", it results in a compatible relationship with the host, colonizes its tissues and reproduces (Lebeda et al. 2008b; de Jong and van den Ackerveken 2009).

As with *Bremia lactucae*, in which the recognition occurs during the interaction between host plasmalemma and the primary infecting structures of the pathogen (primary and secondary vesicules, haustoria) (Lebeda et al. 2008b; de Jong and van den Ackerveken 2009) so is the recognition between *P. cubensis* and its hosts.

The response of resistant genotypes of cucurbit plants to penetration by *P. cubensis* is often characterized by a hypersensitive response (Cohen et al. 1989; Balass et al. 1993; Eckardt 2004; Taler et al. 2004). Cohen et al. (1989) followed the development of *P. cubensis* in suceptible and resistant *C. melo* by using electron microscopy (Fig. 3). These authors found that the pathogen failed to penetrate into the mesophyll cells of the resistant host due to massive accumulation of callose, not only along the host cell walls but also along the inner wall surface of pathogen mycelia. This failure was also accompanied by a massive accumulation of dark, dense material in the cytoplasm of host cells which leads to necrosis. The structural and biochemical responses were studied in the interaction of *Cucumis melo* with *P. cubensis* (Balass et al. 1993). These authors reported that resistance was accompanied by extensive accumulation of phenolics, callose (Fig. 6) and lignin in the infected sites which probably limit the growth of *P. cubensis* mycelium. These changes were accompanied by a rapid increase in peroxidase activity (Balass et al. 1993). Peroxidase activity was used in melon as a marker for resistance to *P. cubensis* (Reuveni et al. 1990; Lebeda and Doležal 1995). No such changes occurred in susceptible genotypes of *C. melo* upon inoculation with *P. cubensis* (Balass et al. 1993).

Resistance of *Cucumis melo* against *P. cubensis* was temperature-dependent (Balass et al. 1993), a phenomenon quite rare in cultivated plants. The resistance was fully expressed at higher temperatures (21–25°C) but was nullified at lower temperatures (12–15°C), perhaps as a result of specific inactivation of transcription of the resistance genes (Pc-1 and Pc-2) at low temperatures (Balass et al. 1993). It should be noted that the effect of temperature on resistance occurred during the first half of the incubation period when the above-mentioned defense mechanisms were launched.

These results gave rise to the idea that resistance was controlled by the metabolic activity of the host. Indeed, in a later study, Taler et al. (2004) showed that resistance in *C. melo* PI124111F, which carries the resistance genes Pc-1 and Pc-2, is metabolic, resulting from enhanced activity of peroxisomal glyoxylate aminotransferase encoded by the genes *At1* and *At2*. When either gene was transformed into a susceptible plant it became resistant (Fig. 7). These authors speculated that Pc-1 and Pc-2 are in fact *At1* and *At2*. Such resistance, which is primarily conferred by enzymatic activity (Lebeda et al. 2001a), was described as "enzymatic resistance" to downy mildew (Eckardt 2004; Taler et al. 2004). In a more recent study, Benjamin et al. (2009) showed that in susceptible genotypes of *C. melo*, *At1* and *At2* are present but not expressed. Over-expression of these glyoxylate-aminotransferase genes in cisgenic melons makes them resistant to downy mildew (Benjamin et al. 2009).

Pathogenesis of *P. cubensis* on cucumber leaves resulted in metabolic changes, including transpiration rate and increased leaf temperature, depending on the stage of pathogen development and disease severity. Spatial and temporal changes in the transpiration rate

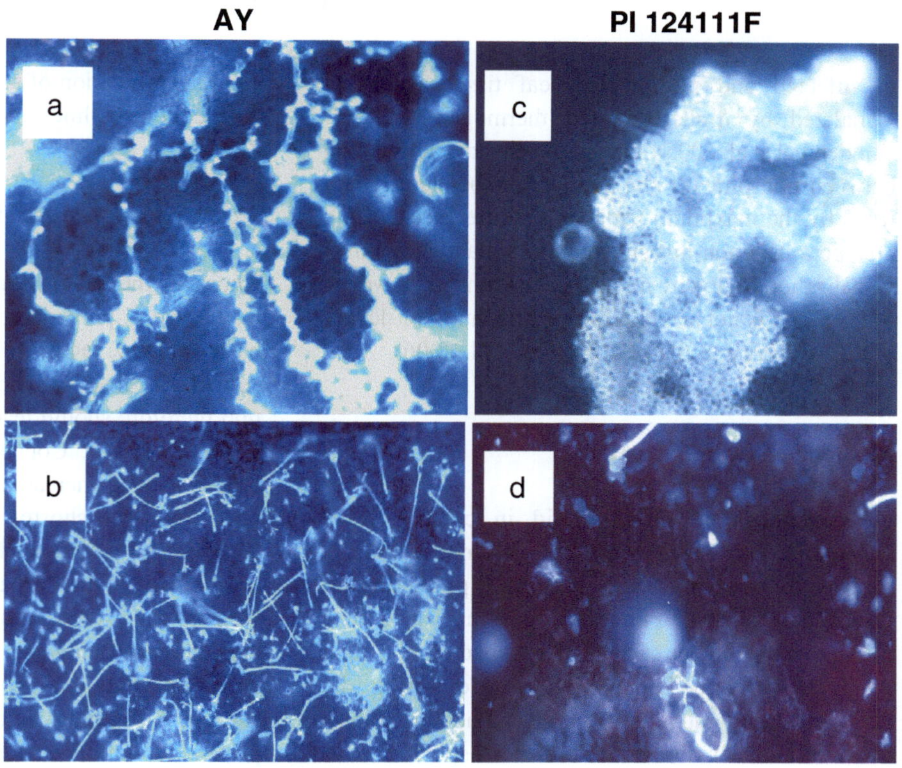

Fig. 6 Epifluorescence micrographs of the development of *Pseudoperonospora cubensis* in *Cucumis melo*: AY (Ananas Yokneam), susceptible and PI124111F, resistant (modified from Cohen et al. 1989). **a** and **c** at 4 days post inoculation (dpi), **b** and **d**, at 7 dpi. Leaves were clarified with ethanol and placed in basic aniline blue and then calcofluor. Note mycelium and sporulation in AY. Note the hypersensitive response (HR) with callose and no sporulation in PI124111F

Fig. 7 Symptomology of *Pseudoperonospora cubensis* on susceptible, resistant and transgenic genotypes of *Cucumis melo* (modified from Taler et al. 2004). Upper panel: symptom of downy mildew on melon leaves of susceptible BU21/3, resistant PI124111F and transgenic line. The resistance gene *At1* was transferred from PI124111F to the transgenic line. Middle panel: epifluorescent micrographs of respective leaves after staining with calcofluor for sporangiophores and basic aniline blue for callose. Bottom panel: natural infection with downy mildew of the susceptible and the transgenic line in the field

of infected and noninfected leaf tissue could be visualized by digital infrared thermography (DIT) (Lindenthal et al. 2005). Due to the negative correlation between transpiration rate and leaf temperature (quantified by the maximum temperature difference, MTD), DIT permits a non-invasive monitoring and an indirect visualization of downy mildew development. However, MTD alone is not suitable for quantification of downy mildew severity in the field (Oerke et al. 2006).

Methods of crop protection

Forecasting is an efficient aid in control of *P. cubensis* (Main et al. 2001). The parameters used in forecasting and their integration in protection of cucurbits against *P. cubensis* are summarized in a few papers (Holmes et al. 2004; Lebeda and Urban 2005; Urban and Lebeda 2006). Monitoring the occurrence and movement of the pathogen enables the prediction of disease outbreaks in specific areas and the application of suitable control measures prior to infection (Holmes et al. 2004; Ojiambo et al 2009; Zhao et al. 2007).

Prevention and agrotechnical aspects

Knowing the biology and ecology of the pathogen may serve in preventing the disease (see above). The basic abiotic factor influencing infection with *P. cubensis* is free leaf moisture. A preventive precaution therefore should lead to elimination of free water from the leaves. Cohen (1977) showed that as the inoculum load increases and temperature approaches optimum, the shorter the leaf wetness duration required for infection. The surface of leaves should not be wet for more than 2–3 h (Cohen 1981). Leaf wetness can be partially controlled in sheltered vegetation by using drip irrigation instead of overhead irrigation, frequent ventilation, and heating before sunrise. Also, under field conditions, drip irrigation is preferable, but dew formation and rain cannot be avoided. Earlier sowing of the crop and decreased plant density can also contribute to the reduction of infection to the extent that it reduces leaf wetness duration (Palti and Cohen 1980). High vegetation density increases the risk of infection as it increases humidity for prolonged periods, stimulates sporulation of *P. cubensis*, and facilitates transfer of sporangia among plants (Lebeda 1990).

Breeding for resistance, classical and transgenic approaches

Availability of sources of resistance and using appropriate methods for testing resistance are among the basic requirements for successful breeding for resistance. There are significant differences in the availability of resistance sources among the most important cucurbits. Most sources are available for *Cucurbita pepo* and *Cucumis melo* (Lebeda 1999; Lebeda et al. 2007a, b; Pitrat 2008; Staub et al. 2008). Recently, great progress has been made in the knowledge of resistance in *Cucurbita* spp. (Lebeda and Křístková 2000; Lebeda and Widrlechner 2004; Lebeda et al. 2006b), although the availability of genotypes with effective and broad levels of resistance is still limited (Ferriol and Picó 2008; Paris 2008). Rather little information is available on resistance against *P. cubensis* in *Citrullus*, *Benincasa*, *Luffa* and *Lagenaria* (Lebeda and Widrlechner 2003). Resistance breeding of watermelon (*Citrullus lanatus*) is still not very well developed (Wehner 2008). The breeding of cucurbits for resistance against *P. cubensis* is further complicated due to the great variability in the pathogen population, pathotypes and races (Lebeda et al. 2006b). In response to recent epidemics, there is an intensified cucumber breeding effort in the USA to develop resistance to downy mildew (Holmes et al. 2006).

Cucumis sativus

Breeding of cucumber for resistance against *P. cubensis* was initiated in Puerto Rico in the 1930s to 1940s by systematic searching for resistance sources (Sitterly 1972). Some resistant/tolerant plants were discovered under field conditions in the materials originating from China and India: Chinese Long from China and Bangalore and accession PI197087 from India (Sitterly 1972; Peterson 1975). These genotypes served, mainly in USA, as the background of cucumber resistance breeding to *P. cubensis* (Staub et al. 2008).

Besides the USA, breeding for resistance also took place in Japan (Ezuka and Komada 1974), Cuba (Pivovarov 1984; Pivovarov and Kudelich 1985), USSR (Medvedeva and Medvedev 1983), and since 1985 also in Czechoslovakia (Lebeda 1990, 1999; Lebeda and Prášil 1994) and in Poland (Doruchowski and Lakowska-Ryk 2000). Unfortunately, no reliable sources of resistance were found in *C. sativus*, and therefore, cucumber cultivars with genetically fixed and efficient resistance were not produced (Lebeda 1991, 1992a; Lebeda and Prášil 1994; Lebeda and Widrlechner 2003; Lebeda et al. 2006b).

The available commercial cultivars do not posses the character of complete incompatibility (resistance). They only allow for a limited level of pathogen sporulation (Lebeda 1999). Resistance in cucumber was reported decades ago (e.g. Cohen 1981), however, in many cultivars (e.g., Palmetto) a relatively rapid breakdown occurred followed by serious infection with *P. cubensis* (Lebeda 1990). One notable exception to this was the durability of resistance in cucumber in the USA. Downy mildew resistant cultivars developed in the late 1969s remained sufficiently resistant to downy mildew that fungicides were not necessary to control the disease until 2004 (Holmes and Thomas 2009). The screening of large sets of *C. sativus* germplasm and cultivars provided no single genotype displaying complete incompatibility to current pathotypes (Lebeda 1992a; Lebeda and Prášil 1994). Differences in field resistance were, nevertheless, found (Table 4), characterized by a delay (7–14 days) in the onset and a slower rate of disease progress under strong infection pressure (Lebeda 1999; Lebeda and Doležal 1995).

Recent achievements in cucumber genome mapping and sequencing (Huang et al. 2009; Ren et al. 2009) provides the new opportunities for research, breeding and development of elite cucumber cultivars with new traits, as well as resistance to diseases and pests.

Cucumis melo

Within the cucurbits, breeding for resistance against *P. cubensis* was most comprehensively elaborated in muskmelon (*Cucumis melo*) (Lebeda et al. 2007a, b; Pitrat 2008). The first resistance research was realized in the 1940s in the USA, where four cultivars (Cuban Castilian, Green Fleshed Rocky Dew, Orange Fleshed Rocky Dew and Smith's Perfect) with high levels of resistance against *P. cubensis* were described (Ivanoff 1944). A highly resistant cultivar (Tainan 2 /PI 321005/) was further released in Taiwan (Sowell and Corley 1974). Currently, the most significant two sources of resistance are the accessions of *C. melo* var. *reticulatus* PI 124111 (Balass et al. 1992, 1993; Cohen 1981; Thomas 1982, 1986; Cohen and Eyal 1987; Kenigsbuch and Cohen 1989; Lebeda 1991, 1999; Lebeda et al. 2007a) and PI 124112 (Lebeda 1991; Kenigsbuch and Cohen 1992a, b; Table 5). Both genotypes originated from Calcutta, India and became the background for the resistance breeding of muskmelon against *P. cubensis* in the USA (Cohen 1981). The cultivars derived from PI 124111 (e.g. the line MR-1) (Table 5) displayed high levels of resistance under natural infection conditions, with attacked plants showing only minute yellow lesions (1–2 mm) with no sporulation (Lebeda and Doležal 1995). However, experiments with artificial inoculations proved also to have race-specific resistance against *P. cubensis* in PI 124111 and MR-1 (Lebeda 1991; Lebeda et al. 2007a).

Resistance of MR-1 was differentially expressed in whole leaves as against leaf discs. Detached whole leaves showed resistance against all tested isolates of *P. cubensis*, whereas the reactions of leaf discs were heterogeneous (some plants displayed resistance while others displayed susceptibility) (Lebeda 1991).

In *C. melo*, there are cultivars with significant levels of field resistance. In Israel, valuable resistance was found under field conditions in three genotypes, including PI 124111, among 19 genotypes tested (Cohen 1981). Nevertheless, the breeding for field resistance in muskmelon is not well developed (Paris 2008).

Attempts to transfer the resistance from *C. melo*, mainly that derived from MR-1, into cucumber was done by using biotechnological approaches (embryo-cultures, protoplast fusion) (Fellner et al. 1996; Lebeda et al. 1996, 1999; Gajdová et al. 2004; Skálová et al. 2004), however, without substantial success.

A successor of PI 124111 that carries resistance to Fusarium wilt races 0, 1 and 2 was named PI 124111 F (Cohen and Eyal 1987). The ultrastructure of resistance against *P. cubensis* in this line revealed (Cohen et al. 1989) intensive callose and lignin deposition in the inoculated sites, which prohibited any growth of the pathogen. Resistance of this line to *P. cubensis* was found to be temperature-dependent: it operates at ≥15°C (Balass et al. 1993). PI 124111 F and its descendent 31/10 (F_{10} from a cross between Hemed and PI 124111 F) carries a specific protein, P45. This protein co-segragated with resistance (Balass et al. 1993) and does not show up in susceptible melons. Taler et al. (2004) discovered that P45 is a peroxisomal glyoxylate aminotransferase. They succeeded in cloning the genes *At1* and *At2* which are responsible for the synthesis of P45 and to transfer them into a susceptible melon under an S35 promoter. The transgenic plants, carrying either *At1* or *At2*, were completely resistant to *P. cubensis*. Recently, Benjamin et al. (2009) showed that *At1* or *At2* occur in susceptible melons but are not

Table 5 Germplasm of *Cucumis melo* with resistance against *P. cubensis* and some other diseases (modified according to Lebeda 1999)

Cucumis melo (accession)	Country of origin	Resistance against disease	Genes of resistance	References
C. melo (PI 124111 /PI 124111F)	India	DM, PM, F	*Pc-1, Pc-2, Pm-3, Pm-6*	Thomas (1986), Cohen and Eyal (1987)
C. melo (PI 124112)	India	DM, PM	*Pc-1, Pc-3, Pm-4, Pm-5*	Pitrat (1990), Cohen (1992a, b), Kenigsbuch and Cohen (1992a)
C. melo var. *acidulus* (PI 200819)	Burma	DM, F	?	Lebeda (1999), Widrlechner (1999, pers. comm.)
C. melo (PI 321005)	Thaiwan	DM, GSB	?	Lebeda (1999), Widrlechner (1999, pers. comm.)
C. melo (line MR-1 derived from PI 124111)	USA	DM, PM, F, A	*Pc-1, Pc-2, Pm-3, Pm-6, Fom-1, Fom-2, Ac*	Thomas (1986), Cohen and Eyal (1987), Pitrat (1990), Lebeda (1991), Cohen (1992a, b), Kenigsbuch and Cohen (1992b)
C. melo var. *agrestis* (CGN 2365)	Kenya	DM	?	Lebeda (1999)

DM, *Pc Pseudoperonospora cubensis*, PM, *Pm Podosphaera xanthii*, F, *Fom Fusarium oxysporum* f. sp. *melonis*, A, *Ac Alternaria cucumerina*, GSB = *Didymella bryoniae* transcribed. When overexpressed in cisgenic plants under a strong promoter plants become resistant (Fig. 7).

In spite of this enormous progress in understanding the resistance of *C. melo* to *P. cubensis*, there is no germplasm or cultivars of *C. melo* with complete resistance to all known pathotypes/races of *P. cubensis* (Lebeda et al. 2006b, 2007a; Pitrat 2008).

Cucurbita spp.

The most important information related the host-pathogen interaction and resistance of *Cucurbita* spp. to *P. cubensis* was summarized above. It is evident that currently there are not enough data about sources of resistance in *Cucurbita* spp. against *P. cubensis*, including characteristics of commercial cultivars. Studies in Japan showed that *C. pepo* cultivar Soumen displayed a high level of field resistance against the isolates of *P. cubensis* from *Cucumis sativus* and *C. melo* (Inaba et al. 1986). Studies in Czechoslovakia showed the occurrence of incomplete resistance and frequent display of race-specific resistance in zucchini and squash cultivars (Lebeda and Křístková 1992, 1993, 2000). Efficient sources of resistance were found in wild and weedy accessions of *Cucurbita* spp., e.g. *Cucurbita foetidissima*, *C. argyrosperma* var. *palmeri* a *C. argyrosperma* var. *sororia* (Lebeda and Widrlechner 2004). However, there is relatively little effort in breeding for resistance in pumpkin and squash (Ferriol and Picó 2008; Paris 2008).

Other cucurbitaceae

Our knowledge on resistance of *Citrullus* spp. to *P. cubensis* is rather limited (Lebeda and Widrlechner 2003). The most important and comprehensive data were reported by Winstead et al. (1957) who found two highly resistant accessions (PI 179660 and PI 179875) amongst a collection of 300 genotypes of *C. lanatus*. A high level of field resistance was reported by Thomas (1970) in cv. Charleston. Currently, breeding of watermelon for resistance to *P. cubensis* is not considered an important goal because of the limited occurrence of the disease (Wehner 2008). This is the main reason why resistant cultivars are not used as a measure in disease control. This conclusion is true also for *Benincasa*, *Luffa* and *Lagenaria* (Lebeda and Widrlechner 2003).

Chemical control

Although chemical control by fungicides may have negative environmental effects and limitations (de Waard et al. 1993), fungicides still constitute the predominant part of the control measures used against oomycetes (Cohen and Coffey 1986; Lebeda and Schwinn 1994; Gisi 2002; Gisi and Sierotzki 2008). According to Gisi (2002) the sales value of fungicides against downy mildews (Table 6) amounted to 1.2 billion SFr in 1996, of which 10% were used to fight mainly *P. cubensis* on cucurbit crops (Urban and Lebeda 2006). Chemical control of *P. cubensis* was done for many decades by contact fungicides, copper formulations at early times, and dithiocarbamates more recently (Palti and Cohen 1980; Cohen 1981). These fungicides prevent zoospore release and cystospore germination. Their application is effective only if done before infection, possibly before sporangial deposition (Urban and Lebeda 2006). Streptomycin, that suppressed tissue colonization by *P. cubensis* (Cohen 1981), is not used in practice.

During the last decades, new, mostly systemic, fungicides have been developed and widely used for disease control (e.g., cymoxanil in 1976, phosetyl-Al in 1977, phenylamides in 1977–1983, propamocarb in 1978, dimethomorph in 1988, cyazofamid in 2001, zoxamide in 2001, mandipropamid in 2005, fluopicolide in 2006). Most systemic fungicides have a specific, single-site mode of action, which means that they are active at one point in one metabolic pathway of the pathogen (Gisi 2002; Urban and Lebeda 2006). The introduction of these systemic fungicides significantly increased the efficiency of plant protection against downy mildews (Cohen and Coffey 1986; Lebeda and Schwinn 1994; Gisi 2002; Cohen et al. 1995; Holmes and Ojiambo 2009; Wang et al. 2009). Systemic fungicides also have some curative effects, as they can stop the development of disease for a certain time after infection. Systemic fungicides quickly translocate in the plant even to the parts not directly treated (Urban and Lebeda 2006). Metalaxyl (and its active enantiomer mefenoxam), oxychloride Cu, propamocarb, prothiocarb and fosetyl-Al are among the most effective substances (Gisi 2002; Tomlin 2003; Urban and Lebeda 2004, 2006, 2007). Fluopicolide is currently the most effective product in the USA (Holmes and Ojiambo 2009).

Systemic fungicides bear a high risk of resistance development in the pathogen. Furthermore, *P. cubensis*

Table 6 Fungicides used against *P. cubensis* and several other oomycete pathogens (modified according to Urban and Lebeda 2006)

Level of systemicity	Cross-resistance group	Common name of compound	Type of activity/ translocation behavior within plants	Biochemical and physiological mode of action
Fully systemic	Phenylamides	Metalaxyl Mefenoxam Benalaxyl Ofurace	Preventive, curative, eradicative/apoplastic, symplastic, translaminar	Inhibition of rRNA synthesis
	Phosphonates	Phosetyl-Al	Preventative, curative/ apoplastic, symplastic	Inhibition of spore germination, retardation of mycelia development and sporulation, induction of host resistance
	Carbamates	Propamocarb Prothiocarb	Preventative, eradicative/ apoplastic	Multisite inhibitor, affect membrane permeability
	Cyano-acetamide oximes	Cymoxanil	Preventative, curative/ apoplastic, symplastic, translaminar	(?)
	Benzamide*	Fluopicolide	Preventative, curative/ apoplastic, symplastic, translaminar	Delocalisation of spectrin-like proteins
Partially systemic	Cinnamic acids	Dimethomorph	Preventative, curative, eradicative/mainly translaminar	Inhibits cell wall synthesis
	Complex III respiration inhibit. (QoI)	Azoxystrobin Fenamidone Trifloxystrobin Kresoxim-methyl Pyraclostrobin	Preventative/translaminar apoplastic (azoxystrobin, fenamid.), "mesostemic" (trifloxystrobin), "quasi-systemic" (kresoxim-methyl)	Inhibit mitochondrial respiration at the enzyme complex III (Qo site)
	Complex III respiration inhibit. (QiI)	Cyazofamid	Preventative, curative, eradicative/translaminar	Inhibit mitochondrial respiration at the enzyme complex III (Qi site)
	Amino acid amide carbamates	Iprovalicarb	Preventative, curative, eradicative/apoplastic, symplastic	Affect cell wall deposition (?)
Non-systemic	Dinitroanilines	Fluazinam	Preventative /-	Inhibit ATP production
	Miscellaneous	Zoxamide	Preventative /-	(?)
	Multisites inhibitors	Inorganic copper fungicides (Cu-oxychloride, Cu-hydroxide)	Preventative /-	Multisite inhibitor
		Organic dithiocarbamate fungicides (e.g. mancozeb)	Preventative /-	Multisite inhibitor
		Chlorothalonil	Preventative /-	Multisite inhibitor
		Folpet	Preventivní /-	Multisite inhibitor
		Other multisites		

(?) not well and/or exactly known

* Anonymous (2006)

is considered by FRAC as one of 10 plant pathogens accepted as showing a high risk of resistance development to fungicides (Pathogen risk list 2005, http://FRAC.info). The introduction of systemic fungicides evoked selection of strains resistant against that fungicide in field pathogen populations (Urban and Lebeda 2006). This occurred first with phenylamide-based products (e.g. Ridomil which contains metalaxyl) against which resistance was developed very shortly after their introduction to the market (Lebeda and Schwinn 1994). The first report of resistance in oomycetes against phenylamides was reported in Israel in 1979, just two years after the introduction of metalaxyl for the control of *P. cubensis* (Reuveni et al. 1980).

During the last 20 years, resistance of *P. cubensis* against other fungicides was also recorded (Table 7) (Urban and Lebeda 2006, 2007; Zhu et al. 2008; Okada and Furukawa 2008; Zhang et al. 2008; Olaya et al. 2009).

Biological control

There is only one report on biological control of downy mildew in cucurbits. Korbel (1990) showed that the treatment of cucumber seeds and spraying the leaves with the mycoparasitic fungus *Pythium oligandrum* (Veselý 1977) delayed the primary infection of the leaves and leaves remained viable longer (Korbel 1990). However, efficiency of protection was relatively low under high infection pressure and thus, this practice is of limited commercial application.

Extracts made from dry leaves of the perennial composite *Inula viscosa* were shown to be effective against several foliar fungal pathogens, including *P. cubensis* (Wang et al. 2004). These extracts are antifungal, inhibiting zoospore release and cystospore germination (Cohen, unpublished). Such extracts are in the process of registration for organic farming. Also the volatile antimicrobial substance allicin (diallylthiosulphinate) from garlic (*Allium sativum*), at concentrations 50–1000 $\mu g ml^{-1}$, reduced the severity of *P. cubensis* on cucumber by approximately 50–100% (Portz et al. 2008).

The non-protein amino acid BABA (β-aminibutyric acid) induces resistance against many diseases in various crops (Cohen 2002a) including *P. cubensis* in cucumber (Ovadia et al. 2000; Walz and Simon 2008). BABA has no direct effect on the pathogen. Rather, it activates host defense (Cohen 2002a; Walz and Simon 2008). BABA was shown to synergize with mancozeb in controlling *P. cubensis* in cucumber (Baider and Cohen 2003).

In the future, it seems that chemically and biologically mediated systemic resistance in cucurbits against *P. cubensis* and some other fungal pathogens (Anand et al. 2007) could be efficient part of Integrated Pest Management (IPM) which is currently implemented for plant disease control to reduce use of fungicides (Lebeda and Schwinn 1994; Urban and Lebeda 2006).

Table 7 Occurrence of strains resistant/tolerant to fungicides in *Pseudoperonospora cubensis* (modified according to Urban and Lebeda 2006; Olaya et al. 2009; Cohen unpublished)

Chemical group/ chemical class	Common name	Countries where resistant/ tolerant strains occurred
Phenylamides	Metalaxyl	Israel (1980)*
		Greece (1981)
		Italy (1985)
		USA (1987)
		USSR (1992)
		Australia (1995)
		Czech Republic (1990 /2004/)
Strobilurins	Azoxystrobin, Kresoxim-methyl, Pyraclostrobin	Japan (1999)
		Taiwan (2001)
		USA (2004)
Phosphonates	Fosetyl-Al	Israel (1984)
		Czech Republic (2004)
Carbamates	Propamocarb	Israel (1984)
Phthalimides	Folpet	Israel (1984)
Dithiocarbamates	Mancozeb	Israel (1984)
Carboxyic acid amides	Dimethomorph	Israel (2006)
	Mandipropamid Iprovalicarb	USA (2007)
	Benthiavalicarb	

* = the year of first described occurrence

Conclusions

Since the publication of the previous reviews (Palti and Cohen 1980; Cohen 1981; Lebeda 1990; Lebeda et al. 2006a, b) extensive knowledge has been gained on the biology and ecology of *P. cubensis*, the variability of interactions between *P. cubensis* and its hosts, sources of resistance and resistance mechanisms, breeding for resistance, and disaese control. One area that has not been well addressed for *P. cubensis* and very little is available in the literature (Sarris et al., 2009) is the genetic diversity of *P. cubensis* and to what extent our understanding of this part of the biology has shaped our effort to effectively manage cucurbit downy mildew. This should be considered as an area that research efforts needs to be concentrated.

Further research should be directed to the following areas:

1) Genome sequencing of *P. cubensis*;
2) Sexual reproduction and strategies of overwintering of *P. cubensis*;
3) Host range of *P. cubensis* on cultivated and wild species of Cucurbitaceae;

4) Development of an internationally-accepted system for pathotype and race determination and denomination;
5) Genetics of pathotype-specific and race-specific resistance/susceptibility;
6) Mechanisms of host resistance: histological, cytological, physiological, biochemical and molecular;
7) Discovery of effective sources of resistance, including their transfer to breeding materials;
8) Understanding the pathogenic variability and resistance to fungicides;
9) Comparative epidemiology and forecasting of downy mildew in cucurbits;
10) IPM of downy mildew in cucurbits including effective fungicides and alternative control measures.

Acknowledgements The authors thank G. J. Holmes (North Carolina, USA) for his valuable critiques of our report. Preparation of this review was supported by grants MSM 6198959215 and QH 71229. Some *P. cubensis* isolates used in this research are deposited in the Czech National Collection of Microorganisms (http://www.vurv.cz/collections/vurv.exe/) at Palacký University in Olomouc, Department of Botany (http://botany.upol.cz).

References

Ackermann, P. (1990). Výskyt a ochrana proti plísni okurkové v Československu (Occurrence and protection against cucurbit downy mildew in Czechoslovakia). In A. Lebeda (Ed.) *Plíseň okurková* (*Cucurbit downy mildew*) (pp. 51–61). Praha, Czechoslovakia: Československá vědecká společnost pro mykologii při ČSAV (Czechoslovak Scientific Society for Mycology by Czechoslovak Academy of Sciences).

Anand, T., Raguchander, T., Karthikeyan, G., Prakasam, V., & Samiyappan, R. (2007). Chemically and biologically mediated systemic resistance in cucumber (*Cucumis sativus* L.) against *Pseudoperonospora cubensis* and *Erysiphe cichoracearum*. *Phytopathologia Mediterranea, 46*, 259–271.

Angelov, D., Georgiev, P., & Krasteva, L. (2000). Two races of *Pseudoperonospora cubensis* on cucumbers in Bulgaria. *Acta Horticulturae, 510*, 81–83.

Anonymous (2006). Pflanzeschutz Nachrichten Bayer, 59, 2–3 (http://www.bayercropscience.com/BCSWeb/ CropProtection.nsf/id/EN1_Journal).

Baider, A., & Cohen, Y. (2003). Synergistic interaction between BABA and Mancozeb in controlling *Phytophthora infestans* in potato and tomato and *Pseudoperonospora cubensis* in cucumber. *Phytoparasitica, 31*, 399–409.

Bains, S. S., & Jhooty, J. S. (1978). Relationships between mineral nutrition of muskmelon and development of downy mildew caused by *Pseudoperonospora cubensis*. *Plant and Soil, 49*, 85–90.

Bains, S. S., & Sharma, N. K. (1986). Differential response of certain cucurbits to isolates of *Pseudoperonospora cubensis* and characteristics of identified races. *Phytophylactica, 18*, 31–33.

Balass, M., Cohen, Y., & Bar-Joseph, M. (1992). Identification of a constitutive 45 kD soluble protein associated with resistance to downy mildew in muskmelon (*Cucumis melo* L.) line PI 124111F. *Physiological and Molecular Plant Pathology, 41*, 387–396.

Balass, M., Cohen, Y., & Bar-Joseph, M. (1993). Temperature-dependent resistance to downy mildew in muskmelon: structural responses. *Physiological and Molecular Plant Pathology, 43*, 11–20.

Bedlan, G. (1989). Erstmaliger Nachweis von Oosporen von *Pseudoperonospora cubensis* (Berk. et Curt.) Rost. an Gewächshausgurken in Őesterreich. *Pflanzenschutzberichte, 3*, 119–120.

Benjamin, I., Kenigsbuch, D., Galperin, M., Abrameto, J. A., & Cohen, Y. (2009). Cisgenic melons over expressing glyoxylate-aminotransferase are resistant to downy mildew. *European Journal of Plant Pathology, 125*, 355–365.

Bjoern, G. K., & Kampmann, H. H. (2000). Screening field Riesenchäel cucumbers for resistance to downy mildew: management of interplot interference problems. *Acta Horticulturae, 510*, 77–80.

Bouwmeester, K., van Poppel, P. M. J. A., & Govers, F. (2009). Genome biology cracks enigmas of oomycete plant pathogens. *Annual Plant Reviews, 34*, 102–133. Oxford, UK: Wiley-Blackwell.

Caten, C. E. (1987). The concept of race in plant pathology. In M. S. Wolfe & C. E. Caten (Eds.), *Populations and plant pathogens: Their dynamics and genetics* (pp. 21–37). Oxford, UK: Blackwell Scientific Publications.

Choi, Y.-J., & Shin, H.-D. (2008). First record of downy mildews caused by *Pseudoperonospora cubensis* on bottle gourd in Korea. *Plant Pathology, 57*, 371.

Choi, Y.-J., Hong, S.-B., & Shin, H.-D. (2005). A re-consideration of *Pseudoperonospora cubensis* and *P. humuli* based on molecular and morphological data. *Mycological Research, 109*, 841–848.

Cohen, Y. (1976). Quantitation of resistance of cucumbers and cantaloups to *Pseudoperonospora cubensis*. *Phytoparasitica, 4*, 25–31.

Cohen, Y. (1977). The combined effects of temperature, leaf wetness, and inoculum concentration on infection of cucumbers with *Pseudoperonospora cubensis*. *Canadian Journal of Botany, 55*, 1478–1487.

Cohen, Y. (1981). Downy mildew of cucurbits. In D. M. Spencer (Ed.), *The downy mildews* (pp. 341–354). London: Academic.

Cohen, Y. (1992a). Genes and mechanisms of resistance againts fungal pathogens breeding lines of muskmelon. In R. W. Doruchowski, E. Kozik, & K. Niemirowicz-Suczytt (Eds.), *Proceedings of Fifth Eucarpia Cucurbitaceae Symposium* (pp. 139–141). Poland: Skierniewice.

Cohen, Y. (1992b). Determination of the sensitivity of *Pseudoperonospora cubensis* to phenylamides. *EPPO Bulletin, 22*, 318–320.

Cohen, Y. (2002a). β-aminobutyric acid-induced resistance against plant pathogens. (Feature Article). *Plant Disease, 86*, 448–457.

Cohen, Y. (2002b). Populations of *Phytophthora infestans* in Israel underwent three major genetic changes during 1983–2000. *Phytopathology, 92*, 300–307.

Cohen, Y., & Coffey, M. D. (1986). Systemic fungicides and the control of Oomycetes. *Annual Review of Phytopathology, 24*, 311–338.

Cohen, Y., & Eyal, H. (1977). Growth and differentiation of sporangia and sporangiophores of *Pseudoperonospora cubensis* on cucumber cotyledons under various combinations of light and temperature. *Physiological Plant Pathology, 10*, 93–103.

Cohen, Y., & Eyal, H. (1987). Downy mildew-, powdery mildew- and Fusarium wilt-resistant muskmelon reeding line PI-124111F. *Phytoparasitica, 15*, 187–195.

Cohen, Y., & Rotem, J. (1971a). Dispersal and viability of sporangia of *Pseudoperonospora cubensis*. *Transactions of British Mycological Society, 57*, 67–74.

Cohen, Y., & Rotem, J. (1971b). Field and growth chamber approach to epidemiology of *Pseudoperonospora cubensis* on cucumbers. *Phytopathology, 61*, 736–737.

Cohen, Y., Perl, M., & Rotem, J. (1971). The effect of darkness and moisture on sporulation of *Pseudoperonospora cubensis* in cucumbers. *Phytopathology, 61*, 594–595.

Cohen, Y., Eyal, H., Hanania, J., & Malik, Z. (1989). Ultrastructure of *Pseudoperonospora cubensis* in muskmelon genotypes susceptible and resistant to downy mildew. *Physiological and Molecular Plant Pathology, 34*, 27–40.

Cohen, Y., Baider, A., & Cohen, B. H. (1995). Dimethomorph activity against oomycete fungal pathogens. *Phytopathology, 85*, 1500–1506.

Cohen, Y., Meron, I., Mor, N., & Zuriel, S. (2003). A new pathotype of *Pseudoperonospora cubensis* causing downy mildew in cucurbits in Israel. *Phytoparasitica, 31*, 458–466.

Colucci, S. J. (2008). Host range, fungicide resistance and management of *Pseudoperonospora cubensis*, causal agent of cucurbit downy mildew. MS Thesis, North Carolina State University, Raleigh, USA

Constantinescu, O. (2000). The fine structure of the sporangium in *Pseudoperonospora humuli* (Chromista, Oomycota, Peronosporales). *Cryptogamic Mycology, 21*, 93–101.

Cooke, O. P. R., & Lees, A. K. (2004). Markers, old and new, for examining *Phytophthora infestans* diversity. *Plant Pathology, 53*, 692–704.

Crute, I. R. (1981). The host specificity of peronosporaceous fungi and the genetics of the relationship between host and parasite. In D. M. Spencer (Ed.), *The downy mildews* (pp. 45–56). London: Academic.

Crute, I. R. (1986). Investigations of gene-for-gene relationships: the need for genetic analyses of both host and parasite. *Plant Pathology, 35*, 15–17.

Day, B., & Hausbeck, M. (2009). Epidemiology of downy mildew: a regional and molecular approach. *Phytopathology, 99*, S172.

de Jong, M., & Van den Ackerveken, G. (2009). Fungal and oomycete biotrophy. *Annual Plant Reviews, 34*, 2009, 102–133. Oxford, UK: Wiley-Blackwell.

de Waard, M. A., Georgopoulos, S. G., Hollomon, D. W., Ishii, H., Leroux, P., Ragsdale, N. N., et al. (1993). Chemical control of plant diseases: Problems and prospects. *Annual Review of Phytopathology, 31*, 403–421.

Dick, M. W. (2001a). The peronosporomycetes. In D. J. McLaughlin, E. G. McLaughlin, & P. A. Lemke (Eds.), *The mycota VII Part A: Systematics and evolution* (pp. 39–72). Berlin: Springer-Verlag.

Dick, M. W. (2001b). *Straminipilous fungi: Systematics of the Peronosporomycetes, including accounts of the marine Straminipilous Protists, the Plasmodiophorids, and similar organisms*. Dordrecht: Kluwer Academic Publishers.

Dick, M. W. (2002a). Towards an understanding of the evolution of the downy mildews. In P. T. N. Spencer-Phillips, U. Gisi, & A. Lebeda (Eds.), *Advances in downy mildew research* (pp. 1–57). Dordrecht: Kluwer Academic Publishers.

Dick, M. W. (2002b). Binomials in the peronosporales, sclerosporales and pythiales. In P. T. N. Spencer-Phillips, U. Gisi, & A. Lebeda (Eds.), *Advances in downy mildew research* (pp. 225–265). Dordrecht: Kluwer Academic.

Dick, M. W. (2002c). The Peronosporomycetes and other flagellate fungi. In D. H. Howard (Ed.), *Fungi pathogenic for humans and animals: Second edition, revised and expanded* (pp. 17–66). Washington: American Society for Microbiology, Marcel Dekker, Inc.

Doruchowski, R. W., & Lakowska-Ryk, E. (2000). F_1 hybrid pickling cucumbers developed for increased yield, earliness and resistance to downy mildew. *Acta Horticulturae, 510*, 45–46.

Eckardt, N. A. (2004). Aminotransferases confer "Enzymatic resistance" to downy mildew in melon. *Plant Cell, 15*, 1–4.

Epinat, C., & Pitrat, M. (1994a). Inheritance of resistance to downy mildew (*Pseudoperonospora cubensis*) in muskmelon (*Cucumis melo*). I. Analysis of a 8×8 diallel table. *Agronomie, 14*, 239–248.

Epinat, C., & Pitrat, M. (1994b). Inheritance of resistance to downy mildew (*Pseudoperonospora cubensis*) in muskmelon (*Cucumis melo*). II. Generation means analysis of 5 genitors. *Agronomie, 14*, 249–257.

Ezuka, A., & Komada, H. (1974). Varietal difference in resistance of cucumber to downy mildew. *Bulletin Tokai-Kinki National Agricultural Experiment Station, 27*, 42–45.

Fellner, M., Binarová, P., & Lebeda, A. (1996). Isolation and fusion of *Cucumis sativus* and *Cucumis melo* protoplasts. In M. L. Gómez-Guillamón, C. Soria, J. Cuartero, J. A. Torés, & R. Fernández-Muňoz (Eds.), *Cucurbits towards 2000. Proceedings of the VIth Eucarpia Meeting on Cucurbit Genetics and Breeding* (pp. 202–209). May 28–30, 1996, Malaga, Spain.

Ferriol, M., & Picó, B. (2008). Pumpkin and winter squash. In J. Prohens & F. Nuez (Eds.), *Vegetables I. Asteraceae, brassicaceae, chenopodiaceae, and cucurbitaceae* (pp. 317–349). New York: Springer.

Forsberg, A. S. (1986). Downy mildew-*Pseudoperonospora cubensis* in Swedish cucumber fields. *Växtskyddsnotiser, 50*, 17–19.

Gajdová, J., Lebeda, A., & Navrátilová, B. (2004). Protoplast cultures of *Cucumis* and *Cucurbita* spp. In A. Lebeda, & H. S. Paris (Eds.) *Progress in cucurbit genetics and breeding research*. Proceedings of Cucurbitaceae 2004, the 8th EUCARPIA Meeting on Cucurbit Genetics and Breeding (pp. 441–454). Olomouc, Czech Republic: Palacký University in Olomouc.

Gent, D. H., Mitchell, M. N., & Holmes, G. J. (2009). Genetic and pathogenic relatedness of *Pseudoperonospora cubensis* and *P. humuli*: Implications for detection and management. *Phytopathology, 99*, S171.

Gisi, U. (2002). Chemical control of downy mildews. In P. T. N. Spencer-Phillips, U. Gisi, & A. Lebeda (Eds.), *Advances in downy mildew research* ((pp, Vol. 1, pp. 119–159). Dordrecht: Kluwer Academic Publishers.

Gisi, U., & Sierotzki, H. (2008). Fungicide modes of action and resistance in downy mildews. *European Journal of Plant Pathology, 122*, 157–167.

Göker, M., Voglmayr, H., Riethmüller, A., Weiß, M., & Oberwinkler, F. (2003). Taxonomic aspects of Peronosporaceae inferred from Bayesian molecular phylogenetics. *Canadian Journal of Botany, 81*, 672–683.

Göker, M., Voglmayr, H., Riethmüller, A., & Oberwinkler, F. (2007). How do obligate parasites evolve? A multi-gene phylogenetic analysis of downy mildews. *Fungal Genetics and Biology, 44*, 105–122.

Hardham, A. R. (2007). Cell biology of plant-oomycete interactions. *Cellular Microbiology, 9*, 31–39.

Holliday, P. (2001). *A dictionary of plant pathology* (2nd ed.). Cambridge: Cambridge University Press.

Holmes, G. J., & Ojiambo, P. (2009). Chemical control of cucurbit downy mildew: a summary of field experiments in the U.S. *Phytopathology, 99*, S171.

Holmes, G. J., & Thomas, C. (2009). The history and re-emergence of cucurbit downy mildew. *Phytopathology, 99*, S171.

Holmes, G. J., Main, C. E., & Keever, Z. T., III. (2004). Cucurbit downy mildew: a unique pathosystem for disease forecasting. In P. T. N. Spencer-Phillips & M. Jeger (Eds.), *Advances in downy mildew research, vol. 2* (pp. 69–80). Dordrecht: Kluwer Academic Publishers.

Holmes, G., Wehner, T., & Thornton, A. (2006). An old enemy re-emerges: downy mildew rears its ugly head on cucumber, impacting growers up and down the Eastern U.S. *American Vegetable Grower*, February, 14–15.

Holub, E. B. (2008). Natural history of *Arabidopsis thaliana* and oomycete symbioses. *European Journal of Plant Pathology, 122*, 91–109.

Horejsi, T., Staub, J. E., & Thomas, C. (2000). Linkage of random amplified polymorphic DNA markers to downy mildew resistance in cucumber (*Cucumis sativus* L.). *Euphytica, 115*, 105–113.

Huang, S., et al. (2009). The genome of the cucumber, *Cucumis sativus* L. *Nature Genetics, 41*, 1275–1281.

Hughes, M. B., & van Halteren, F. (1952). Two biological forms of *Pseudoperonospora cubensis*. *Plant Disease Reporter, 36*, 365–367.

Inaba, T., Morinaka, T., & Hamaya, E. (1986). Physiological races of *Pseudoperonospora cubensis* isolated from cucumber and muskmelon in Japan. *Bulletin National Institute of Agro-Environmental Science, 2*, 35–43.

Ivanoff, S. C. (1944). Resistance of cantaloupes to downy mildew and the melon aphid. *Journal of Heredity, 35*, 35–39.

Iwata, Y. (1941). Specialisation in *Pseudoperonospora cubensis* (Berk. Et Curt.) Rostow. I. Comparative studies on the pathogenicities on the fungi from *Cucumis sativus* L. and *Cucurbita moschata* Duch. *Annales of Phytopathological Society of Japan, 11*, 101–113.

Jamadar, M. M., & Desai, S. A. (1999). Reaction of ridgegourd local cultivars against downy mildew caused by *Pseudoperonospora cubensis* (Berk. et Curt.) Rostow. *Karnataka Journal of Agricultural Sciences, 12*, 204–205.

Jeger, M. J., & Pautasso, M. (2008). Comparative epidemiology of zoosporic plant pathogens. *European Journal of Plant Pathology, 122*, 111–126.

Kanetis, L., Holmes, G. J., & Ojiambo, P. S. (2010). Survival of *Pseudoperonospora cubensis* sporangia exposed to solar radiation. *Plant Pathology, 59*, 313–323.

Kenigsbuch, D., & Cohen, Y. (1989). Inheritance of resistance to downy mildew in a gynoecious muskmelon. *Plant Disease, 73*, 994–996.

Kenigsbuch, D., & Cohen, Y. (1992a). Inheritance of resistance to downy mildew in *Cucumis melo* PI 124112 and commonality of resistance genes with PI 124111F. *Plant Disease, 76*, 615–617.

Kenigsbuch, D., & Cohen, Y. (1992b). Inheritance and allelism of genes for resistance against races 1 and 2 of *Sphaerotheca fuliginea* in muskmelons. *Plant Disease, 76*, 626–629.

Kirkbride, J. H., Jr. (1993). *Biosystematic monograph of the genus Cucumis (Cucurbitaceae)*. Boone: Parkway.

Ko, Y., Chen, C. Y., Liu, C. W., Chen, S. S., Maruthasalam, S., & Lin, C. H. (2008). First report of downy mildew by *Pseudoperonospora cubensis* on chayote (*Sechium edule*) in Taiwan. *Plant Disease, 92*, 1706.

Korbel, E. (1990). Možnost použití biologické ochrany proti plísni okurkové (Possibility of biological control utilisation against cucurbit downy mildew). In A. Lebeda (Ed.), *Plíseň okurková (Cucurbit downy mildew)* (pp. 62–64). Praha, Czechoslovakia: Československá vědecká společnost pro mykologii při ČSAV (Czechoslovak Scientific Society for Mycology by Czechoslovak Academy of Sciences).

Kranz, J. (2003). *Comparative epidemiology of plant diseases*. Berlin: Springer.

Lange, L., Eden, U., & Olson, L. W. (1989a). Zoosporogenesis in *Pseudoperonospora cubensis*. The causal agent of cucurbit downy mildew. *Nordic Journal of Botany, 8*, 497–504.

Lange, L., Eden, U., & Olson, L. W. (1989b). Internal mycelium of *Pseudoperonospora cubensis*. The causal agent of cucurbit downy mildew. *Nordic Journal of Botany, 8*, 505–510.

Lange, L., Eden, U., & Olson, L. W. (1989c). The zoospore of *Pseudoperonospora cubensis*. The causal agent of cucurbit downy mildew. *Nordic Journal of Botany, 8*, 511–516.

Lebeda, A. (1986a). Epidemic occurrence of *Pseudoperonospora cubensis* in Czechoslovakia. *Temperate Downy Mildews Newsletter, 4*, 15–17.

Lebeda, A. (1986b). *Pseudoperonospora cubensis*. In A. Lebeda (Ed.) *Metody testování rezistence zelenin vůči rostlinným patogenům (Methods of vegetable resistance screening against pathogens)* (pp. 81–85). Olomouc, Czechoslovakia: VHJ Sempra, Research and Breeding Institute for Vegetable Crops.

Lebeda, A. (1990). Biologie a ekologie plísně okurkové (Biology and ecology of cucurbit downy mildew). In A. Lebeda (Ed.) *Plíseň okurková (Cucurbit downy mildew)* (pp. 13–45). Praha, Czechoslovakia: Československá vědecká společnost pro mykologii při ČSAV (Czechoslovak Scientific Society for Mycology by Czechoslovak Academy of Sciences).

Lebeda, A. (1991). Resistance in muskmelons to Czechoslovak isolates of *Pseudoperonospora cubensis* from cucumbers. *Scientia Horticulturae, 45*, 255–260.

Lebeda, A. (1992a). Susceptibility of accessions of *Cucumis sativus* to *Pseudoperonospora cubensis*. *Tests of Agrochemicals and Cultivars No. 13 (Annals of Applied Biology, 120 Suppl.)*, 102–103.

Lebeda, A. (1992b). Screening of wild *Cucumis* species against downy mildew (*Pseudoperonospora cubensis*) isolates from cucumbers. *Phytoparasitica, 20*, 203–210.

Lebeda, A. (1999). *Pseudoperonospora cubensis* on *Cucumis* spp. and *Cucurbita* spp.–resistance breeding aspects. *Acta Horticulturae, 492*, 363–370.

Lebeda, A., & Jendrulek, T. (1988). Application of methods of multivariate analysis in comparative epidemiology and research into field resistance. *Journal of Plant Diseases and Protection, 95*, 495–505.

Lebeda, A., & Reinink, K. (1991). Variation in the early development of *Bremia lactucae* on lettuce cultivars with different levels of field resistance. *Plant Pathology, 40*, 232–237.

Lebeda, A., & Křístková, E. (1992). Differences in resistance to *Pseudoperonospora cubensis* in *Cucurbita pepo*. In R. W. Doruchowski, E. Kozik, & K. Niemirowicz-Szczytt (Eds.) *Proceedings of Fifth Eucarpia Cucurbitaceae Symposium* (pp. 175–179). July 27–31, 1992, Skierniewice, Poland: Research Institute for Vegetable Crops

Lebeda, A., & Křístková, E. (1993). Resistance of *Cucurbita pepo* and *Cucurbita moschata* varieties to cucurbit downy mildew (*Pseudoperonospora cubensis*). *Plant Varieties and Seeds, 6*, 109–114.

Lebeda, A., & Prášil, J. (1994). Susceptibility of *Cucumis sativus* cultivars to *Pseudoperonospora cubensis*. *Acta Phytopathologica et Entomologica Hungarica, 29*, 89–94.

Lebeda, A., & Schwinn, F. J. (1994). The downy mildews—an overview of recent research progress. *Journal of Plant Diseases and Protection, 101*, 225–254.

Lebeda, A., & Doležal, K. (1995). Peroxidase isozyme polymorphism as a potential marker for detection of field resistance in *Cucumis sativus* to cucumber downy mildew (*Pseudoperonospora cubensis* (Berk. et Curt.) Rostov.). *Journal of Plant Diseases and Protection, 102*, 467–471.

Lebeda, A., & Křístková, E. (2000). Interactions between morphotypes of *Cucurbita pepo* and obligate biotrophs (*Pseudoperonospora cubensis*, *Erysiphe cichoracearum* and *Sphaerotheca fuliginea*). *Acta Horticulturae, 510*, 219–225.

Lebeda, A., & Gadasová, V. (2002). Pathogenic variation of *Pseudoperonospora cubensis* in the Czech Republic and some other european countries. *Acta Horticulturae, 588*, 137–141.

Lebeda, A., & Widrlechner, M. P. (2003). A set of Cucurbitaceae taxa for differentiation of *P. cubensis* pathotypes. *Journal of Plant Diseases and Protection, 110*, 337–349.

Lebeda, A., & Urban, J. (2004a). Disease impact and pathogenicity variation in Czech populations of *Pseudoperonospora cubensis*. In A. Lebeda, & H. S. Paris (Eds.), *Progress in cucurbit genetics and breeding research. Proceedings of Cucurbitaceae 2004, the 8th EUCARPIA Meeting on Cucurbit Genetics and Breeding* (pp. 267–273). Olomouc, Czech Republic: Palacký University in Olomouc.

Lebeda, A., & Urban, J. (2004b). Distribution, harmfulness and pathogenic variability of cucurbit downy mildew in the Czech Republic. *Acta Fytotechnica et Zootechnica, 7*, 170–173.

Lebeda, A., & Widrlechner, M. P. (2004). Response of wild and weedy *Cucurbita* L. to pathotypes of *Pseudoperonospora cubensis* (Berk. et Curt.) Rostov. (Cucurbit downy mildew). In P. Spencer-Phillips & M. Jeger (Eds.), *Advances in downy mildew research, Vol. 2* (pp. 203–210). Dordrecht: Kluwer Academic Publishers.

Lebeda, A., & Urban, J. (2005). Ochrana okurek a dalších tykvovitých zelenin vůči plísni okurkové (Protection of cucumbers and other cucurbitaceous vegetables to cucurbit downy mildew). In F. Kocourek et al. (Eds.) Metodika pro integrovaný systém ochrany polní zeleniny vůči škodlivým organismům (Methods for integrated system of field vegetables protection to dangerous organisms) (pp. 24–38). Olomouc, Czech Republic: Zelinářská unie Čech a Moravy, Olomouc (Vegetables Grower Union of Bohemia and Moravia).

Lebeda, A., & Urban, J. (2007). Temporal changes in pathogenicity and fungicide resistance in *Pseudoperonospora cubensis* populations. *Acta Horticulturae, 731*, 327–336.

Lebeda, A., Křístková, E., & Kubaláková, M. (1996). Interspecific hybridization of *Cucumis sativus* × *Cucumis melo* as potential way to transfer resistance to *Pseudoperonospora cubensis*. In M. L. Gómez-Guillamón, C. Soria, J. Cuartero, J. A. Torés, & R. Fernández-Muňoz (Eds.) *Cucurbits towards 2000. Proceedings of the VIth Eucarpia Meeting on Cucurbit Genetics and Breeding* (pp. 31–37), May 28–30, 1996, Malaga, Spain.

Lebeda, A., Kubaláková, M., Křístková, E., Navrátilová, B., Doležal, K., Doležel, J., et al. (1999). Morphological and physiological characteristics of plants issued from an interspecific hybridization of *Cucumis sativus* × *Cucumis melo*. *Acta Horticulturae, 492*, 149–155.

Lebeda, A., Luhová, L., Sedlářová, M., & Jančová, D. (2001a). The role of enzymes in plant—fungal pathogens interactions. *Journal of Plant Diseases and Protection, 108*, 89–111.

Lebeda, A., Pink, D. A. C., & Mieslerová, B. (2001b). Host-parasite specificity and defense variability in the *Lactuca* spp.-*Bremia lactucae* pathosystem. *Journal of Plant Pathology, 83*, 25–35.

Lebeda, A., Pink, D. A. C., & Astley, D. (2002). Aspects of the interactions between wild *Lactuca* spp. and related genera and lettuce downy mildew (*Bremia lactucae*). In P. T. N. Spencer-Phillips, U. Gisi, & A. Lebeda (Eds.), *Advances in downy mildew research* (pp. 85–118). Dordrecht: Kluwer Academic Publishers.

Lebeda, A., Štěpánková, J., & Urban, J. (2006a). Plíseň okurky (*Pseudoperonospora cubensis*)—taxonomie, biologie, ekologie, interakce hostitel-patogen a možnosti ochrany (Cucurbit downy mildew (*Pseudoperonospora cubensis*)—taxonomy, biology, ecology, host-pathogen interactions and possibilities of protection). In A. Lebeda, J. Mazáková, & V. Táborský (Eds.) *Protozoa a Chromista. Taxonomie, biologie a hospodářský význam* (*Protozoa and Chromista. Taxonomy, biology and economical impact*) (pp. 47–78). Praha, Czech Republic: Česká fytopatologická společnost (Czech Phytopathological Society).

Lebeda, A., Widrlechner, M. P., & Urban, J. (2006b). Individual and population aspects of interactions between cucurbits and *Pseudoperonospora cubensis*: pathotypes and races. In G. J. Holmes (Ed.), *Proceedings of Cucurbitaceae 2006* (pp. 453–467). Raleigh: Universal.

Lebeda, A., Štěpánková, J., Kršková, M., & Widrlechner, M.P. (2007a) Resistance in *Cucumis melo* germplasm to *Pseudoperonospora cubensis* pathotypes. In A. Lebeda, & P. T. N. Spencer-Phillips (Eds.) *Advances in Downy Mildew Research, Vol. 3. Proceedings of The 2nd International Downy Mildews Symposium* (pp. 157–167.). Olomouc and Kostelec na Hané, Czech Republic: Palacký University in Olomouc and JOLA, v.o.s.

Lebeda, A., Widrlechner, M. P., Staub, J., Ezura, H., Zalapa, J., & Křístková, E. (2007b). Cucurbits (Cucurbitaceae; *Cucumis* spp., *Cucurbita* spp., *Citrullus* spp.), Chapter 8. In R. Singh (Ed.), *Genetic resources, chromosome engineering, and crop improvement series, volume 3—vegetable crops* (pp. 273–377). Boca Raton: CRC.

Lebeda, A., Petrželová, I., & Maryška, Z. (2008a). Structure and variation in the wild-plant pathosystem: *Lactuca serriola-Bremia lactucae*. *European Journal of Plant Pathology, 122*, 127–146.

Lebeda, A., Sedlářová, M., Petřivalský, M., & Prokopová, J. (2008b). Diversity of defence mechanisms in plant-oomycete interactions: a case study of *Lactuca* spp. and *Bremia lactucae*. *European Journal of Plant Pathology, 122*, 71–89.

Limpert, E., Clifford, B., Dreiseitl, A., Johnson, R., Müller, K., Roelfs, A., et al. (1994). Systems of designation of pathotypes of plant pathogens. *Journal of Phytopathology, 140*, 359–362.

Lindenthal, M., Steiner, U., Dehne, H. W., & Oerke, E. C. (2005). Effect of downy mildew development on transpiration of cucumber leaves visualized by digital infrared thermography. *Phytopathology, 95*, 233–240.

Mahrisi, R. P., & Siradhana, B. S. (1984). On the occurrence of oospores of *Pseudoperonospora cubensis* in Rajasthan, India. *Indian Phytopathology, 37*, 323–325.

Mahrisi, R. P., & Siradhana, B. S. (1988). Effect of nutrition on downy mildew disease caused by *Pseudoperonospora cubensis* (Berk. and Curt.) Rostow. on muskmelon. *Annals of Arid Zone, 27*, 153–155.

Main, C. E., Keever, T., Holmes, G. V., & Davis, J. M. (2001). Forecasting long-range transport of downy mildew spores and plant disease epidemics. APS net Feature Story April 25 through May 31. http://www.apsnet.org/ online/future/forecast/

Mauch-Mani, B. (2002). Host resistance to downy mildew diseases. In P. T. N. Spencer-Phillips, U. Gisi, & A. Lebeda (Eds.), *Advances in downy mildew research* (pp. 59–83). Dordrecht: Kluwer Academic.

Medvedeva, N. I., & Medvedev, A. V. (1983). Agronomic and biological assessment of cucumber varieties with potential for breeding for resistance to downy mildew. *Trudy po Prikladnoj Botanike, Genetike i Selekcii, 77*, 25–28.

Michelmore, R. W. (1981). Sexual and asexual sporulation in the downy mildews. In D. M. Spencer (Ed.), *The downy mildews* (pp. 165–181). London: Academic.

Michelmore, R., & Wong, J. (2008). Classical and molecular genetics of *Bremia lactucae*, cause of lettuce downy mildew. *European Journal of Plant Pathology, 122*, 19–30.

Mitchell, M. N., Ocamb, C., & Gent, D. (2009). Addressing the relationship between *Pseudoperonospora cubensis* and *P. humuli* by multigenic characterization and host specificity. *Phytopathology, 99*, S87.

Moravec, J., Lebeda, A., & Křístková, E. (2004). History of growing and breeding of cucurbitaceaous vegetables in Czech Lands. In A. Lebeda, & H. S. Paris (Eds.), *Progress in cucurbit genetics and breeding research. Proceedings of Cucurbitaceae 2004, the 8th EUCARPIA Meeting on Cucurbit Genetics and Breeding* (pp. 21–38). Olomouc, Czech Republic: Palacký University in Olomouc.

Nusbaum, C. J. (1944). The seasonal spread and development of cucurbit downy mildew in the Atlantic coastal states. *Plant Disease, 28*, 82–85.

Nusbaum, C. J. (1948). A summary of cucurbit downy mildew reports from Atlantic coastal states in 1947. *Plant Disease, 32*, 44–48.

Oerke, E. C., Steiner, U., Dehne, H. W., & Lindenthal, M. (2006). Thermal imaging of cucumber leaves affected by downy mildew and environmental conditions. *Journal of Experimental Botany, 57*, 2121–2132.

Ojiambo, P., Kanetis, L., & Holmes, G. (2009). Forecasting long distance movement of *Pseudoperonospora cubensis* and the Cucurbit ipmPIPE. *Phytopathology, 99*, S171.

Okada, K., & Furukawa, M. (2008). Occurrence and countermeasure of fungicide-resistant pathogens in vegetable field of Osaka prefecture. *Journal of Pesticide Science, 33*, 326–329.

Olaya, G., Kuhn, P., Hert, A., Holmes, G., & Colucci, S. (2009). Fungicide resistance in cucurbit downy mildew. *Phytopathology, 99*, S171.

Ovadia, A., Biton, R., & Cohen, Y. (2000). Induced resistance to downy mildew and Fusarium wilt in cucurbits. In N. Katzir, & H. S. Paris (Eds.) "*Proceedings of 7th EUCARPIA Meeting on Cucurbit Genetics and Breeding*". *Acta Horticulture, 510*, 55–59.

Palti, J., & Cohen, Y. (1980). Downy mildew of cucurbits (*Pseudoperonospora cubensis*): The fungus and its hosts, distribution, epidemiology and control. *Phytoparasitica, 8*, 109–147.

Paris, H. S. (2008). Summer squash. In J. Prohens & F. Nuez (Eds.), *Vegetables I. Asteraceae, brassicaceae, chenopodiaceae, and cucurbitaceae* (pp. 351–379). New York: Springer.

Perchepied, L., Bardin, M., Dogimont, C., & Pitrat, M. (2005). Relationship between loci conferring downy mildew and powdery mildew resistance in melon assessed by quantitative trait loci mapping. *Phytopathology, 95*, 556–565.

Perl, M., Cohen, Y., & Rotem, J. (1972). The effect of humidity during darkness on the transfer of assimilates from cucumber leaves to sporangia of *Pseudoperonospora cubensis*. *Physiological Plant Pathology, 2*, 113–122.

Peterson, C. E. (1975). Plant introduction in the improvement of vegetable cultivars. *HortScience, 10*, 575–579.

Petrov, L., Boogert, K., Sheck, L., Baider, A., Rubin, E., & Cohen, Y. (2000). Resistance to downy mildew, *Pseudoperonospora cubensis,* in cucumbers. *Acta Horticulturae, 510*, 203–209.

Pitrat, M. (1990). Gene list for *Cucumis melo* L. *Cucurbit Genetics Cooperative, 13*, 58–68.

Pitrat, M. (2008). Melon. In J. Prohens & F. Nuez (Eds.), *Vegetables I. Asteraceae, brassicaceae, chenopodiaceae, and cucurbitaceae* (pp. 283–315). New York: Springer.

Pivovarov, V. F. (1984). Breeding cucumber for disease resistance using different ecological conditions. *Selekcia i Semenovodstvo USSR, 10*, 20–22 (in Russian).

Pivovarov, V. F., & Kudelich, V. S. (1985). Using the VIR collection in breeding cucumber for downy mildew resistance in the Republic of Cuba. *Trudy po Prikladnoj Botanike, Genetike i Selekcii, 92*, 97–102.

Portz, D., Koch, E., & Slusarenko, A. J. (2008). Effects of garlic (*Allium sativum*) juice containing allicin on *Phytophthora infestans* and downy mildew of cucumber caused by *Pseudoperonospora cubensis*. *European Journal of Plant Pathology, 122*, 197–206.

Ren, Y., et al. (2009). An integrated genetic and cytogenetic map of the cucumber genome. *PloS ONE, 4*, e5795.

Reuveni, M., Eyal, H., & Cohen, Y. (1980). Development of resistance to metalaxyl in *Pseudoperonospora cubensis*. *Plant Disease, 64*, 1108–1109.

Reuveni, R., Shimoni, M., & Karchi, Z. (1990). A rapid assay for monitoring peroxidase-activity in melon as a marker for resistance to *Pseudoperonospora cubensis*. *Journal of Phytopathology, 129*, 333–338.

Riethmüller, A., Voglmayr, A., Göker, M., Weiß, M., & Oberwinkler, F. (2002). Phylogenetic relationships of the downy mildews (Peronosporales) and related groups based on nuclear large subunit ribosomal DNA sequences. *Mycologia, 94*, 834–849.

Robinson, R. W., & Decker-Walters, D. S. (1997). *Cucurbits*. Wallingford: CAB International.

Rod, J. (1990). Principy chemické ochrany (Principles of chemical control). In A. Lebeda (Ed.) *Plíseň okurková (Cucurbit downy mildew)* (pp. 46–50). Praha, Czechoslovakia: Československá vědecká společnost pro mykologii při ČSAV.

Royle, D. J., & Kremheller, H Th. (1981). Downy mildew of the hop. In D. M. Spencer (Ed.), *The downy mildews* (pp. 395–419). London: Academic.

Rubatzky, V. E., & Yamaguchi, M. (1997). *World vegetables. Principles, production and nutritive values* (2nd ed.). New York: Chapman & Hall.

Runge, F., & Thines, M. (2009). A potential perennial host for *Pseudoperonospora cubensis* in temperate regions. *European Journal of Plant Pathology, 123*, 483–486.

Salti, M., Wong, M. Y., Sariah, M., & Masdek, H. N. (2010). First report of *Pseudoperonospora cubensis* causing downy mildew of *Trichosanthes cucumerina* in Malaysia. *Plant Disease, 94*, 642.

Sarris, P.F., Abdelhalim, M., Kitner, M., Skandalis, N., Panopoulos, N.J., Doulis, A.G., & Lebeda, A. (2009). Molecular polymorphism between populations of *Pseudoperonospora cubensis* from Greece and the Czech Republic and their phytopathological and phylogenetic implications. *Plant Pathology* (doi:10.1111/j.1365-3059.2009.02093.x)

Savory, E. A., Tian, M., Erhardt, C., Hausbeck, M., Hammerschmidt, R., & Day, B. (2008). An integrative approach to characterizing the cucumber-*Pseudoperonospora cubensis* interaction. *Phytopathology, 98 (June, Supplement)*, S140.

Shashikumar, K. T., Pitchaimuthu, M., & Rawal, R. D. (2010). Generation mean analysis of resistance to downy mildew in adult muskmelon plants. *Euphytica, 173*, 121–127.

Shetty, N. V., Wehner, T. C., Thomas, C. E., Doruchowski, R. W., & Shetty, K. P. W. (2002). Evidence for downy mildew races in cucumber tested in Asia, Europe, and North America. *Scientia Horticulturae, 94*, 231–239.

Shtienberg, D., Elad, Y., Bornstein, M., Ziv, G., Grava, A., & Cohen, S. (2010). Polyethylene mulch modifies greenhouse microclimate and reduces infection of *Phytophthora infestans* in tomato and *Pseudoperonospora cubensis* in cucumber. *Phytopathology, 100*, 97–104.

Singh, P. P., & Sokhi, S. S. (1989). 1st report of occurrence of oospores of *Pseudoperonospora cubensis* on 2 cucurbitaceous hosts. *Current Science, 58*, 1330–1331.

Singh, R. D., & Singh, J. P. (1998). Improvement and cultivation: *Lagenaria* and *Luffa*. In N. M. Naynar & T. A. More (Eds.), *Cucurbits* (pp. 199–204). Enfield: Science Publishers.

Sitterly, W. R. (1972). Breeding of disease resistance in cucurbits. *Annual Review of Phytopathology, 10*, 471–490.

Skalický, V. (1961). Plíseň okurková-*Peronoplasmopara cubensis*. In J. Benada, & J. Špaček (Eds.) *Zemědělská fytopatologie, díl III-Choroby zeleniny* (*Agricultural phytopathology, vol. III–diseases of vegetable crops*) (pp. 390–393). Praha, Czechoslovakia: Státní zemědělské nakladatelství.

Skalický, V. (1966). Taxonomie der Gattungen der Familie Peronosporaceae. *Preslia, 38*, 117–129.

Skálová, D., Lebeda, A., & Navrátilová, B. (2004). Embryo and ovule cultures in *Cucumis* species and their utilization in interspecific hybridization. In A. Lebeda, & H. S. Paris (Eds.), *Progress in cucurbit genetics and breeding research*. Proceedings of Cucurbitaceae 2004, the 8th EUCARPIA Meeting on Cucurbit Genetics and Breeding (pp. 415–430). Olomouc, Czech Republic: Palacký University in Olomouc.

Skidmore, D. I., & Ingram, D. S. (1985). Conidial morphology and the specialization of *Bremia lactucae* Regel (Peronosporaceae) on hosts in the family Compositae. *Botanical Journal of the Linnean Society, 91*, 503–522.

Slusarenko, A. J., & Schlaich, N. L. (2003). Downy mildew of *Arabidopsis thaliana* caused by *Hyaloperonospora parasitica* (formerly *Peronospora parasitica*). *Molecular Plant Pathology, 4*, 159–170.

Sowell, G., & Corley, W. L. (1974). PI 321005 (Tainan 2), a high quality source of resistance to three cantaloupe diseases. *Plant Disease Reporter, 58*, 899–902.

Staub, J. E., Robbins, M. D., & Wehner, T. C. (2008). Cucumber. In J. Prohens & F. Nuez (Eds.), *Vegetables I. Asteraceae, brassicaceae, chenopodiaceae, and cucurbitaceae* (pp. 241–282). New York: Springer.

Švábová, L., & Lebeda, A. (2005). *In vitro* selection for improved plant resistance to toxin-producing pathogens. *Journal of Phytopathology, 153*, 52–64.

Tahvonen, R. (1985). Downy mildew of cucurbits found for the first time in Finland. *Växtskyddsnotiser, 49*, 42–44.

Taler, D., Galperin, M., Benjamin, I., Cohen, Y., & Kenigsbuch, D. (2004). Plant *eR* genes that encode photorespiratory enzymes confer resistance against disease. *Plant Cell, 16*, 172–184.

Thakur, R. P., & Mathur, K. (2002). Downy mildews of India. *Crop Protection, 21*, 333–345.

Thomas, C. E. (1970). Sporulation of *Pseudoperonospora cubensis* on watermelon. *Plant Disease Reporter, 54*, 108–111.

Thomas, C. E. (1982). Resistance to downy mildew in *Cucumis melo* plant introductions and American cultivars. *Plant Disease, 66*, 500–502.

Thomas, C. E. (1986). Downy and powdery mildew resistant muskmelon breeding line MR-1. *HortScience, 21*, 329.

Thomas, C. E. (1996). Downy mildew. In T. A. Zitter, D. L. Hopkins, & C. E. Thomas (Eds.), *Compendium of cucurbit diseases* (pp. 25–27). St. Paul: American Phytopathological Society Press.

Thomas, C. E., Cohen, Y., Jourdain, E. L., & Eyal, H. (1987a). Use of reaction types to identify downy mildew resistance in muskmelons. *HortScience, 22*, 638–640.

Thomas, C. E., Cohen, Y., Mc Creight, Y. D., Jourdion, E. L., & Cohen, S. (1987b). Inheritance of resistance to downy mildew in *Cucumis melo*. *Plant Disease, 72*, 33–35.

Thomas, C. E., Inaba, T., & Cohen, Y. (1987c). Physiological specialization in *Pseudoperonospora cubensis*. *Phytopathology, 77*, 1621–1624.

Tomlin, C. D. S. (2003). *The pesticide manual* (thirteenthth ed.). Alton: British Crop Protection Council.

Urban, J., & Lebeda, A. (2004). Differential sensitivity to fungicides in Czech populations of *Pseudoperonospora cubensis*. In A. Lebeda, & H. S. Paris (Eds.) *Progress in cucurbit genetics and breeding research*. Proceedings of Cucurbitaceae 2004, the 8th EUCARPIA Meeting on Cucurbit Genetics and Breeding (pp. 275–280). Olomouc, Czech Republic: Palacký University in Olomouc.

Urban, J., & Lebeda, A. (2006). Fungicide resistance in cucurbit downy mildew—methodological, biological and population aspects. *Annals of Applied Biology, 149*, 63–75.

Urban, J., & Lebeda, A. (2007). Variation for fungicide resistance in Czech populations of *Pseudoperonospora cubensis*. *Journal of Phytopathology, 155*, 143–151.

Veselý, D. (1977). Potential biological control of damping-off pathogens in emerging sugar beet by *Pythium oligandrum* Drechsler. *Phytopathologische Zeitschrift, 94*, 113–115.

Voglmayr, H. (2003). Phylogenetic study of *Peronospora* and related genera based on nuclear ribosomal ITS sequences. *Mycological Research, 107*, 1132–1142.

Voglmayr, H. (2008). Progress and challenges in systematics of downy mildews and white blister rusts: new insights from genes and morphology. *European Journal of Plant Pathology, 122*, 3–18.

Voglmayr, H., & Greilhuber, J. (1998). Genome size determination in Peronosporales (Oomycota) by Feulgen Image Analysis. *Fungal Genetics and Biology, 25*, 181–195.

Voglmayr, H., Riethmüller, A., Göker, M., Weiß, M., & Oberwinkler, F. (2004). Phylogenetic relationships of *Plasmopara*, *Bremia* and other genera of downy mildews with pyriform haustoria based on Bayesan analysis of partial LSU rDNA sequence data. *Mycological Research, 108*, 1011–1024.

Walz, A., & Simon, O. (2008). β-aminobutyric acid-induced resistance in cucumber against biotrophic and necrotrophic pathogens. *Journal of Phytopathology, 157*, 356–361.

Wang, W., Ben-Daniel, B. H., & Cohen, Y. (2004). Control of plant diseases by extracts of *Inula viscosa*. *Phytopathology, 94*, 1042–1047.

Wang, H. C., Zhou, M. G., Wang, J. X., Chen, C. J., Li, H. X., & Sun, H. Y. (2009). Biological mode of action of dimethomorph on *Pseudoperonospora cubensis* and its systemic activity in cucumber. *Agricultural Sciences in China, 8*, 172–181.

Waterhouse, G. M., & Brothers, M. P. (1981). The taxonomy of *Pseudoperonospora*. *Mycological Papers, 148*, 1–28.

Wehner, T. C. (2008). Watermelon. In J. Prohens & F. Nuez (Eds.), *Vegetables I. Asteraceae, brassicaceae, chenopodiaceae, and cucurbitaceae* (pp. 381–418). New York: Springer.

Wehner, T. C., & Shetty, N. V. (1997). Downy mildew resistance of the cucumber germplasm collection in North Carolina field tests. *Crop Science, 37*, 1331–1340.

Winstead, N. N., Aycock, R., & Jenkins, J. M. (1957). Reactions of watermelon plant introductions and varieties to downy mildew in North Carolina. *Plant Disease Reporter, 41*, 620–622.

Yang, X. T., Li, M., Zhao, Ch, Zhang, Z., & Hou, Y. L. (2007). Early warning model for cucumber downy mildew in unheated greenhouses. *New Zealand Journal of Agricultural Research, 50*, 1261–1268.

Yeon, I. K. (2007). Incidence of *Pseudoperonospora cubensis* on winter grown oriental melon (*Cucumis melo*) in vinyl houses in Korea. In A. Lebeda, & P. T. N. Spencer-Phillips (Eds.), *Advances in Downy Mildew Research, Vol. 3. Proceedings of The 2nd International Downy Mildews Symposium* (pp. 217–222.). Olomouc and Kostelec na Hané, Czech Republic: Palacký University in Olomouc and JOLA, v.o.s.

Yeon, I. K., Shin, Y. S., Do, H. W., Bae, S. G., & Park, S. D. (2002). Occurrence and chemical control of downy mildew (*Pseudoperonospora cubensis* Rostovzev) infecting oriental melon (*Cucumis melo* L. var. *makuwa* Makino) in plastic greenhouse. *Korean Journal of Horticultural Science & Technology, 20*, 25–28.

Zacha, V., Janýška, A., & Holman, B. (1985). Epifytócia plesne uhorkovej (*Pseudoperonospora cubensis* (Berk. et Curt.) Rost.) v ČSSR v roku 1984. *Sborník ÚVTIZ-Ochrana rostlin, 21*, 226.

Zaker, M., & Ommati, F. (1991). Observation of oospores of *Pseudoperonospora cubensis* on cucumber leaves in Iran. *Iran Journal of Plant Pathology, 27*, 62–63.

Zhang, Yan-ju, Qin, Zhi-wei, & Zhou, Xiu-yan (2006). Study on the over-wintering of cucumber downy mildew in Heilongjiang province of China. Abstracts, 27th International Horticultural Congress & Exhibition, August 13–19, 2006, COEX (Convention & Exhibition), Seoul, Korea; ISHS and KSHS, S11-P-51, p. 376.

Zhang, X., Chen, Y., Zhang, Y. J., & Zhou, M. G. (2008). Occurrence and molecular characterization of azoxystrobin resistance in cucumber downy mildew in Shandong province in China. *Phytoparasitica, 36*, 136–143.

Zhao, C. J., Li, M., Han, X., Zhang, Z., & Wang, Y. (2007). Analysis and monitoring for epidemic system as a basis for cucumber downy mildew warning in greenhouse. *Progress of Information Technology in Agriculture*, pp. 527–532.

Zhou, X. H., Wan, H. J., Qian, C. T., & Chen, J. F. (2008). Development and characterization of *Cucumis sativus-hystrix* introgression lines exhibiting resistance to downy mildew. In M. Pitrat (Ed.), *Cucurbitaceae 2008. Proceedings of the IXth EUCARPIA Meeting on Genetics and Breeding of Cucurbitaceae* (pp. 353–358). Avignon: NRA.

Zhu, S. S., Liu, P. F., Liu, X. L., Li, J. Q., Yuan, S. K., & Sil, N. G. (2008). Assessing the risk of resistance in *Pseudoperonospora cubensis* to the fungicide flumorph in vitro. *Pest Management Science, 64*, 255–261.

Eur J Plant Pathol (2011) 129:61–73
DOI 10.1007/s10658-010-9713-y

Spinach: better management of downy mildew and white rust through genomics

J. C. Correll · B. H. Bluhm · C. Feng ·
K. Lamour · L. J. du Toit · S. T. Koike

Accepted: 20 October 2010 / Published online: 4 December 2010

Abstract Spinach (*Spinacia oleracea*) has become an increasingly important vegetable crop in many parts of the world. Significant changes in production practices, particularly in the U.S. and E.U., have occurred in the past 10–15 years as a result of increased product demand. These changes likely increased the incidence and severity of downy mildew, caused by *Peronospora farinosa* f. sp. *spinaciae*. Recently, progress has been made to define the genetics of resistance to this pathogen and the closely related white rust pathogen, *Albugo occidentalis*. In this paper, we outline the genetic and genomic resources currently available for spinach, draw parallels between spinach diseases and more thoroughly characterized pathosystems, and describe efforts currently underway to develop new genetic and genomic tools to better understand downy mildew and white rust of spinach. Presently, many crucial tools and resources required to define the molecular underpinnings of disease are unavailable for either spinach or its pathogens. New resources and information for spinach genomics would provide a jumpstart for ongoing efforts to define (and deploy) genetic resistance against downy mildew and white rust.

J. C. Correll (✉) · B. H. Bluhm · C. Feng
Department of Plant Pathology, University of Arkansas,
Fayetteville, AR 72701, USA
e-mail: jcorrell@uark.edu

B. H. Bluhm
e-mail: bbluhm@uark.edu

C. Feng
e-mail: cfeng@uark.edu

K. Lamour
Department of Entomology and Plant Pathology,
University of Tennessee,
Knoxville, TN 37996, USA
e-mail: klamour@utk.edu

L. J. du Toit
Washington State University,
Mount Vernon, WA 98273-4768, USA
e-mail: dutoit@wsu.edu

S. T. Koike
University of California Cooperative Extension,
Salinas, CA 93901, USA
e-mail: stkoike@ucdavis.edu

Keywords Disease resistance · Genomics · *Spinacia oleracea*

Abbreviations

Pfs	*Peronospora farinosa* f. sp. *spinaciae*
Ao	*Albugo occidentalis*
RPF	Resistance to *Peronospora farinosa*
NIL	Near Isogenic Line
MAS	Marker Assisted Selection

Spinach production

Spinach (*Spinacia oleracea* L.) is an economically important vegetable crop in many parts of the world. FAO statistics indicate that 20 countries produced approximately 14.5 million MT of spinach in 2008 (http://faostat.fao.org/site/339/default.aspx). Based on FAO statistics, the top ten spinach producing countries are China (12.5 million MT), the U.S. (0.353 million MT), Japan (0.293 million MT), Turkey (0.226 million MT), Indonesia (0.152 million MT), France (0.124 million MT), Italy (0.100 million MT), Korea (0.093 million MT), Belgium (0.090 million MT), and Pakistan (0.082 million MT). Globally, approximately 1 million ha of spinach are grown in Asia and approximately 35,000 ha in each of the E.U., the U.S., Japan, and Turkey (Jan de Visser, *personal communication*). In the U.S., spinach is grown on approximately 35,000 ha, with annual production of approximately 350,000 MT tons valued at over $200 million (USDA-NASS 2008). The U.S. spinach industry has undergone dramatic changes in recent years. Demand for spinach in convenient, pre-cleaned, and pre-packaged units has increased significantly throughout the past decade, thus driving substantial increases in production (Lucier and Plummer 2004; USDA-NASS 2008) (Fig. 1). In the U.S., fresh market spinach production increased over 200% from 1987 to 2007 (USDA-NASS 2008). Per capita consumption of fresh market spinach increased from 0.3 kg/person in 1995 to approximately 1.0 kg/person in 2005 (Irish et al. 2007; USDA-NASS 2008), representing one of the fastest growing rates of per capita consumption among vegetables.

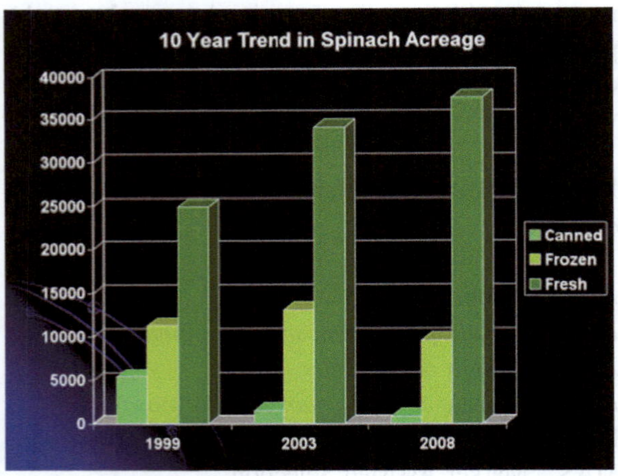

Fig. 1 Spinach production in the U.S. has increased dramatically in the past 10 years

Spinach is a remarkably nutritious vegetable and is increasingly being incorporated into health-conscious diets. Leaves are high in beta carotene and folate, and are also a rich source of vitamin C, calcium, and iron (Ryder 1979; Nonnicke 1989; Dicoteau 2000). Spinach is also an excellent source of antioxidants and has one of the highest ORAC (oxygen radical absorbance capacity) values of any vegetable crop (Prior 2003). In addition, spinach contains high levels of lutein, a carotenoid that prevents age-related macular degeneration (Morelock and Correll 2008).

Seed production

Spinach seed production is limited to areas where environmental conditions include long days to induce flowering and moderate summer temperatures (Metzger and Zeevaart 1985; Morelock and Correll 2008). For U.S. and E.U. spinach production, most seed originates from Denmark (approximately 75% of the world's supply), followed by the Pacific Northwest region of the U.S. (20%) (Daehnfeldt 2007; Foss and Jones 2005). In the Pacific Northwest, approximately 600–1600 ha of spinach seed are grown annually, representing 25%–50% of the U.S. spinach seed supply (Foss and Jones 2005).

Although the genetic basis of sex determination in spinach is complex (Morelock and Correll 2008), spinach is basically dioecious with male and female plants. Hybrid spinach lines were initially developed as a means of combining unique resistances to the downy mildew pathogen (Morelock and Correll 2008). Some open pollinated lines of spinach are still produced, but the vast majority of spinach lines in both Denmark and the U.S. are hybrid cultivars produced by the use of male and female inbred lines each having a different combination of resistances.

Spinach and disease resistance: gaps in our understanding

Several diseases of spinach are economically important, but downy mildew, caused by *Peronospora farinosa* f. sp. *spinaciae* (*Pfs*) (= *P. effusa*) (Brandenberger et al. 1991; Choi et al. 2007), and white rust, caused by

Albugo occidentalis (*Ao*), are of particular concern. Downy mildew is a major production constraint virtually everywhere the crop is grown, whereas white rust is limited to the U.S. (Correll et al. 1994; Morelock and Correll 2008). Although spinach consumption and production are rapidly increasing, spinach has been regarded historically as a minor crop in the U.S and other regions of the world, and the genomic resources available for spinach are limited relative to many other crops. Perhaps not surprisingly, the genetic and molecular basis of resistance to downy mildew and white rust are not well understood in spinach. However, stable genetic resistance to downy mildew and white rust is the lynchpin of effective disease control due to economic and regulatory concerns associated with conventional management strategies (i.e., frequent application of fungicides). To date, durable genetic resistance to either disease has proven to be elusive. Race-specific resistance to *Pfs* has been identified among spinach cultivars and wild relatives, but the underlying genes and signalling pathways are largely undefined. Intriguingly, both pathogens have demonstrated an ability to overcome diverse sources of host resistance, perhaps indicative of diversifying selection and/or intensive reciprocal co-evolution, but the molecular basis for the creation or ascendance of new strains remains a mystery. Despite the lack of information about resistance and virulence at the molecular level, considerable insight can be drawn from the interaction of model species such as *Arabidopsis* with pathogens closely related to *Pfs* and *Ao*. We predict that the steady increase of molecular resources for spinach, *Pfs*, and *Ao* will soon allow long-standing questions about disease interactions to be answered through genomics-based approaches, which in turn will lead to more sustainable management strategies.

In this paper, we outline the genetic and genomic resources currently available for spinach, draw parallels between spinach diseases and more thoroughly characterized pathosystems, and describe efforts currently underway to develop new genetic and genomic tools to better understand downy mildew and white rust of spinach.

Downy mildew and genetics of resistance

Downy mildew, caused by *Peronospora farinosa* f. sp. *spinaciae* (*Pfs*), is perhaps the most significant biotic constraint for spinach production worldwide (Fig. 2) (Correll et al. 1994). Downy mildew occurs everywhere spinach is grown, including areas used exclusively for seed production, and improving resistance is a priority for all breeding programs (Morelock and Correll 2008; Correll et al. 1994). Although the pathogen was first reported in the early 1800's, only three races of the pathogen had been identified before 1990 (Correll et al. 1994). Alarmingly, ten new races of the pathogen were identified between 1990 and 2010, (Fig. 3), and some of the newer races had overcome all known genetic resistance (Irish et al. 2003; 2007). The rapid ascendance of new races is likely a direct result of substantial changes in spinach production during the past decade, including a rapid expansion of planted acreage, 12-month production cycles, high density plantings, reduced usage of rotation crops, and the global movement of the pathogen in seed either as mycelium or as dormant oospores (Inaba and Morinaka 1984; Inaba et al. 1983).

The ability of the pathogen to overcome host resistance is particularly troublesome, as genetic resistance has long been the primary management tool for downy mildew (Morelock and Correll 2008; Correll et al. 1994). Although improving resistance to downy mildew is a critical breeding objective (Morelock and Correll 2008), relatively few studies have been conducted on the genetic control of resistance to the downy mildew pathogen (Irish et al. 2008; Morelock and Correll 2008; Correll et al. 2007).

The emergence of new races of the downy mildew pathogen despite the deployment of new sources of

Fig. 2 Downy mildew on spinach in a controlled greenhouse test

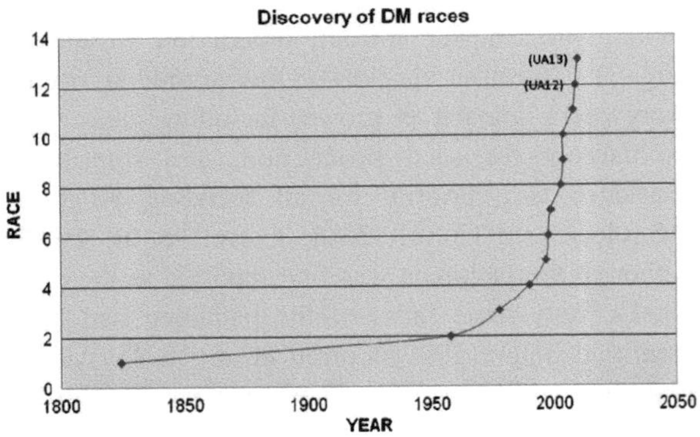

Fig. 3 The observation and documentation of new spinach downy mildew races by year. Note the rapid increase in novel races in the past 10–15 years

resistance is a fascinating case study in diversifying selection. Race 1 was first reported in 1824 (Greville 1824), and over 100 years later, resistance to race 1 was identified in two accessions (PI 140467 and PI 140467) from Iran (Smith 1950). Both sources of resistance manifested as immunity to infection, and the resistance from PI 140467 was conveyed by a single dominant gene (Smith 1950). This source of immunity was incorporated into hybrids 'Califlay' and 'Dixie market,' as well as the parents of 'Early Hybrid 7' and 'Early Hybrid 424' (Ryder 1979). Shortly after race 2 was first described in 1958, resistance was discovered that appeared to be a single dominant gene imparting immunity to race 2 as well as race 1 (Smith et al. 1961; 1962). Later, this resistance was demonstrated to be two closely linked genes rather than a single dominant gene (Eenink 1974, 1976a,b). After race 3 was identified in 1976, resistance was incorporated into hybrids rather quickly. Hybrids 'Mazurka,' 'Polka,' and 'Rhythm' were introduced in 1978, and additional hybrids resistant to race 3 were introduced in subsequent years (Morelock 1999). After race 4 was identified in 1990 (Brandenberger et al. 1991), resistant hybrids 'Bolero' and 'Bossanova' were introduced in 1991 (Morelock 1999). Brandenberger et al. (1992) screened 707 spinach accessions for resistance to race 4 and found nine accessions that were at least partially resistant (9–38% resistance). Accessions CGN09546 (60% of individual plants were resistant) and SP1 82/87 (80% of individual plants were resistant) exhibited the highest levels of resistance to race 4 (Brandenberger et al. 1992). Although inheritance studies were not conducted, seed from the two accessions was increased and distributed to breeders in the private sector. The rapid emergence of races 5, 6, 7, 8, 9, 10, 11, and two additional deviating strains UA12, and UA13 (Correll et al. 1990; 2001; 2003; Correll 1998; Handke et al. 2000; Irish 2004; Irish et al. 2003; 2004; 2007) ignited efforts to broaden host resistance and develop a more thorough understanding of the genetics of resistance to *Pfs*.

Currently, spinach breeding programs rely on pathogenicity tests to identify resistance to downy mildew (Table 1) (Morelock and Correll 2008). However, pathogenicity tests are time consuming, labour intensive, subject to specific environmental conditions, and require the maintenance of reference isolates, which is not trivial for an obligate pathogen. Thus, molecular markers linked to downy mildew resistance loci are urgently needed to provide a more robust means to introgress resistance into advanced spinach breeding material (Michelmore 2003).

More recently, a systematic effort to define downy mildew resistance in spinach led to the identification of a resistance locus (*Pfs1*) and a closely linked molecular marker (Irish et al. 2007; Correll et al. 2007. This locus was identified by developing a near isogenic line (NIL1) using the hybrid Lion as the resistant parent and Viroflay as the susceptible recurrent backcross parent (Irish et al. 2008). The line went through four backcross generations, selection of plants resistant to race 6, and three subsequent selfings, before producing seed for evaluation. The NIL1 line was resistant to multiple races (races 1–7,9,11 and UA13) as anticipated and resistance segregates as a single dominant locus (Irish et al. 2007, Table 2). A second NIL line, NIL2, has resistance to races 1–10. This line is in the advanced stages of development (Correll and Feng *unpublished*).

A total of 6 loci has been hypothesized to control resistance to the 11 known races and 2 additional deviating strains (UA12 and UA13) of the spinach downy mildew pathogen (Correll et al. 2007; Irish et al. 2008) (Table 3). Although the first *P. farinosa* f. sp. *spinaciae* resistance locus described was designated *Pfs-1* (Irish et al. 2008) and other hypothesized loci have been given a "*Pfs*" acronym, this has caused confusion in the spinach community, as the races of the downy mildew pathogen are often designated *Pf1*, or *Pfs1*. Thus, we have adopted the convention used with the Arabidopsis downy mildew pathogen system and have renamed the loci involved in downy mildew

Table 1 Disease reactions of spinach differentials for determining the race identification of isolates of the spinach downy mildew pathogen *Peronospora farinosa* f. sp. *spinaciae*

Differential cultivar	Race[1]												
	1	2	3	4	5	6	7	8	9	10	11	UA12	UA13
Viroflay	+	+	+	+	+	+	+	+	+	+	+	+	+
Resistoflay	–	–	+	+	+	+	+	+	+	+	+	+	+
Califlay	–	+	–	+	–	+	+	–	–	+	–	–	+
Clermont	–	–	–	–	+	+	+	+	+	+	+	+	+
Campania	–	–	–	–	–	+	–	+	+	+	–	+	–
Dolphin	–	–	–	–	–	–	–	+	–	+	–	+	–
Avenger	–	–	–	–	–	–	–	+	–	+	–	+	–
Lion	–	–	–	–	–	–	–	–	–	+	–	–	–
Lazio	–	–	–	–	–	–	–	–	–	–	+	+	+
Whale	–	–	–	(–)	–	(–)	(–)	–	–	+	–	+	+

[1] Races of the downy mildew pathogen as of October 2010. UA12=isolate UA2209 identified in 2009, and UA13=UA0510C indentified in 2010 are deviating isolates but not currently recognized as a bonified race by the International Working Group on Peronospora. Adapted from Irish et al. 2008. +=Susceptible reaction; –=Resistant reaction; (–)=reduced level of infection often referred to as field resistance

resistance of spinach as "*RPF*" loci (for genes involved in the **R**esistance against ***P**eronospora* ***f**arinosa*). Thus, the locus first described as *Pfs-1* will henceforth be referred to as the *RPF1* locus. In addition, two alleles were identified at this locus, *Pfs-1* and *pfs-1*, for the dominant resistance allele and the recessive susceptible allele, respectively. Thus, these two alleles have been renamed *RPF1-1* and *RPF1-2,* corresponding to the dominant and recessive alleles, respectively.

Table 2 Reaction of line NIL1 containing the RPF1 locus introgressed from the hybrid Lion into the susceptible cultivar Viroflay background to different races of *Peronospora farinosa* f. sp. *spinaciae*

Race	Reaction
1	–
2	–
3	–
4	–
5	–
6	–
7	–
8	+
9	nt
10	+
11	–
UA12	+
UA13	–

+=a susceptible reaction

–=a resistant reaction

nt=not tested

UA12 and UA13 are reaction types (see Table 1) caused by strains UA2209 and UA0510C

Although 6 genetically distinct loci have been hypothesized to control resistance to *Pfs*, considerable genetic analysis is needed to verify this hypothesis. In fact, an alternate hypothesis is that there is a single locus with multiple alleles that govern resistance in spinach to *Pfs*. Before naming additional resistance loci, experimental validation through genetic mapping is required.

There is a need to identify additional novel sources of resistance to *Pfs*. However, screening needs to be complemented with studies of the genetics of resistance as one cannot conclude which locus governs resistance in a particular accession to a given race as a locus can control resistance to multiple races (Irish et al. 2008). For example, the *RPF1* locus controls resistance to races 1–7, 9, 11, and UA13 (Irish et al. 2008; Correll et al. *unpublished*) whereas *RPF2* controls resistance to races 1–10 of *Pfs* (Correll et al. *unpublished*).

The Center for Genetic Resources in the Netherlands (CGN) (http://www.cgn.wur.nl/UK/CGN+Plant+Genetic+Resources/Collections/) and the Germplasm Resources Information Network in the U.S. (GRIN) (http://www.ars-grin.gov/) house collections of spinach germplasm, some of which have been screened for responses to the various races of *Pfs*. A consortium of E.U. companies has embarked on an effort through the CGN to collect and screen spinach germplasm for

Table 3 The 6 hypothesized downy mildew pathogen resistance (*RPF*) loci in spinach

	Resistance locus						
	Viroflay	*RPF5*	*RPF4*	*RPF3*	*RPF1*	*RPF2*	*RPF6*
Resistant to races	None	1,2	1-4	1,3,5,8, 9,11,UA12	1-7, 9,11, UA13	1-10	1-5, 7,11, UA13
1		■	■	■	■	■	■
2		■	■		■	■	■
3			■	■	■	■	■
4			■		■	■	■
5				■	■	■	■
6					■	■	
7					■	■	■
8				■		■	
9				■	■	■	
10						■	
11				■	■		■
UA0209D					■	■	■
UA12 (UA2209)				■			
UA13 (UA0510C)					■		■
Super Race?							

Designation	Expected Disease Reaction
	Susceptible
■	Resistant

resistance to various races of *Pfs*, but these data likely will be embargoed until 2015 (Jan deVisser, *personal communication*). The GRIN collection also has acquired new wild spinach germplasm including several collections of *S. tetrandra* and *S. turkistanica*. These lines are particularly difficult to maintain as significant seed dormancy issues must be overcome (Brenner 2009). Lines of several genera related to spinach, namely *Monolepsis, Micromonolepsis,* and *Suckleya* are also in the process of being acquired and should be evaluated for resistance to *Pfs*.

Population diversity of the downy mildew pathogen

Despite the rapid emergence of new races in recent years, little is known about race specificity or genetic diversity in *P. farinosa* f. sp. *spinaciae*. Genomic studies on downy mildew pathogens in general, and *Pfs* in particular, have lagged behind other oomycete pathogens such as *Phytophthora infestans, P. sojae, P. ramorum*, and *P. capsici,* in part due to the inherent difficulties associated with studying obligate plant pathogens. *Pfs* is likely heterothallic and thus requires opposite mating types to produce dormant resting spores (Inaba and Morinaka 1984). Studies with *Pfs* isolates from the U.S. indicate that, although perhaps rare, both mating types may occur as oospores have been observed in inoculation tests with bulk field isolates (Correll et al. *unpublished*). In addition, preliminary analyses of isolates collected from different geographical locations representing diverse races indicated that considerable genetic diversity is present in U. S. populations (Correll et al. *unpublished*). However, a thorough understanding of pathogen diversity is lacking and will ultimately be required to manage downy mildew.

There are two hypotheses regarding the origin of new races of the downy mildew pathogen. One is predicated on the premise that "new" races are already present in natural populations of the pathogen and are ascending due to intense selection via the deployment of certain resistance specificities in commercial spinach production. When the "new" race emerges, it is then moved throughout the world on seed as mycelium, viable sporangia, and/or dormant oospores on seed. Strong circumstantial evidence supports this hypothesis in that one new race was found almost simultaneously in the U.S. and E.U. in 2008. Additional support for this hypothesis is that resistance loci identified decades ago confer resistance against "new" races. For example, the *RPF3* locus, originally identified in Califlay, confers resistance to races 11 and UA12 (Table 3). However, a second possibility is that new isolates are emerging from endemic populations due to rapid, adaptive evolution of the pathogen. Both hypotheses could be tested by examining the molecular diversity of *Pfs* isolates, conducting surveys for the presence of *Pfs* on commercial seed, and identifying genes encoding effectors/avirulence gene products that are recognized by specific spinach resistance loci.

If new races of *Pfs* continue to emerge at the rate witnessed during the past decade, all currently known sources of genetic resistance could be overcome. This stark possibility serves as an impetus for a more thorough understanding of pathogenesis in *Pfs*, especially a molecular-level understanding of what constitutes distinct races of the pathogen. Today, in the era of next-generation DNA sequencing, the question of how new races emerge so rapidly can finally be answered through comparative genomics. Although very little genomic sequence information is currently available for any isolate or race of *Pfs*, the genome of the closely related species *Hyaloperonospora parasitica* (which causes downy mildew of *Arabidopsis thaliana*) has recently been sequenced (http://genome.wustl.edu). This provides a crucial resource for the sequencing of the *Pfs* genome. Once a reference genome sequence of *Pfs* is created, assembled, and annotated, representative isolates from other races can be sequenced, and comparative analyses can highlight likely sources of race-specificity. With this information in hand, race-specific detection with molecular techniques should be possible, and if the number of loci underlying race specificity is relatively limited, it may be possible to screen pathogen populations to predict the emergence of new races years before they reach ascendance and hinder commercial production.

White rust and the genetics of resistance

White rust of spinach, caused by the oomycete pathogen *Albugo occidentalis*, is an economically important disease in the U.S. The white rust pathogen was originally reported on beetberry (*Blitum capitatum*) in Colorado in 1907 (Wilson 1907) and on spinach in Virginia in 1910 (Walker 1952), but took on commercial significance when it was found at several New York vegetable markets on spinach that originated from Texas (Wiant 1937). Although *A. occidentalis* has not been reported outside the United States, white rust remains an economically important disease in Texas, Arkansas, Colorado, and East Coast production areas (Correll et al. 1994; Morelock and Correll 2008). Goode was the first to demonstrate field resistance to white rust in spinach (Goode et al. 1988), and Brandenberger et al. (1994) demonstrated parallels between quantitative resistance to both the white rust and downy mildew pathogens. Resistance to *A. occidentalis* is presumed to be multigenic and quantitative, but has not been studied in a systematic manner. A USDA collection of spinach germplasm was screened for resistance to *A. occidentalis* and, although quantitative differences in resistance were observed, no lines were identified that had complete resistance to the white rust pathogen (Correll *unpublished*).

Even though white rust and downy mildew are both caused by obligate parasites, breeding approaches to improve resistance are quite different. Unlike downy mildew resistance, spinach does not have single gene immunity to white rust; resistance is polygenic (quantitative) although the exact number of genes involved is unknown (Goode et al. 1988; Morelock and Correll 2001, 2003, 2005). The first spinach cultivars with at least some degree of white rust resistance were Crystal, Jewel, and Wintergarden, which were released by the USDA in 1973 (Morelock 1999) but were not widely grown. The partially resistant cultivars Ozarka and Greenvalley were released in 1980 (Bowers and Goode 1980a,b; Morelock 1999). In 1987, the first cultivar with high levels of white rust resistance, Fallgreen, was released (Goode et al. 1988; Morelock 1999). Interestingly, some cultivars that are resistant to white rust also have partial resistance, or "field" resistance, to some races of *Pfs* (Brandenberger et al. 1992; 1994). The genetic or molecular basis of this resistance is not known, but could be exploited to improve overall resistance to *Pfs* by complementing major genes for resistance.

Spinach genetics and genomic resources. What is available and what is needed?

Due to the historical size of spinach production and its prior status as a minor crop, few genomics resources have been developed to improve important horticultural traits. Spinach is a diploid with $2n=12$ chromosomes (Ellis and Janick 1960). The chromosomes of spinach have been identified based on their morphological characteristics, i.e., total length, arm ratio, and presence of satellites (Ellis and Janick 1960). The inheritance of common traits in spinach, such as spiny vs. smooth seed, smooth vs. savoy leaves, and short vs. long petiole, are controlled by single genes (Sneep 1958). Spinach is dioecious, with separate male and female plants, although occasionally monoecious plants can be found that contain both male and female flowers (Morelock and Correll 2008). No sex chromosome has been found in spinach; sex determination is likely controlled by either a sex determination locus or chromosomal segments sharing a mechanism similar to X and Y determination in animals (Bemis and Wilson 1953; Janick and Stevenson 1954).

The genomes of spinach organelles have been sequenced and characterized. Spinach has a tripartite mitochondrial genome that is 327 kb in size (Stern and Palmer 1986). The spinach chloroplast genome has been completely sequenced and is a circular molecule of 150 kb (Schmitz-Linneweber et al. 2001). Sequence homologs have been found among chloroplast, mitcochondrial, and nuclear genomes (Timmis and Scott 1983; Whisson and Scott 1985).

Ito et al. (2000) applied fluorescence in situ hybridization (FISH) and computer-aided karyotyping to identify and characterize spinach chromosomes. Three loci of 45S rDNA were found at the nucleolar organizing region (NOR) of Chromosome 5, and at terminal positions of the short arms of Chromosome 2 and 6. Also, three loci of 5S rDNA were found, one located at the subtelomeric region of the long arm of Chromosome 2, and the other two located at the proximal region of the long arm of Chromosome 5. This was the first physical mapping of genes using chromosome condensation patterns (CP) and FISH in spinach.

A high-density genetic map is essential for mapping new genes or markers, map-based gene cloning, and marker-assisted selection (MAS). Thousands of markers, either morphological or molecular, have been used to construct high-density genetic maps in major crops, such as maize, rice, and wheat. However, the genetic map of spinach is currently far from saturated. Groben and Wricke (1998) developed 13 SSR markers based on the spinach sequence information in Genbank and EMBL databases. Khattak et al. (2006) constructed a genetic map of spinach with 101 amplified fragment length polymorphism (AFLP) markers and nine single sequence repeat (SSR) markers. Seven linkage groups were established, with a total length of 585 cM, and an average distance between the markers of 5.18 cM. The sex determination gene is closely linked to SSR marker SO4 with a distance of 1.9 cM. The lack of a high-density genetic map of spinach results from shortages of (1) appropriate mapping populations, and (2) suitable molecular markers. A good mapping population requires high levels of variation between parents, and the population should be able to be saved for future use or used by other researchers. Furthermore, the markers used for mapping should be reproducible and reliable. Although AFLP markers are a type of PCR-based marker, their development is time consuming and laborious. SSR markers are a better choice for mapping but the current pool of available SSRs is far from sufficient to construct more than a rudimentary genetic map. Spinach sequences deposited into public databases are limited, and many of them correspond to physiological genes that are relatively conserved compared to non-coding regions.

The genome size of spinach has been estimated to be 989 Mb (Arumuganathan and Earle 1991). A spinach bacterial artificial chromosome (BAC) library was recently built (Correll et al. 2009). This library can be used to create a physical map of the spinach genome, to fine map genes of interest, and for map-based gene cloning. We have developed approximately 100 SSRs out of 2000 BAC end sequences (BES) (Feng et al. 2010). A preliminary test of these SSRs showed higher levels of polymorphism among commercial spinach cultivars or hybrids compared to SSRs developed from coding regions.

Although none of the six hypothesized loci underlying resistance to downy mildew (Correll et al. 2007) has been cloned or characterized in spinach, the molecular basis of resistance to downy mildew in Arabidopsis (caused by *Hyaloperonospora arabidopsidis*; formerly *Peronospora parasitica* syn. *Hyaloperonospora parasitica*) is increasingly well understood. To date, 28 distinct resistance genes (designated *recognition of Peronospora parasitica*; *RPP*) have been identified in Arabidopsis (Slusarenko and Schlaich 2003), and several have been cloned and characterized (e.g., *RPP1*, *RPP5*, and *RPP13*; Botella et al. 1998; Bittner-Eddy et al. 2000; Parker et al. 1997). To date, all cloned *RPP* genes are members of the plant disease resistance (*R*) gene family, typified by encoding proteins with an extracellular nucleotide binding site (NBS) fused to a leucine-rich repeat (LRR) with an N-terminal sharing similarity with a Drosophila Toll/mammalian interleukin 1 receptor (TIR) or a coiled-coil domain (CC). Intriguingly, many *RPP* genes are clustered in complex loci known as major recognition gene complexes (MRCs; Holub 1997) and are hypothesized to have originated primarily via duplications and rearrangements (Baumgarten et al. 2003; Meyers et al. 2003). For example, the *RPP5* locus contains seven TIR-NBS-LRR class *R* genes, three related sequences, and two non-*R* genes (Noël et al. 1999). In an evolutionary context, the dynamic nature of MRCs provides Arabidopsis an invaluable resource for rapid co-evolution when threatened by a genetically nimble pathogen. The conveyance of multiple race specificities by *RPF1* in spinach is suggestive of a complex locus analogous to the *RPP* MRCs of Arabidopsis, and such a complex locus for *RPF1* (formerly *Pfs1*) has been hypothesized (Irish et al. 2007).

R-gene mediated resistance to *H. arabidopsidis* is initiated when the product of a given *R* gene recognizes a distinct effector or pathogen-associated molecular pattern (PAMP), which triggers a classic hypersensitive response (HR) of programmed cell death localized at the site of infection. The identification of effector genes from *H. arabidopsidis* that are specifically recognized by characterized *RPP* genes provides unique insight into the Arabidopsis downy mildew pathosystem. For example, the *Arabidopsis thaliana recognized 13* (*ATR13*) effector is recognized by a CC-NBS-LRR class *R* gene encoded by *RPP13* (Allen et al. 2004; Bittner-Eddy et al. 2000). In an additional layer of complexity, *ATR13* also induces resistance against bacterial and viral pathogens (Rentel et al. 2008), which suggests a considerable level of cross talk is interwoven into Arabidopsis downy mildew defense responses. Both *ATR13* and *RPP13* are highly polymorphic in *H. arabidopsidis* and Arabidopsis, respectively, providing evidence of diversifying selection as would be predicted to arise from an ongoing, co-evolutionary battle of pathogenesis (Allen et al. 2004). Whether or not a similar co-evolutionary mechanism at least partially explains the rapid emergence of new races of *Pfs* is a fascinating, yet unanswered, question.

Certain components of *R*-gene mediated signaling pathways are broadly conserved across the plant kingdom. For example, *NPR1* is a central regulator of many, but not all, *R*-gene mediated defenses in Arabidopsis, and represents a point of convergence for pathogen-specific defences and salicylic acid (SA)-dependent systemic acquired resistance (SAR) (Dong 2004). Among characterized *RPP* genes of Arabidopsis, however, downstream signaling pathways appear to be remarkably divergent. For example, *RPP4*-mediated defence requires a suite of well-described signaling components, including *NPR1*, *SID1*, *SID2*, *EDS1*, *PAD4*, and SA (Delaney et al. 1994; Aarts et al. 1998; McDowell et al. 2000; Bittner-Eddy and Beynon 2001; van der Biezen et al. 2002). At the other extreme, *RPP13-ND* functions independently of most signalling components that typify *R*-gene mediated defenses (Bittner-Eddy and Beynon 2001). Where *RPF1* falls in the spectrum of reliance on known resistance signalling intermediaries is currently unknown.

Other factors in addition to *R*-gene mediated resistance play important roles in the suppression of downy mildew in Arabidopsis. For example, resistance to *H. arabidopsidis* is developmentally regulated in ecotype Columbia-0, in that seedlings are highly susceptible, whereas adult plants display substantially greater resistance (McDowell et al. 2005). Intriguingly, downy mildew-resistant (*dmr*) mutants were created by ethyl methane sulphonate mutagenesis in the highly susceptible Landsberg erecta eds1-2 mutant (Parker et al. 1996; Van Damme et al. 2005), providing insight into loci required for susceptibility. Similar approaches could be employed to define the molecular underpinning of resistance conveyed by the *RPF1* locus in spinach.

Genomics-based approaches to elucidate spinach disease interactions: looking to the future

Presently, many of the crucial tools and resources required to define the molecular underpinnings of disease are unavailable for either spinach or its pathogens. New resources and information for spinach genomics would jumpstart ongoing efforts to define (and deploy) genetic resistance against downy mildew and white rust. Most immediately pressing is the need for a detailed physical and genetic map of the spinach genome, which will facilitate positional cloning of genes and development of markers linked to economically important traits. The impact of these resources would be immediate and substantial for public and private breeding efforts. For example, spinach breeding programs currently rely heavily on pathogenicity tests to identify resistance to downy mildew (Morelock and Correll 2008). However, pathogenicity tests are laborious and variable, even in experienced hands. Marker-assisted selection for resistance would circumvent technical difficulties associated with pathogenicity tests (Irish et al. 2007).

After the development of physical and genetic maps, a critical, foundational resource needed for spinach is a sequenced, assembled, and annotated genome. To this end, a spinach BAC library of approximately 75,000 clones was recently constructed from a single plant from line NIL1 containing the *RPF1* downy mildew resistance locus (Correll et al. 2009; Irish et al. 2008; Feng et al. 2010). Given that the genome size of spinach is ~989 Mb, and the library contains 73,728 clones with an average insert size of 183 kb, coverage of the spinach genome should be ~13×. The construction of the BAC library provides a crucial resource for physical mapping, which in turn will facilitate genome sequencing and assembly.

New genomic resources would also significantly augment white rust management, which could indirectly increase downy mildew resistance. The polygenic nature of white rust resistance in spinach necessitates QTL mapping, which will be possible upon the completion of a high-density genetic map. As genes underlying QTL for white rust resistance are identified and characterized, possible interactions between genes conveying resistance to white rust should be explored thoroughly given the known overlap between resistance to white rust and downy mildew (Brandenberger et al. 1992; 1994).

The possibility that *Pfs* is a seed-borne pathogen is cause for considerable concern. However, the impact of seed-borne transmission on the dispersion of races throughout geographically distinct spinach production areas is unknown. Oospores have been demonstrated to be seed-borne and can serve as a primary source of inoculum to initiate seedling infections (Inaba et al. 1983). However, no systematic studies have been conducted to determine the frequency at which oospores contaminate commercial seed produced in either Denmark (where approximately 75% of commercial spinach seed for U.S. and E.U. markets is produced) or the Pacific Northwest (where approximately 15% of commercial spinach seed for U.S. and E.U. markets is produced) (Daehnfeldt 2007; Foss and Jones 2005). Furthermore, no studies other than the initial studies in the early 1980's have examined populations of *Pfs* for the presence of both mating types. To protect spinach production in the immediate future, a rapid, cost-effective, and specific assay to monitor seed transmission would be extremely valuable. The increased availability of sequence information will facilitate the development of real-time PCR assays to detect *Pfs*. Eventually, when race specificity is understood at the molecular level, PCR assays that distinguish races could be developed as a tool for disease management.

Acknowledgements The authors would like to thank the California Spinach Research Committee, Mary Zischke and the California Leafy Greens Research Board, and the many spinach growers and seed companies for their financial support for downy mildew efforts over the years beginning with Bill White Seed Company two decades ago. In addition, we would like to thank Plantum, Netherlands, Naktuinbouw, Netherlands, the International Working Group on Peronospora, and the many spinach breeders and pathologists for helpful discussions on downy mildew of spinach. A special thanks is extended to Jay Schafer, Jan De Visser and Paul Moreno for their continued insight and help on downy mildew of spinach. In addition, we are grateful to the professional staff at De Burgh, in particular Eline, Suzette, Ron, Lars, Jack, Maarten, and Peter. The authors would like to dedicate this review to Dr. Teddy E. Morelock, who passed away in 2009, for his many years of service to both the spinach industry and the many spinach researchers he influenced around the world.

References

Aarts, N., Metz, M., Holub, E., Staskawicz, B. J., Daniels, M. J., & Parker, J. E. (1998). Different requirements for *EDS1* and *NDR1* by disease resistance genes define at least two *R* gene-mediated signaling pathways in *Arabidopsis*. *Proceedings of the National Academy of Sciences USA, 95*, 10306–10311.

Allen, R. L., Bittner-Eddy, P. D., Grenville-Briggs, L. J., Meitz, J. C., Rehmany, A. P., Rose, L. E., et al. (2004). Host-parasite coevolutionary conflict between *Arabidopsis* and downy mildew. *Science, 306*, 1957–1960.

Arumuganathan, K., & Earle, E. D. (1991). Nuclear DNA content of some important plant species. *Plant Molecular Biology Reporter, 9*, 208–218.

Baumgarten, A., Cannon, S., Spangler, R., & May, G. (2003). Genome-level evolution of resistance genes in *Arabidopsis thaliana*. *Genetics, 165*, 309–319.

Bemis, W. P., & Wilson, G. B. (1953). A new hypothesis explaining the genetics of sex determination. *Journal of Heredity, 44*, 41–45.

Bittner-Eddy, P. D., & Beynon, J. L. (2001). *RPP13-Nd* an *Arabidopsis* LZ-NBS-LRR-type resistance gene that is salicylic acid-independent and that does not require disease resistance signaling pathways defined by *EDS1* or *Ndr1*. *Molecular Plant-Microbe Interactions, 14*, 416–421.

Bittner-Eddy, P. D., Crute, I. R., Holub, E. B., & Beynon, J. L. (2000). *RPP13* is a simple locus in *Arabidopsis thaliana* for alleles that specify downy mildew resistance to different avirulence determinants in *Peronospora parasitica* (*At*). *The Plant Journal, 21*, 177–188.

Botella, M. A., Parker, J. E., Frost, L. N., Bittner-Eddy, P. D., Beynon, J. L., Daniels, M. J., et al. (1998). Three genes of the *Arabidopsis RPP1* complex resistance locus recognise distinct *Peronospora parasitica* (*At*) avirulence determinants. *Plant Cell, 10*, 1847–1860.

Bowers, J. L., & Goode, M. J. (1980a). Spinach varieties. *Proceedings of the Arkansas State Horticultural Society, 101*, 64.

Bowers, J. L., & Goode, J. J. (1980b). Ozarka and Greenvalley: new disease resistant spinach cultivars. *Arkansas Farm Research, 32*(2), 6.

Brandenberger, L. P., Correll, J. C., & Morelock, T. E. (1991). Identification of and cultivar reactions to a new race (race 4) of *Peronospora farinosa* f. sp. *spinaciae* on spinach in the United States. *Plant Disease, 75*, 630–634.

Brandenberger, L. P., Morelock, T. E., & Correll, J. C. (1992). Evaluation of spinach germplasm for resistance to a new race (Race 4) of *Peronospora farinosa* f. sp. *spinaciae*. *HortScience, 27*, 1118–1119.

Brandenberger, L. P., Correll, J. C., Morelock, T. E., & McNew, R. W. (1994). Characterization of resistance of spinach to white rust (*Albugo occidentalis*) and downy mildew (*Peronospora farinosa* f. sp. *spinaciae*). *Phytopathology, 84*, 431–437.

Brenner, D. M. (2009). The U.S. spinach germplasm collection. (Paper presented at The T.E. Morelock International Spinach Conference, Fayetteville, AR). From: http://spinach.uark.edu/Spinach%20Conference%202009%20Program.pdf, November.

Choi, Y. J., Hong, S. B., & Shin, H. D. (2007). Re-consideration of *Peronospora effusa* infecting *Spinacia oleracea* as distinct species, *Peronospora effusa*. *Mycological Research III*, 381–391.

Correll, J. C. (1998). Review of the biology of *Peronospora farinosa* f. sp. *spinaciae*. In: *Crop protection compendium*. Wallingford, U.K.: CABI).

Correll, J. C., Koike, S. T., Brandenberger, L. P., Black, M. C., & Morelock, T. E. (1990). A new race (Race 4) of downy mildew on spinach. *California Agriculture, 44*, 14–15.

Correll, J. C., Morelock, T. E., Black, M. C., Koike, S. T., Brandenberger, L. P., & Dainello, F. J. (1994). Economically important diseases of spinach. *Plant Disease, 78*, 653–660.

Correll, J. C., Irish, B. M., Koike, S. T., & Morelock, T. E. (2001). Update on downy mildew (*Peronospora farinosa* f. sp. *spinaciae*) of spinach in the United States. Paper presented at the National Spinach Conference, Fayetteville, AR USA, November.

Correll, J. C., Irish, B. M., Koike, S. T., Shafer, J., & Morelock, T. E. (2003). Update on downy mildew of spinach. Paper presented at the National Spinach Conference, Fayetteville, AR USA, November

Correll, J. C., Feng, C., Irish, B. M., Koike, S. T., Morelock, T. E., Bentley, T. C. et al. (2007). Spinach downy mildew: overview of races and the development of molecular markers linked to major resistance genes. In: A. Lebeda, & P. T. N. Spencer-Phillips (Eds.), *Advances in downy mildew research, 3* (pp. 135–142).

Correll, J. C., Feng, C. Kudrna, D. A., Ammiraju, J., Raju, & Wing, R. A. (2009). Construction of a spinach (*Spinacae oleracea*) BAC library. *Plant and Animal Genomes XVII.* p. 604.

Daehnfeldt, L. (2007). Danish production of vegetable seed—Is it only spinach? Retrived June 18, 2007 from http://www.daehnfeldt.com/index.asp?NyhedsId=38.

Delaney, T. P., Uknes, S., Vernooij, B., Friedrich, L., Weymann, K., Negrotto, D., et al. (1994). A central role of salicylic acid in plant disease resistance. *Science, 266*, 1247–1249.

Dicoteau, D. R. (2000). *Vegetable crops*. New Jersey: Prentice Hall.

Dong, X. (2004). NPR1, all things considered. *Current Opinion in Plant Biology, 7*, 547–552.

Eenink, A. H. (1974). Linkage in *Spinacia oleracea* L. between the locus for resistance to *Peronospora spinaciae* Laub. and the locus for tolerance for cucumber virus 1. *Euphytica, 23*, 485–487.

Eenink, A. H. (1976a). Linkage of *Spinacia oleracea* L. of two race-specific genes for resistance to downy mildew *Peronospora farinosa* f. sp. *spinaciae* Byford. *Euphytica., 25*, 713–715.

Eenink, A. H. (1976b). Resistance in spinach to downy mildew. In: *Proceedings of Eucarpia Meeting*. Leafy Vegetables, pp. 53–54. Wageningen, Holland.

Ellis, J. R., & Janick, J. (1960). The chromosomes of *Spinacia oleracea*. *American Journal of Botany, 47*, 210–214.

Feng, C., Correll, J. C., & Bluhm, B. H. (2010). A spinach BAC library for marker development, gene discovery, and functional genomics. *Phytopathology, 100*, S35.

Foss, C. R., & Jones, L. J. (2005). Crop profile for spinach seed in Washington. From: http://www.ipmcenters.org/cropprofiles/docs/waspinachseed.html.

Goode, M. J., Morelock, T. E., & Bowers, J. L. (1988). 'Fall Green' spinach. *HortScience, 23*(5), 931.

Greville, R. K. (1824). *Flora Edinensis*. Edinburgh: Blackwood & Strand.

Groben, R., & Wricke, G. (1998). Occurrence of microsatellites in spinach sequences from computer datebases and development of polymorphic SSR markers. *Plant Breeding, 117*, 271–274.

Handke, S., Seehaus, H., & Radies, M. (2000). Detection of a linkage group of the four *turkestanica*. *Gartenbauswissenschaft., 65*, 73–78.

Holub, E. B. (1997). Organization of resistance genes in *Arabidopsis*. In I. R. Crute & E. B. Holub (Eds.), *The gene-for-gene relationship in plant-parasite interactions* (pp. 5–26). Wallingford: CABI.

Inaba, T., & Morinaka, T. (1984). Heterothallism in *Peronospora effusa*. *Phytopathology, 74*, 214–216.

Inaba, T., Tkahashi, K., & Morinaka, T. (1983). Seed transmission of spinach downy mildew. *Plant Disease, 67*, 1139–1141.

Irish, B. M. (2004). New races of the downy mildew pathogen of spinach, identification of molecular markers for disease resistance, and molecular diversity of spinach germplasm. Dissertation, University of Arkansas.

Irish, B. M., Correll, J. C., Koike, S. T., Schafer, J., & Morelock, T. E. (2003). Identification and cultivar reaction to three new races of the spinach downy mildew pathogen from the United States and Europe. *Plant Disease, 87*, 567–572.

Irish, B. M., Correll, J. C., Raid, R. N., & Morelock, T. E. (2004). First report of *Peronospora farinosa* f. sp. *spinaciae* (race 5) of spinach in Florida. *Plant Disease, 88*, 84.

Irish, B. M., Correll, J. C., Koike, S. T., & Morelock, T. E. (2007). Three new races of the spinach downy mildew pathogen identified by a modified set of spinach differentials. *Plant Disease, 91*, 1392–1396.

Irish, B. M., Correll, J. C., Feng, C., Bentley, T., & de los Reyes, B. G. (2008). Characterization of a resistance locus (Pfs-1) to the spinach downy mildew pathogen (*Peronospora farinosa* f. sp. *spinaciae*) & development of a molecular marker linked to Pfs-1. *Phytopathology, 98*, 894–900.

Ito, M., Ohmido, N., Akiyama, Y., Fukui, K., & Koba, T. (2000). Characterization of spinach chromosomes by condensation patterns and physical mapping of 5S and 45S rDNAs by FISH. *Journal of the American Society for Horticultural Science, 125*(1), 59–62.

Janick, J., & Stevenson, E. C. (1954). A genetic study of the heterogametic nature of the staminate plant in spinach (*Spinacia oleracea L.*). *Proceedings of the American Society for Horticultural Science, 63*, 444–446.

Khattak, J. Z. K., Torp, A. M., & Andersen, S. B. (2006). A genetic linkage map of *Spinacia oleracea* and localization of a sex determination locus. *Euphytica, 148*, 311–318.

Lucier, G., & Plummer, C. (2004). Vegetable consumption expected to rise in 2004. Vegetables and melons outlook, Electronic Outlook Report from the Economic Research Service, United States Department of Agriculture VGS-302 April 21, 2004. From http://www.ers.usda.gov/publications/vgs/Apr04/vgs302.pdf.

McDowell, J. M., Cuzick, A., Can, C., Beynon, J., Dangl, J. L., & Holub, E. B. (2000). Downy mildew (*Peronospora parasitica*) resistance genes in *Arabidopsis* vary in functional requirements for *NDR1*, *EDS1*, *NPR1* and salicylic acid accumulation. *Plant Journal, 22*, 523–529.

McDowell, J. M., Williams, S. G., Funderburg, N. T., Eulgem, T., & Dangl, J. L. (2005). Genetic analysis of developmentally regulated resistance to downy mildew (*Hyaloperonospora parasitica*) in *Arabidopsis thaliana*. *Molecular Plant-Microbe Interactions, 18*, 1226–1234.

Metzger, J. D., & Zeevaart, J. A. D. (1985). *Spinacia oleracea*. In A. H. Halevy (Ed.), *Handbook of flowering, vol. IV* (pp. 384–392). Boca Raton: CRC.

Meyers, B. C., Kozik, A., Griego, A., Kuang, H., & Michelmore, R. W. (2003). Genome-wide analysis of NBS-LRR-encoding genes in *Arabidopsis*. *Plant Cell, 15*, 809–834.

Michelmore, R. W. (2003). The impact zone: genomics and breeding for durable disease resistance. *Current Opinion in Plant Biology, 6*, 397–404.

Morelock, T. E. (1999). Spinach: variety test and description. *Hortscience, 34*(6), 987–988.

Morelock, T. E., & Correll, J. C. (2001). *Spinach breeding in Arkansas*. Paper presented at the National Spinach Conference, Fayetteville, AR USA.

Morelock, T. E., & Correll, J. C. (2003). *Spinach breeding in the mid-south*. Paper presented at the National Spinach Conference, AR USA, November.

Morelock, T. E., & Correll, J. C. (2005). *Spinach breeding in the mid-south*. Paper presented at the National Spinach Conference., Fayetteville, AR USA, November.

Morelock, T. E., & Correll, J. C. (2008). Spinach breeding. In J. Prohens & F. Nuez (Eds.), *Vegetables I* (pp. 183–212). New York: Springer.

Noël, L., Moores, T. L., Van der Biezen, E. A., Parniske, M., Daniels, M. J., Parker, J. E., et al. (1999). Pronounced intra specific haplotype divergence at the *RPP5* complex disease resistance locus of *Arabidopsis*. *Plant Cell, 11*, 2099–2111.

Nonnicke, I. L. (1989). *Vegetable production*. Van Nostrand Runhold.

Parker, J. E., Holub, E. B., Frost, L. N., Falk, A., Gunn, N. D., & Daniels, M. J. (1996). Characterization of *eds1*, a mutation in *Arabidopsis* suppressing resistance to *Peronospora parasitica* specified by several different *RPP* genes. *Plant Cell, 8*, 2033–2046.

Parker, J. E., Coleman, M. J., Szabò, V., Frost, L. N., Schmidt, R., van der Biezen, E. A., et al. (1997). The *Arabidopsis* downy mildew resistance gene *RPP5* shares similarity to the Toll and interleukin-1 receptors with *N* and *L6*. *Plant Cell, 9*, 879–894.

Prior, R. L. (2003). Spinach as a source of antioxidant phytochemicals with potential health effects. Paper presented at the National Spinach Conference, Fayetteville, AR. USA, November.

Rentel, M. C., Leonelli, L., Dahlbeck, D., Zhao, B., & Staskawicz, B. J. (2008). Recognition of the *Hyaloperonospora parasitica* effector ATR13 triggers resistance against oomycete, bacterial, and viral pathogens. *Proceedings of the National Academy of Sciences USA, 105*, 1091–1096.

Ryder, E. J. (1979). *Leafy salad vegetables*. New York: AVI.

Schmitz-Linneweber, C., Maier, R. M., Alcaraz, J., Cottet, A., Herrmann, R Gl, & Mache, R. (2001). The plastid chromosome of spinach (*Spinacia oleracea*): complete nucleotide sequence and gene organization. *Plant Molecular Biology, 45*, 307–315.

Slusarenko, A. J., & Schlaich, N. L. (2003). Downy mildew of *Arabidopsis thaliana* caused by *Hyaloperonospora parasitica* (formerly *Peronospora parasitica*). *Molecular Plant Pathology, 4*, 159–170.

Smith, P. G. (1950). Downy mildew immunity in spinach. *Phytopathology, 40*, 65–68.

Smith, P. G., Webb, R. E., Millett, A. M., & Luhn, C. H. (1961). Downy mildew on spinach. *California Agriculture, 15*, 5.

Smith, P. G., Webb, R. E., & Luhn, C. H. (1962). Immunity to race 2 of spinach downy mildew. *Phytopathology, 52*, 597–599.

Sneep, J. (1958). The present position of spinach breeding. *Euphytica, 7*, 1–8.

Stern, D. B., & Palmer, J. D. (1986). Tripartite mitochondrial genome of spinach: physical structure, mitochondrial gene mapping, and locations of transposed chloroplast DNA sequences. *Nucleic Acids Research, 14*, 5651–5666.

Timmis, J. N., & Scott, N. S. (1983). Sequence homology between spinach nuclear and chloroplast genomes. *Nature, 305*, 65–67.

United States Department of Agriculture – National Agricultural Statistics Service. (2008). From http://www.nass.usda.gov/QuickStats/index2.jsp

Van Damme, M., Andel, A., Huibers, R. P., Panstruga, R., Weisbeek, P. J., et al. (2005). Identification of *Arabidopsis* loci required for susceptibility to the downy mildew pathogen *Hyaloperonospora parasitica. Molecular Plant-Microbe Interactions, 18*, 583–592.

van der Biezen, E. A., Freddie, C. T., Kahn, K., Parker, J. E., & Jones, J. D. (2002). Arabidopsis *RPP4* is a member of the *RPP5* multigene family of TIR-NB-LRR genes and confers downy mildew resistance through multiple signaling components. *Plant Journal, 29*, 439–451.

Walker, J. C. (1952). *Diseases of vegetable crops*. New York: McGraw Hill.

Whisson, D. L., & Scott, N. S. (1985). Nuclear and mitochondrial DNA have sequence homology with a chloroplast gene. *Plant Molecular Biology, 4*, 267–273.

Wiant, J. S. (1937). White rust on Texas spinach. *Plant Disease Reporter, 21*, 114–115.

Wilson, G. W. (1907). Studies in North American *Peronosporales*. I. The genus *Albugo. Torrey Botanical Club Bulletin*, 3461–3485.

Eur J Plant Pathol (2011) 129:75–88
DOI 10.1007/s10658-010-9683-0

Advances in sunflower downy mildew research

Ferenc Viranyi · Otmar Spring

Accepted: 6 September 2010 / Published online: 16 September 2010

Abstract This review summarises the progress in research on sunflower downy mildew as reported in publications of the past 10 to 15 years, the period since the last comprehensive review on *Plasmopara halstedii*. Particular attention is paid to subjects that showed much progress and may be of particular interest to sunflower pathologists, mycologists or molecular biologists. Accordingly, single sections are devoted to the problems of taxonomic and phylogenetic aspects, host specificity, the host—pathogen interaction including resistance phenomena, as well as epidemiology and disease management. Reflecting the progress achieved in some fields and illuminating the deficits in others should stimulate the reader's interest in this very significant pathosystem.

Keywords *Plasmopara halstedii* · Taxonomy · Host range · Resistance · Phenotypic and genotypic diversity · Epidemiology

Introduction

Downy mildew of sunflower, incited by *Plasmopara halstedii* (Farl.) Berlese et de Toni, is considered one of the most severe pathogenic threats of this crop worldwide. Since the pathogen shows high phenotypic diversity, particularly in virulence and fungicide sensitivity, effective disease control depends on profound knowledge of the biology of the pathogen, its physiological capacities and requirements as well as the molecular mechanisms involved in the interaction with the host and environment. In a previous review, Gulya and co-workers (1997) summarized in detail the state of knowledge in the biology, host—pathogen relationships and disease control of sunflower downy mildew.

However, since then significant progress has been made in research and a number of publications have presented new findings and ideas. There is a need, therefore, to outline the advances in research over the last decade with particular emphasis on the taxonomic position, host specialisation and genetics of this oomycete, coupled with resistance phenomena, epidemiology and disease management. Accordingly, in this review we have chosen to elaborate in detail the highlights of this pathosystem, with the aim of providing up-to-date information to those interested in this devastating disease.

F. Viranyi
Plant Protection Institute, Szent István University,
Gödöllö 2103, Hungary

O. Spring (✉)
Institute of Botany, University of Hohenheim,
Garbenstr. 30,
Stuttgart 70593, Germany
e-mail: spring@uni-hohenheim.de

Historical and phylogenetic aspects of sunflower downy mildew taxonomy

The taxonomic situation of *Plasmopara halstedii* has previously been illuminated in various reviews (e.g., Zimmer and Hoes 1978; Sackston 1981; Viranyi 1992; Gulya et al. 1997). However, the scientific information contributed from East European scientists to this field particularly within the first two decades after World War II and summarized by Novotelnova (1966) appears to be inadequately considered. Hence, there is an ongoing discussion on the correct classification and the species delineation since the first description of the pathogen.

Farlow (1883) described the downy mildew pathogen *Peronospora halstedii* on the basis of pathogen samples found on *Eupatorium purpureum, Ambrosia artemisiifolia, Bidens frondosa, Rudbeckia laciniata, Silphium terebinthaceum* and the perennial sunflower species *Helianthus strumosus, H. tuberosus and H. doronicoides* (*H. mollis X giganteus*). It appears noteworthy that he first referred to a sample Halsted found on *Eupatorium purpureum* near the Bussey Institution in 1876 and it can be deduced that he had this specimen (Farlow Herbarium #79717) in mind when he stated that a sample found on *H. tuberosus* by Ellis showed "...the most marked deviation from the type...". When Schröter (1886) had separated *Plasmopara* from *Peronospora* due to the germination by means of zoospores instead of forming germ tubes, Berlese and de Toni (1888) transferred the taxon to the new genus under the name *Plasmopara halstedii* (Farlow) Berlese & de Toni. In their description, the host range comprised seven genera, including four perennial species of *Helianthus*, but again no annuals. Based on subtle morphological differences, Wilson (1907) described the genus *Rhysotheca* in which *Plasmopara halstedii* had been embedded, but the new genus never gained broad acceptance. Over the subsequent years, numerous collections of downy mildews on species of Asteraceae were made in North America and classified as *Plasmopara halstedii*, due to morphological similarities of sporangiophores and sporangia, although high variability in size and shape was recognized early (Stevens 1913). The first specimens documenting the occurrence of downy mildew on *Helianthus annuus* were dated from the 1890s and in the 1920s it became a serious threat to sunflower production in many states of the U.S. (Henry and Gilbert 1924; Young and Morris 1927).

The pathogen was considered "...to occur on almost any composite ..." (Nishimura 1922). This was in contrast to Gäumann's narrow host-based species concept (1923) and to the situation found in most species of the Peronosporaceae according to the current classification. Frequently *Plasmopara halstedii* was regarded as a species complex (Leppik 1966). Clear evidence for the conspecificity of the pathogen originally described by Farlow with isolates found on cultivated sunflower *H. annuus* is still lacking. In Eastern Europe, where sunflower downy mildew became a serious threat in the middle of the 20th century, intensive cross infection studies with pathogen inoculum from oilseed sunflower were performed to elucidate the host range of sunflower downy mildew (reviewed by Novotelnova 1966).

Field experiments never showed disease symptoms on host plants other than *H. annuus* and a few other annual species of the genus *Helianthus*. This encouraged Novotelnova to reclassify the species complex (Novotelnova 1962, 1964, 1966). She retained Farlow's name *Plasmopara halstedii* for pathogens found on Eupatorieae, described several new species on hosts from other tribes and separated the downy mildew of sunflower as *Plasmopara helianthi* Novot. Within the latter she further differentiated the specialized forms *helianthi, perennis* and *patens* based on natural and artificial infection studies with annual or perennial *Helianthus* species and on slight morphometric differences in sporangia, respectively. However, her species concept never gained broad acceptance and sunflower downy mildew is still mostly known as *Plasmopara halstedii.*

Fifty years later it appears possible and necessary to test the conspecifity for Farlow's *P. halstedii* and Novotelnova's *P. helianthi*. Molecular genetic tools have fundamentally influenced our understanding of the oomycete phylogeny and changed the classification at all taxonomic levels (reviewed by Voglmayr 2008, Spring and Thines 2004). Moreover, on the basis of phylogenetic studies, new phenotypic characters were searched for and evaluated for reliability (Spring and Thines 2004; Thines 2006).

Plasmopara in the sense of Schröter (1886) was recognised as a polyphyletic entity which, in past years, has been separated into a core genus (harbouring *P. halstedii*) and several new genera (for references see Voglmayr and Constantinescu 2008). A broad survey on molecular genetic data from downy mildews on Asteraceae is still lacking due to technical problems in getting reliable DNA sequence information from old desiccated material compiled in herbarium collections. However, new methodologies to overcome such restrictions are in progress (Telle and Thines 2008). The first data obtained from herbarium specimens of downy mildew infected Asteraceae (Thines, personal communication) clearly indicate that Farlow's suggestion to separate *P. halstedii* on Tubuliflorae from *P. ganglioniformis* (= *Bremia lactucae*) on Liguliflorae is insufficient as a species concept for downy mildew pathogens of this plant family. Recent phylogenetic results gradually resolve the large downy mildew complex living on Asteraceae and confirm separate entities on the generic (e.g. *Bremia, Novotelnova, Paraperonospora, Plasmopara, Protobremia*) and species level (Voglmayr and Constantinescu 2008). Narrow species concepts such as those found in *Novotelnova scorzonerae*, *Paraperonospora leptosperma*, *Plasmopara angustiterminalis* or *Plasmopara* on *Flaveria bidentis* are supported by the new characters analysed (Spring et al. 2003; Thines 2006; Komjati et al. 2007; Voglmayr and Constantinescu 2008). In this context, recent studies on downy mildew of *Xanthium strumarium* (Komjáti et al. 2007) in Hungary and on a naturalized population of the perennial sunflower hybrid *Helianthus* x *laetiflorus* (Spring et al. 2003) revealed that the former is correctly classified as *P. angustiterminalis* Novotelnova while the latter was indistinct from downy mildew samples collected from annual oilseed sunflower. The fact that both isolates readily infected *H. annuus* seedlings when sporangia were used as inoculum is questioning Novotelnova's (1966) concept of differentiating specific forms of *P. helianthi* according to infection patterns on annual or perennial sunflower species.

Host range and distribution of sunflower downy mildew

The unresolved taxonomic situation is directly linked with questions on the true host range of sunflower downy mildew, its geographic origin and current distribution. While former assumptions of a broad host range of *P. halstedii* were based on morphological similarities of downy mildew pathogens found on Asteraceae (Farlow 1883; Nishimura 1922; Leppik 1966), infection studies performed by East European pathologists with inoculum from sunflower downy mildew revealed a very narrow host range for this pathogen (reviewed by Novotelnova 1966). In tests with more than 70 potential hosts under field and artificial conditions, only the annual *H. lenticularis* (a western variety of wild *H. annuus*) and *H. debilis* as well as the perennial sunflower species *H. maximiliani*, *H. grosseserratus* and *H. tomentosus* (syn. *H. tuberosus*) and the hybrid *Heliathus* x *multiflorus* were susceptible (Christova and Mitov 1960; Pustovoit 1963, 1964; Novotelnova 1966). This list was expanded by Leppik (1966) and Virányi (1984) who succeeded to infect *H. divaricatus* and *H. argophyllus*, respectively, with zoosporangia gained from oilseed sunflower. To date, infection of species from other genera of Asteraceae has not been achieved with such inoculum.

It must be emphasized that previous infection tests were predominantly carried out using only inoculum from cultivated *H. annuus*. Experiments testing the susceptibility of cultivated *H. annuus* to inocula from other Asteraceae have been reported to a much lesser extent. This is due to the problem that such experiments continuously require fresh inoculum and that the culture of pathogen strains isolated from wild species on their original hosts is difficult to establish. Infection of *H. annuus* succeeded with zoosporangia from a pathogen on *Helianthus* x *laetiflorus*, found on a naturalized colony of the perennial host in Germany (Spring et al. 2003). The pathogen survived in the rhizome of its host, leading to yearly systemic infection of shoots and leaves. Based on morphological, chemical and molecular genetic characters, the pathogen was indistinct from strains of downy mildew found on cultivated annual sunflower. Interestingly, its infection behaviour on sunflower differential lines did not allow the assignment of a certain pathotype, because several of the commonly used sunflower differential lines (Tourvieille et al. 2000) died from damping-off before sporulation occurred. This may suggest that the pathogen was introduced with its perennial host rather than being spread from local sunflower fields.

Possible host jumps of downy mildew pathogens of other Asteracean genera to *H. annuus* were reported from *Ambrosia artemisiifolia* (Walcz et al. 2000) and from *Xanthium strumarium* (Virányi 1984; Komjáti et al. 2007) in Hungary and from *Iva xanthifolia* in the U.S. (Gulya 2002a). In all three cases, successful cross-infection under laboratory conditions was achieved using zoosporangia of the wild hosts for whole seedling inoculation (Cohen and Sackston 1973) of sunflower. However, natural infection of cultivated sunflower with sporangia of downy mildew from wild Asteracean host species has not yet been proven under field conditions. In the case of the reported *X. strumarium* incidence, the pathogen could be maintained on certain sunflower genotypes (e.g. the differential line QHP 1), but killed others by early damping-off. Molecular genetic and other characters undoubtedly identified this pathogen as *P. angustiterminalis* Novot. (Komjáti et al. 2007; 2008), a species phylogenetically close to *P. halstedii* (Voglmayr et al. 2004). Such molecular genetic investigations are lacking for the pathogens found on *Ambrosia* and *Iva*, but could still be performed if herbarium specimens are available.

The fact that two species of *Plasmopara* potentially infect the same host is of particular interest, because it could provide the possibility for hybridization as previously shown for oomycetes (Adhikari et al. 2003; Schardl and Craven 2003; Spring and Zipper 2006; Érsek and Nagy 2008).

As reviewed earlier by Sackston (1981), the origin of the sunflower downy mildew pathogen is seen in North America, where its host *H. annuus* is one of the species of *Helianthus* with the widest geographic distribution (Heiser 1969; Rogers et al. 1982). On wild sunflower, downy mildew infection is remarkably rare (the US National Fungus Collection, BPI, stores specimens of 11 *Helianthus* species infected with downy mildew, while the genus is comprised of 49 species) and epidemics on wild populations have never been reported. This makes wild *H. annuus* a potential source for disease resistance and raises the question of how and from where did sunflower downy mildew start its world-wide spread.

Although nutritional use of sunflower is known from Indian tribes in North America since ca. 5,000 years (for review see Putt 1997) and sunflower varieties selected by the Havasupai and Hopi Indians are assumed to be the potential origin of the modern cultivated crop (Heiser 1976; Gulya 1992), intensive sunflower agriculture did not originate in the U.S., but was initiated in the late 19th century by Russian emigrants. It is noteworthy that downy mildew was not known from Europe until the 1940s when *P. halstedii* occurred in former Yugoslavia and later in other East European countries (for review see Sackston 1992), while first reports of sunflower downy mildew in North American sunflower fields are from the early 20th century (Young and Morris 1927; Bisby et al. 1938). This makes very unlikely the host jump from a downy mildew pathogen of wild European Asteraceae (e.g. *X. strumarium*) to cultivated sunflower and supports Sackston's (1992) theory of the North American origin of sunflower downy mildew. Most likely the export of unnoticed contaminated seeds (Meliala et al. 2000; Ioos et al. 2007) gave rise to the worldwide distribution of sunflower downy mildew.

Consecutive development of numerous pathotypes appears to be the natural result of sunflower breeding and subsequent selection of compatible pathogen genotypes rather than the result of independent acquisition from natural populations endemic to different geographic regions of the world. On the other hand, introduced populations of the pathogen may have subsequently evolved in different areas of the world to form the numerous pathotypes. Detailed molecular genetic studies on this topic are still lacking, but a recent investigation on French isolates from 14 pathotypes indicates that multiple introductions of different pathotypes in combination with selection pressure exerted by host resistance genes may have accelerated the evolution of pathotype diversity in France (Delmotte et al. 2008).

In this context it is noteworthy, that partial sequencing of a RNA mycovirus from sunflower downy mildew revealed identical nucleotide sequences in all *Plasmopara halstedii* strains of eight pathotypes originating from eight countries in three continents (Heller-Dohmen et al. 2008). This suggests contamination of the sunflower downy mildew pathogen with the virus before its worldwide distribution and pathotype segregation.

Genetic diversity—origin and molecular genetic aspects

Numerous attempts have been made in the past to differentiate sunflower downy mildew pathotypes by means of morphological and molecular genetic characters, but infection of sunflower differential lines is still the only practical test for pathotype classification. Nevertheless, molecular genetic investigations revealed a high level of genetic variability which is the basis for pathotyping.

The early applied techniques such as Restriction Fragment Length Polymorphism (RFLP) (Borovkov and McClean 1993), RAPD (Borovkova et al. 1992; Roeckel-Drevet et al. 1997), and Inter Simple Sequence Repeats (ISSR) (Intelmann and Spring

2002) analysis indicated genetic diversity between the samples investigated, but were insufficient to differentiate populations on the intraspecific level. Sequencing of the large subunit of ribosomal DNA was helpful to resolve the phylogeny of the Peronosporaceae including the genus *Plasmopara* (Riethmüller et al. 2002; Voglmayr et al. 2004) but was inappropriate to resolve intraspecific groups of *P. halstedii*. For the latter, the less conserved internal transcribed genomic region (ITS) appeared to be a more promising tool. In particular, the ITS2 region of *P. halstedii* attracted interest due to its enormous size in comparison to other Peronosporaceae (Thines et al. 2005a). This length polymorphism is caused by inserted repetitive elements, representing a synapomorphy of downy mildews with pyriform haustoria (Thines et al. 2005b). It is not only useful for interspecific phylogenetic studies, but is promising for investigations at the population level (Thines 2007a, b). Based on differences in partial sequence analysis of the nuclear ITS region, Spring et al. (2006) finally succeeded to differentiate between certain pathotypes of sunflower downy mildew originating from different geographic regions.

More recently, molecular genetic studies based on single nucleotide polymorphisms (SNPs) and size variation (insertions/deletions) in expressed sequence tags (ESTs) were performed using sunflower downy mildew isolates from France and Russia (Giresse et al. 2007). The authors found high genetic variability and the technique seems to be appropriate for high-throughput genotyping. Their current results showed a low level of heterozygosity, hence indicating that *P. halstedii* is a selfing species. This is concordant with the homothallic gametangiogamy found in single zoospore infection of sunflower with the pathogen (Spring 2000). On the other hand, homothally raises questions on the processes generating the observed genetic and phenotypic variability in sunflower downy mildew (Delmotte et al. 2008). Besides sexual recombination by occasional outcrossing, numerous mechanisms such as mitotic recombination, gene conversion, transposable elements or dispensable chromosomes were found to account for the genetic variability in other oomycetes (Chamnanpunt et al. 2001; Kamoun 2003). In addition, zoospore fusion was found to create asexual hybrids in *Phytophthora* species (Érsek et al. 1995; Érsek and Nagy 2008). Dual infections of sunflower with downy mildew strains differing in pathogenicity and fungicide tolerance have recently resulted in mitotically derived recombinant offspring, hence giving evidence for asexual genetic recombination in *P. halstedii* (Spring and Zipper 2006). More recent experiments showed that such unusual genetic recombination may even involve different species of *Plasmopara*, as long as they are able to infect the same host tissue (Hammer 2009). Although the mode of gene transfer has not yet been clarified, anastomosis appears to be a possible way to transfer nuclei, mitochondria or just DNA fragments between hyphae of two individuals (Hammer and Spring 2006). For unequivocal proof of gene transfer via anastomosis, visual staining for microscopic identification of the participating mycelia is required, but so far, attempts to transform sunflower downy mildew have resulted in transient expression only (Hammer et al. 2007). Similar experiences have been reported for experiments to transform another biotrophic pathogen of the genus, *P. viticola* (Dubresson et al. 2008). In this case, *gfp* expression in grape downy mildew lasted over four generations.

Pathogen-host interaction, resistance phenomena

The interaction between sunflower and *Plasmopara halstedii* can be described with a typical gene-for-gene relationship where the different pathotypes each having specific virulence character will meet sunflower genotypes with or without effective resistance gene(s) against them. Although genetic resistance in sunflower to *P. halstedii* has long been described, and a series of Pl genes identified and incorporated into downy mildew resistant sunflower cultivars, there remained many questions to be answered in order to effectively control this devastating sunflower disease over the long term. In fact, researchers in different countries are working hard to understand better the mechanism, as well as the genetic and molecular basis of resistance, searching for new genes or gene clusters conferring resistance, selecting for such genes using molecular markers in breeding programmes and looking for alternative ways to confer resistance in sunflower.

Expression of resistance

P. halstedii infection of sunflower usually takes place in below ground plant parts by directly penetrating root and/or hypocotyl tissues resulting in systemic downy mildew infection. It had already been shown that chemical stimuli secreted by the roots attract zoospores that move towards the root surface, thus enhancing successful penetration (Delanoe 1972). Unfortunately, there are no data available so far as to whether or not any difference exists in the reactions of either compatible or incompatible interactions regarding this phenomenon nor have such signals been identified.

Postpenetration events in sunflower may vary depending on the pathogen–host interaction, i.e. the type of resistance (Mouzeyar et al. 1993; Virányi and Gulya 1996). Accordingly, the host may strictly stop pathogen development at or near the penetration site, or alternatively, it may allow colonization of root, hypocotyl or even cotyledon tissues. However, in most cases, resistance is associated with restricted or intensive host cell death. In other words, sunflower reacts to infection with a hypersensitive reaction (HR) a well-known resistant host response to pathogen attack (Virányi 1980; Mouzeyar et al. 1993). However, the genetic background of this phenomenon was not known. Recently, Herbette et al. (2003) suggested the involvement of GPX (glutathione peroxidase) enzymes in the hypersensitive response in sunflower and Radwan et al. (2005) using Real Time Polymerase Chain Reaction (RT-PCR) analysis showed that resistance was associated with the activation of a hsr203J-like gene, a molecular marker of HR in tobacco. Activation of this gene was specifically observed during the incompatible interaction and coincided with cell collapse in sunflower hypocotyls. Interestingly, this reaction is tissue specific and did not occur in leaves. It may, therefore, explain, at least in part, the observed types of latent and/or local infections of sunflower. The former allows the pathogen to survive and even sexually reproduce in roots (Heller et al. 1997) and/or hypocotyls (Virányi 1980) of virtually resistant host genotypes, while the latter prohibited pathotype differentiation on leaf discs (Rozynek and Spring 2001) and may give rise to floral infection which contributes to pathogen dissemination through apparently healthy seeds (Meliala et al. 2000; Spring 2001).

In a recent study, Chaki et al. (2009) investigated the role of reactive oxygen and nitrogen species (ROS, RNS) in sunflower susceptible or resistant to *P. halstedii* by using confocal laser scanning microscopy (CLSM) and biochemical analyses. They analyzed the superoxide radical production, hydrogen peroxide content, L-arginine-dependent nitric oxide synthase and S-nitrosoglutathione reductase activities, and their location. Their results showed that the pathogen induced nitrosative stress in the susceptible but not in the resistant sunflower.

Resistance genes, gene clusters

Considerable efforts had been undertaken since the 1990s to locate resistance genes in the sunflower genome and to use the knowledge in plant breeding. In an earlier study, Vear et al. (1998) using RFLP markers pointed out that Pl4 appears not linked or allelic with Pl2 and that Pl5 is not located in the cluster including Pl1, Pl2 and Pl6. Similarly, Brahm et al. (1998) could construct linkage maps that represented the genomic region carrying the respective resistance loci. By homology cloning, Gentzbittel et al. (1998) obtained probes by which they demonstrated that at least three NBS-like loci were located on linkage group 1, i.e. in the region where downy mildew resistance loci had been described. More recently, Radwan et al. (2004) in France and Dußle et al. (2004) in Germany achieved considerable results with PCR markers for the Pl5/Pl8 locus from complete CC-NBS-LLR sequences and with the localization of the Pl_{ARG} gene using SSR markers, respectively. Furthermore, Pankovic et al. (2007) developed two co-dominant cleaved amplified polymorphic sequence (CAPS) markers that completely co-segregated with the Pl6 gene conferring resistance to pathotype 730. They concluded that CAPS markers will facilitate efficient marker-assisted selection for sunflower resistance to downy mildew pathotype 730.

Sources and inheritance of resistance

The frequency of Pl genes in wild sunflower species is high, especially in the annual species, some of these having resistance for all known pathotypes. Whereas it may occur that a single species has several genes to control a single pathotype (Seiler 1998), Terzic et al. (2007) included in their study 29 populations of five wild annual sunflowers that were crossed with cultivated ones to find out new sources of resistance to downy mildew. Their results suggested that the annual wild sunflower species, particularly *H. annuus* and *H. argophyllus,* could be used to obtain such resistance. Inheritance studies carried out in Spain (Molinero-Ruiz et al. 2003) and in Serbia (Pankovic et al. 2005) documented that resistance to pathotypes 310 and 703, or to pathotype 730 were controlled by single dominant genes.

Quantitative or non-race-specific resistance

Genetic resistance of sunflower to downy mildew, conferred by major gene(s) is effective but will be overcome by the appearance of new pathogenic forms of *P. halstedii* possessing an altered virulence profile. Within a new breeding programme recently started in France, Tourvieille de Labrouhe et al. (2005) showed that pyramiding, alternation or mixtures of a few major genes appeared to increase the duration of effective control of downy mildew. As an alternative, resistance of sunflower which does not depend on the host/pathogen interaction has been studied by Vear et al. (2006) in large scale trials for 3 years. A significant level of partial resistance was shown in cultivated sunflower lines that do not have Pl genes effective against the pathotypes present. This resistance appeared independent of pathogen virulence phenotype and its heredity was under additive control. This type of quantitative resistance was also tested by Serre et al. (2008) under growth chamber conditions and the importance of plant age and environment interaction during host resistance response was pointed out. Partial resistance to *P. halstedii* in high oleic sunflower hybrids has been reported as well by Baldini et al. (2006) in Italy. Quite recently, in summarizing their 4-year-study, Tourvieille de Laboruhe et al. (2009) suggested that it should be possible to select for non-race-specific downy mildew resistance and to include it in modern varieties. However, since this type of resistance is partial, it may be necessary to combine it with major gene resistance.

Induced resistance

Apart from genetic and durable (quantitative/non-race-specific/multiple) resistance, chemically induced resistance might be considered as an additional and useful element in the integrated management against sunflower downy mildew. The phenomenon has already been known for years (reviewed by Sticher et al. 1997) and the plant activator benzothiadiazole (BION 50 WP as commercial product) has been used successfully against powdery mildew of wheat. In addition, researchers from Hungary and Italy also reported the positive effect of this compound in depressing downy mildew symptoms under field and greenhouse conditions, respectively (Tosi et al. 1999; Bán et al. 2004); and a remarkable increase in induced resistance against broomrape attack by *Orobanche cumana* was evident due to benzothiadiazole treatment of sunflower in Germany (Sauerborn et al. 2002). Furthermore, microscopical observations on treated and inoculated sunflower tissues revealed well-defined and characteristic alterations as host responses to infection and such changes resembled those occurring with Pl-gene mediated resistant plants following infection (Bán et al. 2004). Investigations in different laboratories have been focusing on the biochemical and molecular genetic background of this induced resistance mechanism. For example, Serrano et al. (2007) in Spain reported an increased peroxidase and chitinase activity in BTH-treated and inoculated susceptible and resistant sunflower seedlings. In Hungary, investigations have been made on the expression of genes allegedly associated with resistance response by measuring the accumulation of the transcripts of glutathion-S-transferase (GST), defensin (PDF), and catalase (CAT) following treatments with benzothiadiazole (BTH), dichloroisonicotinic acid (INA), or beta-aminobutyric acid (BABA) and/or *P. halstedii* inoculation. Preliminary results revealed significant increases in the activity of these genes in the susceptible but not in the resistant sunflowers, whereas the results with the partially resistant (HLI-type) sunflower were contradictory (Körösi and Virányi 2008). Apart from the plant activators, mentioned above, Nandeeshkumar et al. (2008a) reported about an enhanced and early activation of defence-related responses to downy mildew in a susceptible sunflower cultivar treated with chitosan. Using Northern hybridization analysis they revealed an increased level of transcripts for five known defence response genes. Furthermore, in a separate experiment, they also gained positive results with plant growth promoting rhizobacteria that were associated with an enhanced activation of catalase, phenylalanine ammonia lyase, peroxidase, polyphenoloxydase and chitinase (NandeeshKumar et al. 2008b). In addition, essential oil of a plant species *Bupluerum gibraltarium* was assumed to act as an activator of plant defence in sunflower attacked by *P. halstedii* (Fernandez-Ocana et al. 2004). Perazzolli et al. (2008) demonstrated that besides benzothiadiazole, *Trichoderma harzianum* T39 strain effectively protected susceptible grapevine cultivars against downy mildew (*Plasmopara viticola*) under greenhouse conditions. Although no such experiments have yet been performed for the sunflower–*P. halstedii* pathosystem, the results with grapevine downy mildew might be of interest. They found differences between the two biotic and chemical inducers in terms of their activation time, persistence of the effect and systemic protection, suggesting that different pathways other than that of salicylic-acid dependent BTH elicitation are activated by *T. harzianum* in grapevine. It is still worth mentioning that plant activators (chemical resistance inducers) usually do not cause any retardation in plant growth or development of the treated crop plant but such unexpected side-effects might occur under specific conditions and/or with specific pathogen–host relations (LaMondia 2008).

Epidemiology and disease management

Disease incidence and severity, population dynamics

The incidence and severity of downy mildewed sunflower plants in the field has been reported by earlier investigators to range between traces to near 50% or even up to 80%, primarily depending on inoculum potential, the cultivar used, and the environmental conditions (for reference see Sackston 1981). Additionally, the time of evaluation can be crucial for estimating the degree of infection severity, as Gulya (2006) reported from the 2005 North Dakota (USA) sunflower crop situation. Based on field observations he concluded that late season surveys generally underestimate the impact of early season downy mildew, primarily because a portion of affected plants die early in the season and are not considered in autumn surveys. *Plasmopara halstedii* is present worldwide except for Australia where it has not been detected on sunflower so far. An up-to-date record of its global distribution was recently published by Gulya (2007) who listed as many as 35 pathotypes (virulence phenotypes). He found a considerable variation, by continent, on the number of pathotypes, Asia and South America having the fewest number of pathotypes identified followed by Africa, Europe and North America. The countries with the highest numbers of pathotypes are Canada, France and the USA. There could be various explanations for the appearance of new virulent forms (Delmotte et al. 2008), but the increase in pathotypes during the last decade is probably due to the introduction of new cultivars with different genetic pedigrees and to the intensity of production, particularly in Europe. Though it is known that local spread of this pathogen within a field is mainly by soil particles (e.g. during tillage) and wind-borne sporangia from adjacent infected sunflower plants, its introduction into new countries and continents can only be by seed.

Factors affecting disease development

P. halstedii as an oomycete producing motile zoospores, is considered to exclusively depend on the presence of free water to complete its life cycle. In this context, the most susceptible stage of host development is between germination and emergence (Meliala et al. 2000). As an example, severe downy mildew outbreaks were observed in 2005 in some North Dakota fields where unusually high amounts of precipitation occurred during June, immediately following planting (Gulya 2006). Similarly, Tourvieille et al. (2008a) investigated the incidence of spring rainfall on the severity of primary downy mildew attack of sunflower in field trials with staggered sowing dates. Disease risk appeared greatest if there was heavy rainfall when sunflower seedlings were at their most susceptible stage, whereas heavy rainfall before sowing or after emergence had no effect on the percentage of diseased plants. Furthermore, Göre (2009) recorded a serious outbreak of sunflower downy mildew in Turkey during the spring of 2007 and 2008. He thought that low temperature and extensive spring rains encouraged the disease, resulting in approximately 85% yield loss and lower quality of sunflower production in the Marmara region of Trace. Interestingly, Baldini et al. (2006), aiming at developing a prediction model, studied the main climatic factors affecting development and spread of downy mildew in high-oleic sunflower. By calculating the accumulated rainfall and average temperatures for different sowing date conditions, they did not find water availability as a limiting factor between accumulated rainfall and percentage of infection. In contrast, air temperature clearly had an effect on the percentage of disease, the most favourable mean air temperatures being between 10–15°C during the 5 days after sowing.

Besides environmental conditions, disease intensity may also be influenced by the aggressiveness of the pathogen population. Recently, French researchers measured this important trait of *P. halstedii* by using pathotypes 100 and 710, as well as two sunflower lines differing in their level of quantitative resistance (Sakr et al. 2009). Based on two criteria, i.e., the latent period and sporulation density, they successfully differentiated between the two pathogen strains in terms of their aggressiveness.

Soilborne vs. airborne inoculum

Apart from the soilborne nature of *P. halstedii,* since oospores of this pathogen give rise to systemically infected plants (Spring and Zipper 2000), local lesions on leaves as a result of infection by wind-borne sporangia may occasionally occur in some areas. The first report of a massive appearance of this disease symptom was given as early as the 1970s by Zimmer (1972) in the United States, whereas in Europe such secondary local infections could hardly be seen until the late1990s. In a field survey in 1995 in Hungary, an unusual disease appearance was observed: numerous plants with angular leaf spots, some of which were located close to the main veins or near the end of petiole, were observed (Rátainé Vida 1996). From such leaves the pathogen tended to spread along the veins reaching the petiole. Subsequent microscopic observations gave evidence of a close correlation between the presence of hyphae in leaf and corresponding petiole (Virányi, unpublished). In fact, such secondary systemic spread of the pathogen has become more common in different sunflower cultivars under extremely favourable (wet and cool) weather conditions. Similar to Hungary, a high potential for transition from local to systemic infection was found by Spring (2008) in Germany as well. Though localized lesions are usually not considered to be of economic importance, in the case of seed production, seeds from such plants may carry-over the pathogen to the next season and to various locations (Spring 2008).

Detection, risk assessment

Considering the risk that seed transmission of *P. halstedii* poses to unexpected propagation of new pathogenic variants or fungicide-tolerant strains into other regions, it is necessary to guarantee the absence of the pathogen in seed shipments. With the rapid improvement of detection technology, it became more probable to find a method that could be sufficient and productive in monitoring sunflower seed samples for latent infection. Various methods have been used to develop diagnostic tools for the identification of this pathogen in sunflower plant tissue and/or seed samples. Roeckel-Drevet et al. (1999) generated an oligonucleotide primer for the selective amplification of *P. halstedii,* Bouterige et al. (2000) employed an enzyme linked immunosorbent assay (ELISA) test, and Spring and Haas (2004) were successful with fatty acid patterns as markers of infected vs. healthy sunflower seeds. Recently, Ioos et al. (2007) developed a method for the direct detection of the pathogen in seed samples based on the ribosomal large sub unit DNA. Though their test was able to detect a near 1:50 ratio of seed contamination, this was not practical, probably because of the problem of representative sampling. Furthermore, from the practical point of view it might be of interest for sunflower growers to know in advance, i.e. before sowing what pathotype (s) exist(s) and to what extent in a particular field. For this reason, Gulya (2004) and Tourvieille et al. (2008b) developed a bioassay using soil samples to assess downy mildew risk at the field level. Under different geographical and climatic conditions, their results revealed a good correlation between the soil infestation measured by the soil bioassay and the presence of infected plants in previous years.

Chemical control

The possibility of effective fungicidal control against sunflower downy mildew is closely associated with the introduction and use of metalaxyl in the late 1970s. This systemic, phenylamide-type compound was considered to be very effective against downy mildew of sunflower for almost two decades. However, the first tolerant strains could be identified in Hungary (Oros and Virányi 1984) and in Turkey (Delen et al. 1985), and expanded field populations tolerant to this fungicide have been described from France (Lafon et al. 1996), the United States (Gulya et al. 1999), Spain (Molinero-Ruiz et al. 2000) and Germany (Spring et al. (2006). The widespread occurrence of *P. halstedii* populations insensitive to metalaxyl/mefenoxam (the latter being a pure stereoisomer of the former) prompted researchers to look for alternative compounds in order to combat such fungicide tolerance. For example, in the United States, as many as 30 fungicides have been tested as seed treatment for the efficacy against sunflower downy mildew and even the most effective compound, fenamidone, allowed 3–16 % infection in field trials (Gulya 2002b). No report is available so far for the authors regarding the appearance of *P. halstedii* tolerance to fungicides other than phenylamides.

Outlook

In this review we tried to outline the recent progress in research and development of sunflower downy mildew involving various aspects of the pathogen's taxonomy and biology, host resistance, and the interaction between host and pathogen. Although our aim was to give an account of the latest experimental results published following the review by Gulya and co-workers in 1997, we often realized that this time-limit might hinder our general philosophy: besides providing an up-to-date and sufficient information package of the progress achieved in downy mildew research over the last decade, the continuous character of research activity was kept in mind as well. This makes it necessary to enhance some former aspects and to re-evaluate previous results in light of new experimental possibilities, particularly as contributed by molecular genetic techniques.

By reading through a number of scientific reports in journals and other periodicals we also realized that questions remained unanswered and some new questions arised. For example, future studies should 1) elucidate the final taxonomic status and the exact host range of this oomycete pathogen, and 2) clarify the origin of genetic variation within the species pathogenic on sunflower, and the local/non-local appearance of virulence phenotypes. Furthermore, until a reliable molecular method of pathotype identification is firmly established, the traditional methodology based on the interpretation of reactions on sunflower differential lines should be improved and the number of differentials possibly expanded. This includes the establishment of a valid system for the cultivation and distribution of seed samples required for pathotyping new isolates. New sources of resistance to downy mildew, including non-race-specific or durable resistance, need to be identified and special attention should also be paid to alternative types of resistance such as the use of abiotic/biotic resistance inducers. Finally, continuing use of chemical control will be required, but its efficacy should be improved by the discovery of new molecules and/or mixtures of various compounds in order to combat fungicide tolerance.

References

Adhikari, T. B., Liu, J. Q., Mathur, S., Wu, C. X., & Rimmer, S. R. (2003). Genetic and molecular analyses in crosses of race 2 and race 7 of *Albugo candida*. *Phytopathology, 93*, 959–965.

Baldini, M., Danuso, F., Turi, M., Sandra, M., & Raranciuc, S. (2006). Main factors influencing downy mildew (*Plasmopara halstedii*) infection in high-oleic sunflower hybrids in Northern Italy. *Crop Protection, 27*, 590–599.

Bán, R., Virányi, F., & Komjáti, H. (2004). Benzothiadiazole-induced resistance to *Plasmopara halstedii* (Farl.) Berl. et de Toni in sunflower. In: P. Spencer-Phillips & M. Jeger (eds.) Advances in Downy Mildew Research Vol 2 (pp. 265–273), Kluwer Academic Publishers

Berlese, A. N., & De Toni, J. B. (1888). Phycomycetae. In P.A. Saccardo (ed.), *Sylloge Fungorum Omnium Hucusque Cognitorum*, Vol 7 (pp. 181–322).

Bisby, G. R., Buller, A. H. R., Dearness, J., Fraser, W. P., & Russell, R. C. (1938). *The fungi of Manitoba and Saskatchewan*. Ottawa: Nature Research Council.

Borovkov, A. Y., & McClean, P. E. (1993). A tandemly repeated sequence from the *Plasmopara halstedii* genome. *Gene, 123*, 127–130.

Borovkova, I. G., Borovkov, A. Y., McClean, P. E., Gulya, T. J. & Vick, B. A. (1992). Restriction fragment length polymorphisms and RAPD markers in DNA of *Plasmopara halstedii*, the downy mildew fungus of sunflower. *Proceedings of the 13th International Sunflower Conference* (pp. 1420–1425) Pisa, Italy.

Bouterige, S., Robert, R., Bouchara, J. P., Marot, L. A., Molinero, V., & Senet, J. M. (2000). Production and characterization of two monoclonal antibodies specific for *Plasmopara halstedii*. *Applied and Environmental Microbiology, 66*, 3277–3282.

Brahm, L., Röcher, T., Horn, R., Prüfe, M., Köhler, H., & Friedt, W. (1998). Mapping Downy mildew resistance in sunflower. In: ISA Symposium III (pp. 103–110). Sunflower Downy Mildew, Fargo, ND, 13-14 January 1998.

Chaki, M., Fernandez-Ocana, A. M., Valderrama, R., Carreras, A., Esteban, F. J., Luque, F., et al. (2009). Involvement of reactive nitrogen and oxygen species (RNS and ROS) in sunflower-mildew interaction. *Plant & Cell Physiology, 50*, 665–679.

Chamnanpunt, J., Shan, W.-X., & Tyler, B. M. (2001). High frequency mitotic gene conversion in genetic hybrids of

the oomycete *Phytophthora sojae*. *Proceedings of the National Academy of Sciences of the United States of America, 98*, 14530–14535.
Christova, E., & Mitov, N. (1960). Mnata po slinchogle—Plasmopara halstedii (Farl) Berl. et de Toni v Bulgaria i sredstva za borba c neya. *Bolgarsk. Akad. Nauk. otd. S-kh. Nauk. Tsentr. N. Issl-Inst. zashch, rast. Nauch. Tr. III, 381–97* [cited from Novotelnova 1966].
Cohen, Y., & Sackston, W. E. (1973). Factors affecting infection of sunflower by *Plasmopara halstedii*. *Canadian Journal of Botany, 51*, 15–22.
Delanoe, D. (1972). Biologie et epidemiologie du mildiou du tournesol (*Plasmopara helianthi* Novot.). *CETIOM Informations Techniques, 29*, 1–49.
Delen, N., Onogur, E., & Yilfiz, M. (1985). Sensitivity levels to metalaxyl in six *Plasmopara helianthi* Novot. isolates. *J. Turkish Phytopath, 14*, 31.
Delmotte, F., Giresse, X., Richard-Cervera, S., M'Baya, J., Vear, F., Tourvieille, J., et al. (2008). Single nucleotide polymorphisms reveal multiple introductions into France of *Plasmopara halstedii*, the plant pathogen causing sunflower downy mildew. *Infection, Genetics and Evolution, 8*, 534–540.
Dubresson, R., Kravchuk, Z., Neuhaus, J. M., & Mauch-Mani, B. (2008). Optimisation and comparison of transient expression methods to express the green fluorescent protein in the obligate biotrophic oomycete *Plasmopara viticola*. *Vitis, 47*, 235–240.
Duβle, C. M., Hahn, V., Knapp, S. J., & Bauer, E. (2004). Pl_{Arg} from *Helianthus argophyllus* is unlinked to other known downy mildew resistance genes in sunflower. *Theoretical and Applied Genetics, 109*, 1083–1086.
Érsek, T., & Nagy, Z. A. (2008). Species hybrids in the genus Phytohthora with emphasis on the alder pathogen *Phytophthora alni*: a review. *European Journal of Plant Pathology, 122*, 31–39.
Érsek, T., English, J. T., & Scholz, J. E. (1995). Creation of species hybrids of Phytophthora with modified host ranges by zoospore fusion. *Phytopathology, 85*, 1343–1347.
Farlow, W. G. (1883). Note on some species in the third and eleventh centuries of Ellis's North American Fungi. *Proceedings of the American Academy of Arts and Sciences, 18*(10), 71–73.
Fernandez-Ocana, A. M., Gomez-Rodriguez, M. V., Velasco-Negueruela, A., Camacho-Simarro, A. M., Fernandez-Lopez, C., & Altarejos, J. (2004). In vivo antifungal activity of the essential oil of *Buplueurum gibraltarium* against *Plasmopara halstedii* in sunflower. *Journal of Agricultural and Food Chemistry, 52*, 6414–6417.
Gäumann, E. (1923). Beiträge zu einer Monographie der Gattung *Peronospora* Corda. *Beiträge zur Kryptogamenflora der Schweiz, 5*, 1–360.
Gentzbittel, L., Mouzeyar, S., Badaoui, S., Vear, F., Tourvieille de Labrouhe, D., & Nicolas, P. (1998). Cloning of molecular markers for disease resistance in sunflower *Helianthus annuus* L. In: *ISA Symposium III. Sunflower Downy Mildew* (pp. 94–102). *Fargo* ND, 13-14 January 1998.
Giresse, X., De Labrouhe, D. T., & Richard-Cervera, S. (2007). Twelve polymorphic expressed sequence tags-derived markers for *Plasmopara halstedii*, the causal agent of sunflower downy mildew. *Molecular Ecology Notes, 7*, 1363–1365.
Göre, M. E. (2009). Epidemic outbreaks of downy mildew caused by *Plasmopara halstedii* on sunflower in Thrace, part of the Marmara region of Turkey. *Plant Pathology, 58*, 396.
Gulya, T. J. (1992). Native American variety may provide sunflower crop with crucial resistance. *NPGS Network News, 8*, 29–30.
Gulya, T. J. (2002a). First report of the cross-infectivity of *Plasmopara halstedii* from marshelder to sunflower. *Plant Disease, 86*, 919.
Gulya, T. J. (2002b).Efficacy of single and two-way fungicide seed treatments for the control of metalaxyl-resistant strains of *Plasmopara halstedii* (sunflower downy mildew). In: *Proceedings of Brighton Crop Protection Conference, Brighton* (pp. 575–580). UK, November 18–22.
Gulya, T. J. (2004). A seedling bioassay to detect the presence of *Plasmopara halstedii* in soil. In P. T. N. Spencer-Phillips & M. Jeger (Eds.), *Advances in Downy Mildew research* (Vol. 2, pp. 233–240). Dordrecht: Kluwer Academic Publishers.
Gulya, T. J. (2006). The downy mildew situation on the 2005 North Dakota sunflower crop. In: *Proceedings of the NSA Sunflower Research Workshop 2006, Fargo, ND.*
Gulya, T. J. (2007) Distribution of *Plasmopara halstedii* races from sunflower around the world. *Advances in Downy Mildew Research* Vol 3, Palacky University and JOLA Publishers (pp. 121–134).
Gulya, T., Rashid, K. Y., & Masirevic, S. M. (1997). Sunflower diseases. In A. A. Schneider (Ed.), *Sunflower Technology and Production* (pp. 263–380). Madison: American Society of Agronomy.
Gulya, T., Draper, M., Harbour, J., Holen, C., Knodel, J., & Lamey, A. et al. (1999). Metalaxyl resistance in sunflower downy mildew in North America. In: *Proceedings of the 21st NSA Sunflower Research Workshop* (pp. 1118–123). *Bismarck,* ND.
Hammer, T. R. (2009). Untersuchungen zum asexuellen Gentransfer bei biotrophen Oomyceten anhand der Fallbeispiele *Plasmopara halstedii* und *Peronospora tabacina*. Dissertation, University of Hohenheim (pp. 1–128).
Hammer, T. R. & Spring, O. (2006). Kernaustauschvorgänge durch Anastomosen bei *Plasmopara halstedii*. In: *Proceedings of the 55th Deutsche Pflanzenschutztagung*, Göttingen, p. 206.
Hammer, T. R., Thines, M., & Spring, O. (2007). Transient expression of gfp in the obligate biotrophic oomycete *Plasmopara halstedii* using electroporation and a mechanoperforation method. *Plant Pathology, 56*, 177–182.
Heiser, C. B. (1969). The North American sunflowers. *Memoirs of the Torrey Botanical Club, 22*, 1–218.
Heiser, C. B. (1976). *The sunflower*. Norman: University of Oklahoma Press.
Heller, A., Rozynek, B., & Spring, O. (1997). Cytological and physiological reasons for the latent type of infection in sunflower caused by *Plasmopara halstedii*. *Journal of Phytopathology, 145*, 441–445.
Heller-Dohmen, M., Göpfert, J. C., Hammerschmidt, R., & Spring, O. (2008). Different pathotypes of the sunflower downy mildew pathogen *Plasmopara halstedii* all contain isometric virions. *Molecular Plant Pathology, 9*, 777–786.
Henry, A. W., & Gilbert, H. C. (1924). Important fungous diseases of the common sunflower. *Minnesota Studies Plant Science, 5*, 285–305.

Herbette, S., Lenne, C., de Labrouhe, D. T., Drevet, J. R., & Roeckel-Drevet, P. (2003). Transcripts of sunflower antioxidant scavangers of the SOD and GPX families accumulate differentially in response to downy mildew infection, phytohormones, reactive oxygen species, nitric oxyde, protein kinase and phosphatase. *Physiologia Plantarum, 119*, 418–428.

Intelmann, F., & Spring, O. (2002). Analysis of total DNA by minisatellite and simple-sequence repeat primers for the use of population studies in *Plasmopara halstedii*. *Canadian Journal of Microbiology, 48*, 555–559.

Ioos, R., Laugustin, L., Rose, S., Tourvieille, J., & Tourvieille de Labrouhe, D. (2007). Development of a PCR test to detect the downy mildew causal agent *Plasmopara halstedii* in sunflower seeds. *Plant Pathology, 56*, 209–218.

Kamoun, S. (2003). Molecular genetics of pathogenic oomycetes. *Eukaryotic Cell, 2*, 191–199.

Komjáti, H., Walcz, I., Virányi, F., Zipper, R., Thines, M., & Spring, O. (2007). Characteristics of a *Plasmopara angustiterminalis* isolate from *Xanthium strumarium*. *European Journal of Plant Pathology, 119*, 421–428.

Komjáti, H., Bakonyi, J., Spring, O., & Virányi, F. (2008). Isozyme analysis of *Plasmopara halstedii* using cellulose-acetate gel electrophoresis. *Plant Pathology, 57*, 57–63.

Körösi, K., & Virányi, F. (2008). Molecular changes in downy mildew-infected sunflower triggered by resistance inducers. In: Velasco L. (ed.) *Proceedings of the 17th International Sunflower Conference* Vol 1, (pp. 157–161). *Cordoba,* Spain, 8–12 June 2008.

Lafon, S., Penaud, A., Walser, P., De Guenin, M.-C. H., Molinero, V., Mestres, R., et al. (1996). Le mildiou du tournesol toujours sous sourveillance. *Phytoma, 484*, 35–36.

LaMondia, J. A. (2008). Efficacy of fungicides and a systemic acquired resistance activator (acibenzolar-S-methyl) against tobacco blue mold. *Crop Protection, 28*, 72–76.

Leppik, E. E. (1966). Origin and specialization of *Plasmopara halstedii* complex in the Compositae. *FAO Plant Protection Bulletin, 14*, 72–76.

Meliala, C., Vear, F., & Tourvieille de Labrouhe, D. (2000). Relation between date of infection of sunflower downy mildew (*Plasmopara halstedii*) and symptoms development. *Helia, 23*, 35–44.

Molinero-Ruiz, M. L., Dominguez, J., Melero-Vara, J. M., & Gulya, T. J. (2000). Tolerance to metalaxyl in Spanish isolates of *Plasmopara halstedii*. In: *Proceedings of 15th International Sunflower, Conference, Toulouse, France,* 12–15 June, Vol. 2, G11–G15.

Molinero-Ruiz, M. L., Melero-Vara, J. M., & Dominguez, J. (2003). Inheritance of resistance to two races of sunflower downy mildew (*Plasmopara halstedii*) in two *Helianthus annuus* L. lines. *Euphytica, 131*, 47–51.

Mouzeyar, S., Delabrouhe, D. T., & Vear, F. (1993). Histopathological studies of resistance of sunflower (*Helianthus annuus* L.) to downy mildew (*Plasmopara halstedii*). *Journal of Phytopathology, 139*, 289–297.

NandeeshKumar, P., Sudisha, J., Ramachandra, K. K., Prakash, H. S., Niranjana, S. R., & Shekar, S. H. (2008). Chitosan induced resistance to downy mildew in sunflower caused by *Plasmopara halstedii*. *Physiological and Molecular Plant Pathology, 72*, 188–194.

NandeeshKumar, P., RamachandraKini, K., Prakash, H. S., Niranjana, S. R., & Shekar Shetty, H. (2008). Induction of resistance against downy mildew on sunflower by rhizobacteria. *Physiological and Molecular Plant Pathology, 72*, 188–194.

Nishimura, M. (1922). Studies in *Plasmopara halstedii*. *Journal of the College of Agriculture, Hokkaido Imperial University, 11*, 185–210.

Novotelnova, N. S. (1962). *Plasmopara halstedii* (Farl.) Berl. et de Toni as a collective species (Basis for taxonomical subdivision of genus *Plasmopara* on Compositae). *Bot Zhurnal, 47*, 970–981.

Novotelnova, N. S. (1964). Polymorphism of peronosporous fungi from genus *Plasmopara*. *Bot. Zhurnal, 50*, 301–312.

Novotelnova, N. S. (1966). Downy mildew of sunflower. Izdatelstvo "Nauka", Moskva-Leningrad. Translated from Russian, Indian National Scientific Documentation, Hillside Road New Delhi.

Oros, G., & Virányi, F. (1984). Resistance of *Plasmopara halstedii* to metalaxyl in the greenhouse. *Temperate Downy Mildews Newsletter, 3*, 22–23.

Pankovic, D., Radovanovic, N., Jocic, S., & Skoric, D. (2005). Inheritance of resistance to sunflower downy mildew conferred by *Pl6* gene, investigated by whole seed inoculation technique and PCR markers. In: *10th FAO Consultation Meeting* (p. 10). *Novi Sad,* Yugoslavia, 17–20 July 2005.

Pankovic, D., Radovanovic, N., Jocic, S., Satovic, Z., & Skoric, D. (2007). Development of co-dominant amplified polymorphic sequence markers for resistance of sunflower to downy mildew race 730. *Plant Breeding, 126*, 440–444.

Perazzolli, M., Dagostin, S., Ferrari, A., Elad, Y., & Pertot, I. (2008). Induction of systemic resistance against *Plasmopara viticola* in grapevine by *Trichoderma harzianum* T39 and benzothiadiazole. *Biological Control, 47*, 228–234.

Pustovoit, G. V. (1963). Selection of sunflower for group immunity as a method of interspecific hybridization. *Trudy za 1912–1962 gg. Vniimk, Selkhozgiz, M.: 75–92.* [cited from Novotelnova 1966].

Pustovoit, G. V. (1964). Selection of sunflower for group immunity as a method of interspecific hybridization. *Dokl, Vaskhnil, 1*, 2–7 [cited from Novotelnova 1966].

Putt, E. D. (1997). Early history of sunflower. In A. A. Schneider (Ed.), *Sunflower technology and production* (pp. 1–19). Madison: American Society of Agronomy.

Radwan, O., Bouzidi, M. F., Nicolas, P., & Mouzeyar, S. (2004). Development of PCR markers for the *Pl5/Pl8* locus for resistance to *Plasmopara halstedii* in sunflower. Helianthus annuus L. from complete CC-NBS-LRR sequences. *Theoretical and Applied Genetics, 109*, 176–185.

Radwan, O., Mouzeyar, S., Venisse, J. S., Nicolas, P., & Bouzidi, M. F. (2005). Resistance of sunflower to the biotrophic oomycete *Plasmopara halstedii* is associated with a delayed hypersensitive response within the hypocotyl. *Journal of Experimental Botany, 56*, 2683–2693.

Rátainé Vida, R. (1996). Re-appearance of sunflower downy mildew. *Növényvédelem (Crop Protection), 32*, 533–535 (in Hungarian).

Riethmüller, A., Göker, M., Weiß, M., Oberwinkler, F., & Voglmayr, H. (2002). Phylogenetic relationships of the downy mildews (Peronosporales) and related groups based on nuclear large subunit ribosomal DANN sequences. *Mycologia, 94*, 834–849.

Roeckel-Drevet, P., Coelho, V., Tourvieille, J., Nicolas, P., & Tourvieille deLabrouhe, D. (1997). Lack of genetic variability in french identified races of *Plasmopara halstedii*, the cause of downy mildew in sunflower *Helianthus annuus*. *Canadian Journal of Microbiology, 43*, 260–263.

Roeckel-Drevet, P., Tourvieille, J., Drevet, J. R., Says-Lesage, V., Nicolas, P., & Tourvieille de Laboruhe, D. (1999). Development of a polymerase chain reaction diagnostic test for the detection of the biotrophic pathogen *Plasmopara halstedii* in sunflower. *Canadian Journal of Microbiology, 45*, 797–803.

Rogers, C. E., Thompson, T. E., & Seiler, G. J. (1982). *Sunflower species of the United States*. Bismarck: National Sunflower Association.

Rozynek, B., & Spring, O. (2001). Leaf disc inoculation, a fast and precise test for the screening of metalaxyl tolerance in sunflower downy mildew. *Journal of Phytopathology, 149*, 309–312.

Sackston, W. E. (1981). Downy mildew of sunflower. In D. M. Spencer (Ed.), *The downy mildews* (pp. 545–575). London: Academic.

Sackston, W. E. (1992). On a treadmill: breeding sunflowers for resistance to diseases. *Annual Review of Phytopathology, 30*, 529–551.

Sakr, N., Ducher, M., Tourvieille, J., Walser, P., Vear, F., & Tourvieille de Labrouhe, D. (2009). A method to measure aggressiveness of *Plasmopara halstedii* (Sunflower Downy Mildew). *Journal of Phytopathology, 157*, 133–136.

Sauerborn, J., Buschmann, H., Ghiasvand Ghiasi, K., & Kogel, K. H. (2002). Benzothiadiazole activates resistance in sunflower (*Helianthus annuus*) to the root-parasitic weed *Orobanche cumana*. *Phytopathology, 92*, 59–64.

Schardl, C. L., & Craven, K. D. (2003). Interspecific hybridization in plant-associated fungi and oomycetes: a review. *Molecular Ecology, 12*, 2861–2873.

Schröter, J. (1886). Fam. Peronosporacei. In F. Cohn (Ed.), *Kryptogamen-Flora von Schlesien* (pp. 228–252). Breslau: J. U. Kern.

Seiler, G. (1998). Wild sunflower as a potential source of resistance to downy mildew. In: *ISA Symposium III. Sunflower Downy Mildew* (pp. 113–121). *Fargo*, ND, 13-14 january 1998.

Serrano, A. R., Del Castillo, J. L., Novo, J. J., Fernandez Ocana, A., & Gomez Rodriguez, M. V. (2007). Chitinase and peroxydase activities in sunflower hypocotyls: effects of BTH and inoculation with *Plasmopara halstedii*. *Biologia Plantarum, 51*, 149–152.

Serre, F., Walser, P., Roche, S., Vear, F., & Tourvieille de Labrouhe, D. (2008). Research on a growth chamber test to measure quantitative resistance to sunflower downy mildew. In: Velasco L. (ed.) *Proceedings of the 17th International Sunflower Conference* Vol. 1 (pp. 109–114). *Cordoba*, Spain, 8–12 June 2008.

Spring, O. (2000). Homothallic sexual reproduction in *Plasmopara halstedii*, the downy mildew of sunflower. *Helia, 32*, 19–26.

Spring, O. (2001). Non-systemic infections of sunflower with *Plasmopara halstedii* and their putative role in the distribution of the pathogen. *Journal of Plant Diseases and Protection, 108*, 329–336.

Spring, O. (2008). Transition of secondary to systemic infection of sunflower with *Plasmopara halstedii*—An underestimated factor in the epidemiology of the pathogen. *Fungal Ecology, 2*, 75–80.

Spring, O., & Haas, K. (2004). Eicosapentaenoic acid, a possible marker for downy mildew contamination in sunflower seeds. In: P. Spencer Phillips & M. Jeger (eds.) Advances in Downy Mildew Research Vol 2 (pp. 241–248). Kluwer Academic Publishers.

Spring, O., & Thines, T. (2004). On the necessity of new characters for classification and systematics of biotrophic Peronosporomycetes. *Planta, 219*, 910–914.

Spring, O., & Zipper, R. (2000). Isolation of oospores of sunflower downy mildew, *Plasmopara halstedii*, and microscopical studies on oospore germination. *Journal of Phytopathology, 148*, 227–231.

Spring, O., & Zipper, R. (2006). Evidence for asexual genetic recombination in sunflower downy mildew, *Plasmopara halstedii*. *Mycological Research, 110*, 657–663.

Spring, O., Voglmayr, H., Riethmüller, A., & Oberwinkler, F. (2003). Characterization of a *Plasmopara* isolate from *Helianthus* x *laetiflorus* based on cross infection, morphological, fatty acids and molecular phylogenetic data. *Mycological Progress, 2*, 163–170.

Spring, O., Bachofer, M., Thines, M., Riethmüller, A., Göker, M., & Oberwinkler, F. (2006). Intraspecific relationship of *Plasmopara halstedii* isolates differing in pathogenicity and geographic origin based on ITS sequence data. *European Journal of Plant Pathology, 114*, 309–315.

Stevens, F. L. (1913). *The fungi which cause plant disease* (pp. 91–92). New York: Macmillan.

Sticher, L., Mauch-Mani, B., & Métraux, J. P. (1997). Systemic acquired resistance. *Annual Review of Phytopathology, 35*, 235–270.

Telle, S., & Thines, M. (2008). Amplification of cox2 (620 bp) from 2 mg of up to 129 years old herbarium specimens, comparing 19 extraction methods and 15 polymerases. *PLoS ONE, 3*(10), e3584. doi:10.1371/journal.pone.0003584.

Terzic, S., Dedic, B., Atlagic, J., & Masirevic, S. (2007). Transferring of *Plasmopara* resistance from annual wild into cultivated sunflower. In: *ISA First Symposium on Sunflower Industrial Use, Udine,* Italy, 11–13 September 2006, (Abstr.)

Thines, M. (2006). Evaluation of characters available from herbarium vouchers for the phylogeny of the downy mildew genera (Chromista, Peronosporales), with focus on scanning electron microscopy. *Mycotaxon, 97*, 195–218.

Thines, M. (2007a). Characterisation and phylogeny of repetitive elements giving rise to exceptional length of ITS2 in several downy mildew genera (Peronosporaceae). *Fungal Genetics and Biology, 44*, 199–207.

Thines, M. (2007b). Repeats in the ITS2 of *Plasmopara* species and their relevance for phylogenetic studies. *Advances in Downy Mildew Research, 3*, 31–35.

Thines, M., Komjati, H., & Spring, O. (2005a). Exceptional length of ITS in *Plasmopara halstedii is due* to multiple repetitions in the ITS-2 region. *European Journal of Plant Pathology, 112*, 395–398.

Thines, M., Komjati, H. & Spring, O. (2005b). Unusual insertions of phylogenetic relevance in the nuclear ITS region of some Peronosporaceae (Oomycetes). *Proceedings of the 23 rd Fungal Genetic Conference, Fungal Genetics Newsletter 52*, 72.

Tosi, L., Luigetti, R., & Zazzerini, A. (1999). Benzothiadiazole induces resistance to *Plasmopara helianthi* in sunflower plants. *Journal of Phytopathology, 147*, 365–370.

Tourvieille de Labrouhe, D., Mestries, E., & Walser, P. (2005). Quelles perspectives pour la lutte génétique vis-a-vis du mildiou du tournesol? *OCL, 12*(2), 85–93.

Tourvieille, L. D., Gulya, T. J., Masirevic, S., Penaud, A., Rashid. K. Y., & Viranyi, F. (2000). New nomenclature of races of *Plasmopara halstedii* (sunflower downy mildew). In: *Proceedings of the 15th International Sunflower Conference* (Vol. 2, I: pp. 61–65). Toulouse.

Tourvieille de Labrouhe, D., Walser, P., Serre, F., Roche, S. & Vear, F. (2008a). Relations between spring rainfall and infection of sunflower by *Plasmopara halstedii* (downy mildew). In: Velasco L. (ed.) *Proceedings of the 17th International Sunflower Conference* Vol. 1, (pp. 97–102). *Cordoba*, Spain, 8–12 June 2008.

Tourvieille de Labrouhe, D., Penaud, A., Walser, P., Mestries, E., Moinard, J., & Serre, F., et al. (2008b). Determining the sunflower downy mildew risk by soil analysis. In: L. Velasco (ed.) *Proceedings of the 17th International Sunflower Conference* Vol 1, (pp. 169–174). *Cordoba*, Spain, 8–12 June 2008.

Tourvieille de Laboruhe, D., Serre, F., Walser, P., Roche, S., & Vear, F. (2009). Nonrace specific resistance to downy mildew (*Plasmopara halstedii*) in sunflower (*Helianthus annuus*). *Euphytica, 164*, 433–444.

Vear, F., Milon, J., Philippon, J., Mouzeyar, S., Nicolas, P. & Tourvieille de Laboruhe, D. (1998). Analysis and location of sunflower downy mildew resistance genes. In: *ISA Symposium III. Sunflower Downy Mildew* (pp. 84–93). *Fargo*, ND, 13-14 January 1998.

Vear, F., Serre, F., Roche, S., Walser, P. & Tourvieille de Labrouhe, D. (2006). Recent research on downy mildew resistance useful for breeding industrial—use sunflowers. In: *ISA 1st Symposium on Sunflower Industrial Use* (pp. 1–8). *Udine*, Italy, 11–13 September 2006.

Virányi, F. (1980). Hypersensitive tissue necrosis on *Plasmopara*-resistant sunflower seedlings. *International Sunflower Association Newsletter, 4*, 11–19.

Virányi, F. (1984). Recent research on the downy mildew of sunflower in Hungary. *Helia, 7*, 35–38.

Virányi, F. (1992). Downy mildew of sunflower. In H. S. Chaube et al. (Eds.), *Plant diseases of international importance* (Prentice Hall, Vol. 2, pp. 328–344). NJ: Eaglewood Cliffs.

Virányi F., & Gulya T. J. (1996). Expression of resistance in the *Plasmopara halstedii*—sunflower pathosystem. In: *ISA Symposium I. Disease Tolerance in Sunflower* (pp. 14–21). *Beijing*, 13 June 1996.

Voglmayr, H. (2008). Progress and challenges in systematics of downy mildews and white blister rusts: New insights from genes and morphology. *European Journal of Plant Pathology, 122*, 3–18.

Voglmayr, H., & Constantinescu, O. (2008). Revision and reclassification of three *Plasmopara* species based on morphological and molecular phylogenetic data. *Mycological Research, 112*, 487–501.

Voglmayr, H., Riethmüller, A., Göker, M., Weiß, M., & Oberwinkler, F. (2004). Phylogenetic relationships of *Plasmopara*, *Bremia* and other genera of the downy mildew pathogens with pyriform haustoria based on Bayesian analysis of partial LSU rDNA sequence data. *Mycological Research, 108*, 1011–1024.

Walcz, I., Bogár, K., & Virányi, F. (2000). Study on an *Ambrosia* isolate of *Plasmopara halstedii*. *Helia, 23*, 19–24.

Wilson, G. W. (1907). Studies in North American Peronosporales II. Phytophthorae and Rhysothecae. *Bulletin of the Torrey Botanical Club, 34*, 387–416.

Young, P. A., & Morris, H. E. (1927). *Plasmopara* downy mildew of cultivated sunflowers. *American Journal of Botany, 14*, 551–553.

Zimmer, D. E. (1972). Field evaluation of sunflower for downy mildew resistance. *Plant Disease Reporter, 56*, 428–431.

Zimmer, D. E., & Hoes, J. A. (1978). Diseases. In J. F. Carter (Ed.), *Sunflower science and technology, agronomy 19* (p. 225). Madison: American Society of Agronomy.

Eur J Plant Pathol (2011) 129:89–100
DOI 10.1007/s10658-010-9664-3

Histological responses of host and non-host plants to *Hyaloperonospora parasitica*

Hua Li · Xintian Ge · Shiue Han · Krishnapillai Sivasithamparam · Martin John Barbetti

Accepted: 9 July 2010 / Published online: 17 September 2010

Abstract Differences in *Hyaloperonospora parasitica* development and plant tissue responses were compared for 10 cruciferous hosts (including both resistant and susceptible genotypes), 3 leguminous and 1 graminaceous non-host species. Cotyledons, or true leaves in the case of *Triticum aestivum* and *Pisum sativum,* were studied at 2, 8, 24 h and 3, 5, 7 days post inoculation (dpi). The high levels of zoosporangial germination observed on all species tested, as well as on glass slides, suggested that inhibition of germination did not play a significant role in distinguishing host *versus* non-host resistance. During the early stages of infection, at spore germination and host penetration, there was no evidence of a clear-cut difference between *Brassica* host species which displayed a hypersensitive, partially resistant or susceptible reaction compared with non-host species. Haustoria formation was the key infection phase for the establishment of biotrophy. Across all tested species, haustoria were initiated inside the epidermal cells. However, there were significant differences in the frequency and timing of haustorial formation and the final size of haustoria among the tested species at early infection stage. Fully developed haustoria were never observed in *Raphanus raphanistrum, Triticum aestivum, Lupinus angustifolius* nor *Trifolium subterraneum*. Instead, the haustorium development appears to abort in the penetrated epidermal cells of these species. Although haustoria were formed in the epidermal and mesophyll cells of *Sinapsis alba* and *Pisum sativum,* subsequent hyphal growth and/or continued haustoria formation were rare or few, respectively. Hypersensitive reaction was the key resistance response observed among the host and non-host resistant species tested. It is noteworthy that, in the initial stages of pathogenesis, there was no differentiating point that separated the non-host species from those that were hosts.

Keywords Biotrophy · Brassicaceae · Downy mildew · Fabaceae · Hypersensitive response · Host-pathogen interaction · Resistance

H. Li (✉) · X. Ge · S. Han · K. Sivasithamparam · M. J. Barbetti
School of Plant Biology, Faculty of Natural and Agricultural Sciences, The University of Western Australia, 35 Stirling Highway, Crawley, WA 6009, Australia
e-mail: hli@cyllene.uwa.edu.au

M. J. Barbetti
Department of Agriculture and Food Western Australia, 3 Baron-Hay Court, South Perth, South Perth, WA 6151, Australia

Introduction

Oomycete pathogens such as species of *Phytophthora, Pythium, Albugo* and the various downy-mildew genera, form a unique branch of eukaryotic plant pathogens with an independent evolutionary history (Kamoun et al. 1999). Among these oomycetes, downy-mildew species cause some of the most destructive plant diseases in the world (Viennot-

Bourgin 1981). *Hyaloperonospora* (formerly *Peronospora*) *parasitica*, the causal agent of downy mildew of cruciferous vegetables and field crops, is endemic across the oilseed rape-growing regions in Australia (Howlett et al. 1999; Ge et al. 2008), Europe (Paul et al. 1998), China and Japan (Satou and Fukumoto 1996) and, in particular, on the Indian subcontinent (Nashaat et al. 2004). While young *Brassica* plants are most susceptible to *H. parasitica* (Channon 1981; Silué et al. 1996), infection at a later stage of plant development can also result in yield and quality reductions (Channon 1981). As fungicides are often expensive and lead to the development of pathogen resistance to such chemicals, identification and deployment of host resistance has increasingly become the most efficient and cost-effective means for managing this disease. Host resistance has been studied from many perspectives in the *Arabidopsis-H. parasitica* pathosystem (Mauch-Mani and Slusarenko 1993; 1996; Donofrio and Delaney 2001) including the role of signalling pathways. For example, McDowell et al. (2000) reported that the *RPP7* gene initiated host resistance through a novel signalling pathway that functions independently of salicylic acid accumulation or jasmonic acid response components.

In nature, plants are exposed to many potential pathogens. For this reason, plants possess different constitutive and inducible mechanisms to recognize and defend themselves against pathogens. While a wide range of pathogens may attempt to penetrate plant tissues, only a limited number of these invasion interactions are successful, suggesting that plants can exhibit a 'general' host resistance, defined by Heath (1980) and Jahnen and Hahlbrock (1988) as 'non-host' resistance. Non-host resistance is highly effective and durable, and hence it is often suggested that the mechanisms of non-host resistance could be exploited to generate disease resistant crop plants. Several components of non-host disease resistance have been identified (Zhou et al. 1999; Mellersh and Heath 2003; Yun et al. 2003). Heath (2000) and Holub and Cooper (2004) pointed out that in spite of tremendous progress in plant science, non-host resistance is still poorly understood in contrast with host resistance. It is still not clear why a pathogen can be highly virulent on one plant species while having no effect on others. Understanding non-host resistance is key to deciphering the complex plant defence mechanisms against pathogens.

An understanding of the mechanisms of host and non-host resistance to *H. parasitica* would not only provide information about the way in which host-parasite specificity is determined but it also offers potential for future more precise and long-lasting control of this disease through development of new cultivars with more effective and durable host resistance. The determination of both the factors responsible for non-host resistance and the racial status of this pathogen could together lead to identification of physiological and molecular markers for use in securing effective resistance to this disease against the pathotypes/races of this pathogen.

In this paper we investigate the early infection events of hypersensitive, partially resistant or susceptible *Brassica* and other crucifer host species by *H. parasitica*. These responses are compared with those of the pathogen on three legumes and a cereal species included as non-cruciferous non-host plants, in order to provide additional understanding of the interactions involved in the determination of host-*H. parasitica* specificity.

Materials and methods

Plant materials

Plant materials used in this study, including 10 cruciferous hosts (including both resistant and susceptible genotypes), viz. two *B. napus*, one each of *B. oleracea* var. *italica, B. nigra, B. juncea*, *B. rapa*, *B. carinata*, *Eruca vesicaria* ssp. *sativa*, *Raphanus raphanistrum* and *Sinapis alba*; 3 leguminous, viz. *Lupinus angustifolius, Pisum sativum* and *Trifolium subterraneum*, and 1 graminaceous, viz. *Triticum aestivum* (Table 1). All cruciferous plants used in this study were considered as hosts, while three leguminous and one graminaceous hosts were considered to be non-host species, based on previous reports.

Fifteen seeds of each genotype were sown into 150×70×70 mm pot, 6 pots per genotype. After emergence, the seedlings were thinned to 10 plants/pot. Samples for histological study were taken randomly from different plants from different pots. A total of 24 plants per genotype (6 replicates with 4 plants in each replicate) were assessed for disease severity at 7 dpi.

Table 1 Plant material tested in this study and its reaction to *Hyaloperonospora parasitica* (isolate UWA DM 53) expressed as the severity of downy mildew disease 7 days post inoculation, on cotyledons/true leaves (0-9 scale was applied, where 0 = no symptoms of downy mildew; 9 = heavy sporulation, cotyledon collapsed; for details see Williams 1985)

Species	Genotype /cultivar	Host/non-host	Reaction to host	Disease severity
Brassica napus	Thunder TT	host	S	8.33
B. oleracea var. *italica*	Shogun	host	S	6.97
B. nigra	P23845	host	PR	3.63
B. napus	Pioneer 46Y78	host	PR	3.43
B. juncea	Dune	host	HR	1.13
Eruca vesicaria ssp. *sativa*	MJB 1A	host	HR	0.67
B. rapa	Xi Shui Bai	host	HR	0.57
B. carinata	ATC94011	host	HR	0.53
Sinapis alba	Concerta	host	NR	0
Raphanus raphanistrum	Wild radish	host	NR	0
Pisum sativum	Dunwar	non-host	NR	0
Lupinus angustifolius	Tanjil	non-host	NR	0
Trifolium subterraneum	Woogenellup	non-host	NR	0
Triticum aestivum	Gamenya	non-host	NR	0
Significance $P<0.01$	LSD			0.44

Disease severity data were analysed using two-way analysis of variance (ANOVA)

Significancy of mean of disease severity differences within species or genotype were assessed by test for least significant difference (LSD)

Reaction to host within parenthesis indicates: *S* susceptible; *PR* partially resistant; *HR* hypersensitive response; *NR* no response

Pathogen isolate and inoculum preparation

A single zoosporangial isolate of *H. parasitica* (UWA DM 53) originating from *B. napus* was used for this study. Inoculum was prepared from a culture of this isolate maintained on cotyledons of oilseed rape (*B. napus cv.* Thunder TT). Cotyledons that supported abundant sporulation by the pathogen were collected and placed in 50 ml of distilled water in a 200 ml glass flask. The flask was shaken vigorously to dislodge the zoosporangia. The resulting zoosporangial suspension was filtered through a single layer of cheesecloth and the concentration determined with a haemocytometer counting chamber and adjusted to 1×10^5 zoosporangia ml^{-1}.

Inoculation

The experiment was conducted in a phytotron growth cabinet where air temperature was maintained at 15 (night)-20 (day) °C, photoperiod of 12/12 h, light intensity was 150 $\mu E\ m^{-2}\ s^{-1}$ and plants were watered daily with deionised water. After 10 d, a 10 µl droplet of zoosporangial suspension was deposited on the each lobe of 10 day old cotyledons of each test genotype, except for wheat and pea where the suspension was applied to the upper leaf surface of 10 day old first true leaf. A 10 µl droplet of water was deposited at the same inoculation sites on control plants. Inoculated and control plants were placed into plastic boxes with sealed lids in order to maintain high humidity post-inoculation for a period of 24 h. Subsequently, the lids were removed and plants watered daily. At 6 dpi, the plants in the plastic boxes were again covered by lids for a further 24 h to maintain high humidity.

Disease severity assessment

Disease severity on cotyledons/true leaves was assessed at 7 dpi using the 0–9 scale as defined by Williams (1985); where 0 = no symptoms or signs of *H. parasitica*; 1 = minute scattered necrotic flecks under the inoculum drop, no sporulation; 2 = larger necrotic flecks under the inoculum drop, no sporulation; 3 = very sparse sporulation, one to a few conidiophores, necrotic flecking but often with tissue necrosis evident; 5 = sparse sporulation, tissue necrosis; 7 = abundant sporulation, tissue necrosis and chlorosis may be present; and, 9 = heavy sporulation, cotyledon collapsed.

Histological studies

Epifluorescence microscopy

Plants inoculated with zoosporangia were sampled 0, 2, 8 h and 1, 3, 5 and 7 dpi. Four leaf segments taken from different plants at each time-point were decolorized in 10% KOH over night and stained with aniline blue in 0.07 M K_2HPO_4 for UV light observation as described by Heath (1974). A zoosporangium was considered as germinated if it produced either a germ tube of at least half of the diameter of the zoosporangium or a germ tube with a primary appressorium. A total of 200 zoosporangia per leaf segment were scored. The total number of zoosporangia, germinated zoosporangia, appressoria, and haustoria were recorded, as was the presence of any cell necrosis, at the penetration sites and in cells. The samples were examined and photographed using a Olympus BX 51 microscope photograph system with an excitation filter (G365) and an emission filter (LP 420) inserted into a beam of incident light from a mercury vapor lamp.

Bright-field microscopy

Infected cotyledons/true leaves were sampled 0, 2, 8 h and 1, 3, 5 and 7 dpi. Four cotyledon/true leaf segments (taken from different plants) per time-point were cleared in 95% ethanol over night and stained with 1% cotton blue and whole, wet mounts examined and photographed using same microscopy photograph system as above with bright field lighting. Drops of zoosporangial suspension were deposited on glass slides in moist chambers as comparisons. One hundred zoosporangia were assessed at five random fields of view on each of five inoculation sites per cultivar. The number of zoosporangia, germinated zoosporangia, appressoria, and haustoria were recorded, as was the presence of any cell necrosis, at the penetration sites and within cells.

Scanning electron microscopy

Plants were sampled at 2 and 4 hpi for this experiment. The infected region of the tissue was cut and fixed in 2.5% glutaraldehyde in 0.05 M phosphate buffer pH 7 at room temperature. Samples were further processed by using PELCO BioWave Microwave Processor. Tissues were vacuum infiltrated for 6 min (2 min on / 2 min off / 2 min on) at 80 W. Specimens were then washed with 0.05 M phosphate buffer for 40 s at 80 W, dehydrated in a series of ethanol solutions ranging from 50% to 100% (dry ethanol) and twice with 100 % dry acetone again for 40 s at 250 W. Samples were then critically point dried (PELCO Critical Point Dryer) using liquid carbon dioxide, sputter coated with gold and examined and photographed using the JEOL 6400 SEM.

Statistical analysis

Disease severity data were analysed using GenStat® Release 9.1 (Lawes Agricular Trust, Rothamsted Experimental Station). The data were analysed using two-way analysis of variance (ANOVA). Significant differences between mean disease severities within species or genotype were assessed by testing for least significant difference (LSD).

Results

Disease intensity

Disease severities across the tested genotypes differed significantly ($P<0.01$) (Table 1). The highest score (8.3) for disease was recorded on the highly susceptible *B. napus* cv. Thunder TT, where the infected cotyledons showed heavy sporulation and collapsed. The partially resistant *B. napus* cv. Pioneer 46Y78 had a disease score of 3.4, while *B. oleracea* and *B. nigra* had disease scores 7 and 3.6, respectively. However, other crucifer species tested, including *B. carinata*, *B. rapa*, *B. juncea* and *E. vesicaria* ssp. *sativa*, all showed only a hypersensitive response on the inoculated cotyledons. No symptoms were evident on host cruciferous species *S. alba* or *R. raphanistrum*, nor on any of the non-host genotypes tested.

Infection progress

A summary of the development of *H. parasitica* on the various host and non-host plants is shown in

Table 2. At 2 h post inoculation (hpi), zoosporangia germinated equally well (>76%) on all host and non-host species tested, but zoosporangial germination rates were lower on glass slides (43%). Appressoria formed at the tip of the germ-tube on all host, non-host species and even on the glass slides. Appressoria formation resulted in the direct penetration of epidermal cells for *Brassica* and other cruciferous species (Fig. 1a) and also for *P. sativum*. Hyphae penetrated mostly between anticlinalwalls of epidermal cells, but infrequently into the lumen of an epidermal cell. On the cotyledons of *L. angustifolius* and *Trifolium subterraneum*, penetrations were direct and/or indirect (i.e., through stomata). However, there was no penetration of the leaves of the monocotyledon species (*Triticum aestivum*). Compared with the *Brassica* and other cruciferous host plants, there was a significantly lower number of penetrations by the pathogen on all non-host plants, with the exception of *P. sativum* and the cruciferous host *R. raphanistrum*. Following penetration, haustoria formation was observed to varying degrees on all of the host species, but not observed on *B. juncea* and *B. carinata* at this stage. Often the first haustorium was formed in one of the epidermal cells adjoining to where the hypha had penetrated and, frequently, one haustorium was observed in each of the epidermal cells proximal to the penetrating hypha (Fig. 1b). Haustoria also formed in the non-host species *P. sativum* and *L. angustifolius*, but were not formed in *Trifolium subterraneum* or *Triticum aestivum*. Infected cells in *B. rapa* developed a distinctive fluorescence under UV light (Fig. 1c).

At 8 hpi, on all host species, except for *R. raphanistrum*, as well as on *P. sativum*, the penetrating hypha lay in between the anticlinal cell walls of two neighbouring epidermal cells within which one, two or three haustoria were formed. The penetrating hypha usually reached as far as the immediately adjacent mesophyll cell and haustoria was sometimes formed in this mesophyll cell, as well as in the epidermal cells. There were, however, no significant differences among these species in relation to the number and size of the haustoria formed. Infected cells developed a distinctive fluorescence in *B. carinata*, *B. rapa*, *B. oleracea*, *B. nigra* and *S. alba*, and of these the fluorescence was most intense in *S. alba* and *B. carinata*. There was significantly fewer haustoria and less depth of penetration in the host *R. raphanistrum* and in the three non-hosts *L. angustifolius, Trifolium subterraneum* and *Triticum aestivum*, compared with species described above. Further, the haustoria in *R. raphanistrum, L. angustifolius, Trifolium subterraneum* and *Triticum aestivum* were relatively small, infrequent and were formed only in the epidermal cells. On all host species except *R. raphanistrum,* and on the non-host *P. sativum*, 100% of zoosporangia germinated, and in all cases, penetration was direct. In contrast, there were only 46, 12, 9 and 7% penetrated directly, and 0, 25, 65 and 0% penetrated through stomata, for *R. raphanistrum, L. angustifolius, Trifolium subterraneum* (Fig. 1d) and *Triticum aestivum*, respectively. On the host *R. raphanistrum* and non-host species *Triticum aestivum* and *Trifolium subterraneum* (Fig. 1d), as well as on the glass slide, germ-tubes were longer than for all the other species tested.

At 24 hpi, there were significant differences in the extent of tissue colonisation by the pathogen among the different hosts. In the highly susceptible *B. napus* cv. Thunder TT, hyphae grew past approximately 5–7 cells forming more than 10 haustoria inside the proximal mesophyll cells. On the partially resistant *B. napus* cv. Pioneer and on *B. rapa*, hyphae of the pathogen grew past approximately 3-4 cells and formed roughly 5-6 haustoria inside the proximal mesophyll cells. On all other host species and on the non-host species *P. sativum,* there were about 3-4 haustoria formed inside the initially-penetrated epidermal cells. In contrast, in *R. raphanistrum, Triticum aestivum, Trifolium subterraneum* and *L. angustifolius*, further invasion was restricted to the first epidermal cell penetrated. In these cells, the small haustoria (approximately 3 μm wide ×7 μm long) were evident at 8 hpi. At 24 hpi however, the appressoria on the surface of epidermal cell expanded to about 15–20 (wide)× 30–40 (long) μm with no evidence of the original haustorium. The haustorium development appeared to have aborted/collapsed at this point. Instead, a newly emerging hypha was seen to extend from the appressorium growing out, without any evidence of penetration, up to 180 μm on the surface of the epidermis (Fig. 1e). Infected cells of *E. vesicaria* ssp. *sativa* and *S. alba* showed a distinctive fluorescence. Hypersensitive cell death was readily observed among the species which showed distinctive fluorescence and also in *R. raphanistrum*. The mesophyll cell containing the first (oldest) haustorium to be produced, close to the penetrated

Table 2 Development of *Hyaloperonospora parasitica* isolate UWA DM 53 on host and non-host cotyledons/true leaves tested

Host		2 hpi						8 hpi					24 hpi		
Species	Genotype/cultivar	% germ	appre.	penet.	haust.	% germ	% appre.	% direct./stomatal penetration	haust.	flu.	appre.	penet. mesophyll cell	haust.	cd	callose
B. napus	Thunder TT	99	+	+	0-1	100	100	100/0	2-3	–	+	+	10–15	–	–
B. oleracea	Shogun	99	+	+	0-1	100	100	100/0	2-3	+	+	+	3-4	–	–
B. nigra	P23845	99	+	+	0–2	100	100	100/0	2-3	+	+	+	3–5	–	–
B. napus	Pioneer 46Y78	99	+	+	1-2	100	100	100/0	2-3	–	+	+	5-6	–	–
B. juncea	Dune	98	+	+	0	100	100	100/0	2	–	+	+	3–5	+	–
E. vesicaria ssp. *sativa*	MJB 1A	99	+	+	1-2	100	100	100/0	2	–	+	+	2-3	+	–
B. rapa	Xi Shui Bai	99	+	+	0-1	100	100	100/0	2-3	+	+	+	5 (3–8)	+	+
B. carinata	ATC94011	99	+	+	0	100	100	100/0	2-3	+	+	+	3–5	+	+
S. alba	Concerta	98	+	+	0-1	100	100	100/0	2-3	+	+	+	3–5	+	–
R. raphanistrum	Wild radish	76	+	+	0	82	35	46/0	0-1	–	+	–	0	+	–
P. sativum	Dunwar	90	+	+	0-1	100	100	100/0	2	–	+	+	2-3	+	+
L. angustifolius	Tanjil	89	+	+	1-2	92	74	12/25	0-1	–	+	+	0-1	–	–
Trifolium subterraneum	Woogenellup	89	+	+	0	93	30	9/65	0-1	–	+	+	0-1	–	–
Triticum aestivum	Gamenya	88	+	–	0	86	29	7/0	0	–	+	+	0-1	–	–
Glass slides		43	+	–	–	72	+	–	–	–	100	–	–	–	–

appre. appressorium, *haust.* haustoria, *germ* germination, *cd* cell death associated with HR, *flu* fluorescence

\+ present, – absent

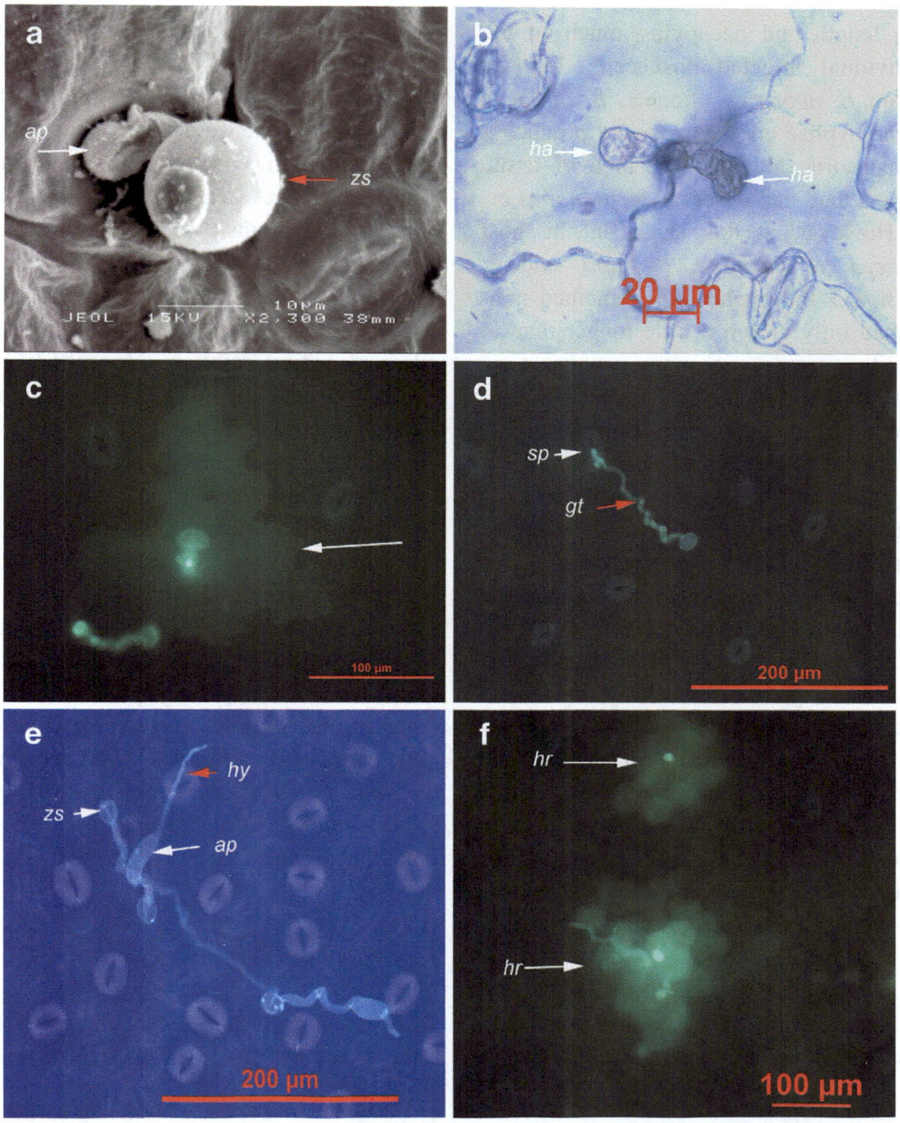

Fig. 1 Formation of infection structures of *Hyaloperonospora parasitica* on host and non-host plant. Samples for Fig. 1. **a**, scanning electron microscopy image; **b,** fresh cotyledon/leaf cleared in boiling 95% ethanol, stained with 1% cotton blue; **c, d, e** and **f**, fresh cotyledon/leaf cleared in 10% KOH and stained with 0.05% Aniline Blue in 0.067 M K2HPO4 (pH 9). **a,** direct penetration (*white arrow*) of epidermal cell of *Brassica napus* cv Thunder TT by a germ tube from a zoosporangium (*red arrow*). **b**, one haustorium (*arrows*) was observed in each of the epidermal cells proximal to the penetrating hypha of *B. rapa* after 4 hpi. **c**, infected cells in *B. rapa* developed a distinctive fluorescence (*arrow*) under uv light 4 hpi. **d**, indirect penetration through stomata (*white arrow*) of *Trifolium subterraneum* from a long germ tube (red arrow). **e**, failure of penetration, the appressorium getting larger (*white arrow*) with a hypha growing out of the appressorium, on *Lupinus angustifolius*. **f**, typical hypersensitive reaction (*arrows*) on *B. carinata* 3 dpi. *ap*: appressorium, *gt*: germ tube, *sp*: stomatal penetration, *ha*: haustorium, *hy*: hypha, *zs*: zoosporangia

epidermal cell, was the first to show a hypersensitive reaction. Callose deposition at the point of penetration was observed in the host species *B. carinata* and *B. rapa* and in the non-host *P. sativum*.

At 3 dpi, there were significant differences in the colonisation development of the pathogen across the different host species. In the highly susceptible *B. napus* cv. Thunder TT, the pathogen produced intercellular hyphae that ramified around many cells with multiple branches and haustoria, eventually colonizing much of the cotyledon tissues. By this stage, haustoria were broad and conspicuous, reaching 20 µm in length and occupying much of the lumen of individual infected host cells. On the partially resistant *B. napus* cv. Pioneer, *B. oleracea* and *B. nigra*, the hyphae grew extensively producing fully formed haustoria, although the area of tissue colonised was less than that observed for *B. napus* cv. Thunder TT. However, at this stage, in the host species *B. juncea*, *B. carinata*, *B. rapa,* and *E. vesicaria* ssp. *sativa,* the hyphae still remained non-branched and had only grown a little longer than at 24 hpi. In these host species, not many new haustoria developed from the intercellular hyphae and penetrated cells showed a typical hypersensitive reaction (HR) (Fig. 1f), and a great of number of cells displayed a HR, including those adjacent to the penetrated cells (Fig. 2a). Intensive callose deposition at the point of penetration was observed on *B. carinata* and *S. alba*. On *S. alba* and *P. sativum*, there were little difference compared with what was observed at 24 hpi in relation to either the length of the hyphae or the number of haustoria present, but fully developed haustoria were observed in the proximal mesophyll cells (2b). On the host *R. raphanistrum* and on the other three non-hosts and on the glass slides, at this point of time (3dpi), appressoria were significantly larger than those observed at 24 hpi, with hyphae growing out from the appressoria. However, these hyphae rarely penetrated or produced huastoria in these particular species.

At 5 dpi, in the highly susceptible *B. napus* cv. Thunder TT, hyphae with many branches and many haustoria were observed, and the whole cotyledon was colonised and hyphae had proceeded to grow into the petiole and towards the stem. On the partially resistant *B. napus* cv. Pioneer, and on *B. oleracea* and *B. nigra,* hyphae were seen to have colonised most of the cotyledon with branches and many haustoria present. There were, however, no visible external disease symptoms evident to the naked eye on these three species at this point of time. On *B. juncea*, *B. carinata*, *B. rapa*, *S. alba, E. vesicaria* ssp. *sativa* and the non-host *P. sativum,* there were lesions typical of hypersensitive reaction. No further progression of infection structures was evident in either the host *R. raphanistrum* or in the three non-host species.

At 7 dpi, the sporangiophores were fully developed on *B. napus*, *B. oleracea* and *B. nigra*, and hyphae were also observed growing within the petiole and towards the stem. However, most severe disease symptoms were observed on *B. napus* cv. Thunder TT. Hypersensitive reaction lesions were clearly evident on the host species *B. juncea*, *B. carinata*, *B. rapa*, and *E. vesicaria* ssp. *sativa*. No sporangiophores were visible on the plants that showed hypersensitive reactions. Visible hypersensitive reactions were not evident on *S. alba* or *P. sativum.* No disease symptoms were observed on the host species *S. alba*, *R. raphanistrum* nor on any of the non-host species studied.

Discussion

This study showed clear separation of the reactions of the tested species to *H. parasitica* into four groups.

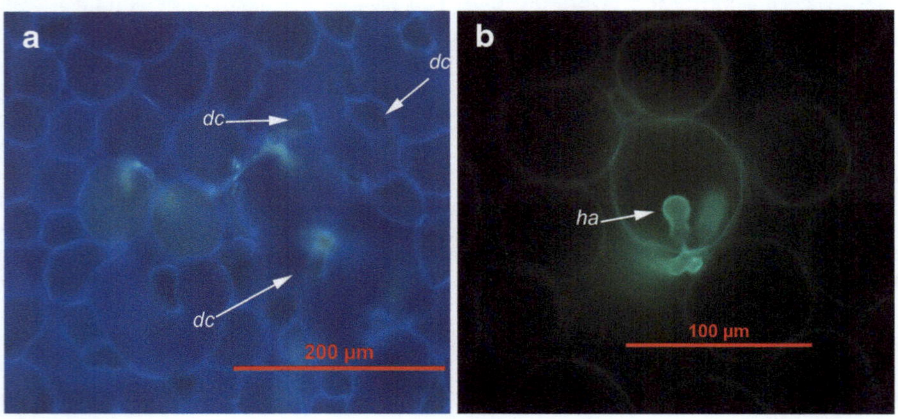

Fig. 2 Formation of infection structures of *Hyaloperonospora parasitica* on host plant. All samples cleared in 10% KOH and stained with 0.05% Aniline Blue in 0.067 M K2HPO4 (pH 9). **a**, additional cells dead adjacent to the penetrated cells (*arrows*) on *B. carinata* 3 dpi. **b**, a fully developed haustorium in a *Sinapis alba* mesophyll cell 3 dpi. *dc*: dead cells associate HR, *ha*: haustorium

Group 1, comprised of the highly susceptible hosts, represented by *B. napus* cv. Thunder TT and *B. oleracea*, with highest disease scores. Group 2, comprised of susceptible hosts, including *B. napus* cv. Pioneer, and *B. nigra* that responded with moderate disease scores. This group of hosts reacted to *H. parasitica* similarly to Group1, but overall disease severity was lower and disease progression was slower. Group 3, showed HR to *H. parasitica*, represented by host species *B. juncea*, *B. carinata*, *B. rapa*, and *E. vesicaria* ssp. *sativa*. Group 4, included host crucifer species *S. alba*, *R. raphanistrum* and all non-host species, all of which showed no disease symptoms. *Hyaloperonospora* has only recently been separated from *Peronospora* (Constantinescu and Fatehi 2002). *H. parasitica* contains isolates from diverse species of Brassicaeae, and also of other closely related families such as Resedaceae and Capparidaceae and, remarkably, even the unrelated Cistaceae (Constantinescu and Fatehi 2002). While the pathogenicity of *H. parasitica* varies at both the host generic and the species levels, the basis for its particular host range still remains un-clear. For example, *S. alba* and *R. raphanistrum* have been reported as host plants for *H. parasitica* (Dickinson and Greenhalgh 1977). However, in our study, *R. raphanistrum* had a similar response to *H. parasitica* as the non-host legumes *Trifolium subterraneum* and *L. angustifolius*, while *S. alba* responded similarly to the non-host legume *P. sativum*. Whether the reaction of *S. alba* or *R. raphanistrum* to this isolate of *H. parasitica* is truly pathotype specific still remains to be clarified. Based on infection progression and disease severity from our study, there were no clear cut-offs at the anatomical level which separate species that are hosts of *H. parasitica* from species that are non-hosts to *H. parasitica*.

The high levels of zoosporangial germination on all species tested as well as on glass slides suggest that inhibition of germination does not play a significant role in the expression of host and/or non-host resistance. A similar observation was made by Heath (1974) in relation to non-host resistance to cowpea rust (*Uromyces phaseoli* var. *vignae*). In our study, appressoria were formed on all tested species as well as on glass slides by 8 hpi. However, there were significant differences in terms of both the number and the size of appressoria among the host and non-host species. On all host species and on the non-host *P. sativum*, zoosporangia sometimes formed germ tubes and occasionally failed to do so, but, in all cases where direct penetration was initiated, apressoria in the form of swollen discs were formed. In contrast, while zoosporangia formed germ tubes on *R. raphanistrum*, *L. angustifolius*, *Trifolium subterraneum*, *Triticum aestivum* and on glass slides, apressoria, however, were either absent or only in low numbers but with an average size greater than that those on other plant species tested. For *L. angustifolius* and *Trifolium subterraneum*, penetration occurred directly or through stomata, but for *Triticum aestivum* there was only very low occurrence (7%) of direct penetration. In *R. raphanistrum*, *L. angustifolius*, *Trifolium subterraneum*, *Triticum aestivum*, the attempted penetration appears to have been initiated. However, no further developments were observed. The appressoria on the surface of epidermal cell were expanded and a new hypha extended from the appressoria growing out on the surface of the epidermis. Such restriction in progression of penetration was previously reported for bean rust on bean by Wynn (1976). Wynn (1976) observed when penetration failed with the first appressoria over stomata, hypha continued to grow toward a second stomata to form another appressoria. The branched infection hyphae were unable to penetrate the plant and remained on the surface. In our study, the formation of larger appressoria and the growth of a hypha growing out of each of the appressoria on glass slides and on these four species, together suggest that at least the development processes of the pathogen up to this point may be primarily determined by factors other than specific signals from plants. Little is known about the signalling associated with or required for appressorium formation on plant surfaces. Although surface hydrophobicity and hardness can be sufficient to stimulate appressorium development in some pathogens (e.g., Carzaniga et al. 2001; Tucker and Talbot 2001), certain pathogens are known to require specific signals to develop infection structures (e.g., the sensing of cutin by *Colletotrichum* spp. (Dickman et al. 2003), or sensing the height of the guard cell lip by *Uromyces appendiculatus* (Allen et al. 1991). In our study, *H. parasitica* was able to penetrate both host and non-host plants, suggesting that the penetration stage is not critical in determining the success or failure of pathogenesis. However, there were significant differences in relation to the frequency and timing of penetration among the tested species. Interestingly, biotrophy requires an impressive series of cellular interactions between the plant and the parasite (Lebeda et al. 2001; Mendgen and Hahn 2002).

Haustoria were formed either inside the epidermal and/or the mesophyll cells across all tested species at different stages post inoculation. However, there were significant differences in terms of the frequency and timing of haustorial formation and in the final size of haustoria among the tested species. Fully developed haustoria were never observed in *R. raphanistrum*, *Triticum aestivum*, *L. angustifolius* or *Trifolium subterraneum*. Instead, the haustorium development appears to cease, collapse or abort in penetrated epidermal and/or the mesophyll cells of these species. Although haustoria formed in the epidermal and/or mesophyll cells of *S. alba* and *P. sativum*, further hyphal growth was rare and only a few additional haustoria developed. Taken together, this suggests that there is some inhibition to haustorium formation on these particular species. Overall, our findings suggest that *H. parasitica* is able to initiate an infection of plant tissue up to the stage at or immediately following first formation of one or more haustoria, which then induces a rapid activation of defence response(s) in infected cells or tissues. This results in the cessation of further progress in infection. This is in agreement with Lebeda et al. (2008) that plant resistance is mostly related to recognition (pre-haustorial or post-haustorial) limiting pathogen intercellular hypha growth and haustoria formation. The restriction of haustoria formation may be a crucial limiting factor in host–pathogen communication and play an important role in expression of resistance (Lebeda et al. 2006).

Callose deposition was detectable after KOH treatment and aniline blue staining of plant tissues. In *B. carinata*, the rapid accumulation of callose at the point of penetration of the cell wall of infected cotyledons was observed as early as 8 hpi. After 24 h, callose deposits were easily visible under fluorescence at the point of penetration of hosts *B. carinata*, *B. rapa*, and in the non-host *P. sativum*. No callose deposition was evident in un-inoculated control plants nor in inoculated *R. raphanistrum*. Similar observations were made by Gindro et al. (2003) in and around the stomatal cells of grapevine infected by *P. viticola*; by Diez-Navajas et al. (2008) in non-host infected by *P. viticola*; by Sedlářová and Lebeda (2001) for lettuce (*Lactuca* spp.) infected by *B. lactucae*; and by Heintz and Blaich (1990) around haustoria of *Uncinula necator* in grapevines resistant to powdery mildew.

Hypersensitive reaction was found to be a typical response to infection by *H. parasitica* in resistant hosts and in the non-host *P. sativum*. In host species such as *B. rapa*, *B. carinata*, *B. juncea*, *B. nigra*, *B. oleracea* or *S. alba*, a significant proportion of penetrated cells showed a yellow fluorescence by 8 hpi, and for *E. vesicaria* ssp. *sativa* and in non-host *P. sativum* at 24 hpi. The yellow fluorescence observed in these species was strongly associated with a hypersensitive reaction. Similar observations were made by Dai et al. (1995) in grapevine infected by *Plasmopara viticola* and by others (Bennett et al. 1996; Lebeda et al. 2008) in lettuce cells challenged by *Bremia lactucae*. This fluorescence may have been caused by the accumulation of phenolics (Bennett et al. 1996) and of lignin-like deposits produced by peroxidase mediated cross-linking (Bolwell 1993; Fry 1986; Graham and Graham 1991; Scalbert et al. 1985). In partially resistant species like *Vitis rupestris* (Dai et al. 1995; Unger et al. 2007) as well as in the partially resistant grapevine cultivar Solaris (Gindro et al. 2003), necrosis appeared as necrotic spots only several days post infection by *P. viticola*. The hypersensitive response is an effective defence mechanism against both biotrophs and non-biotrophs alike because of the accompanying up-regulation of a multitude of 'defence' responses that produce a strongly antimicrobial environment in and around the dead cells (Koch and Slusarenko 1990; Goodman and Novacky 1994; Hua Li et al. 2004; 2007). Interestingly, in our study, the hypersensitive reaction-like events (e.g., early enhancement of yellow fluorescence, and cell death around the penetrated cells) were observed in *S. alba* and *P. sativum* despite the absence of a visible hypersensitive response. At all stages of the infection processes we observed, susceptible responses were characterized by the absence of those reactions seen in incompatible interactions. Hypersensitive responses in the resistant host species, and also in one non-host, typically occurred during the early stages of infection. This is in contrast to the rapid colonisation of plant tissue, pathogen reproduction and secondary infection cycles that were observed on susceptible *B. napus* cv. Thunder TT.

In conclusion, the comparison of infection processes on different cruciferous host plants (resistant and susceptible genotypes) and on leguminous and graminaceous non-host species by *H. parasitica* indicated the critical stages during infection which determine the outcome of pathogenesis. We propose that each stage beyond initial penetration could represent a 'switching point' (Heath 1974) at which the plant response(s) determines the subsequent progress or limitation of the infection resulting in compatible or incompatible interactions. Even though both the host and non-host resistance responses appear to have certain similarities, it remains to be determined if these responses involve the same or different signal transduction pathways. Although the interactions of *H. parasitica* with *Arabidopsis thaliana* have been used as model for studies on signaling pathways it is clear that such studies should first be conducted on the pathogen interaction with closely related host species as well as non-host plants, both at phenotypic as well as histological levels.

Acknowledgment This research was supported by a funding from both the Australian Research Council and from the Department of Agriculture and Food Western Australia.

References

Allen, E. A., Hazen, B. E., Hoch, H. C., Kwon, Y., Leinhos, G. M. E., Staples, R. C., et al. (1991). Appressorium formation in response to topographical signals by 27 rust species. *Phytopathology, 81*, 323–331.

Bennett, M., Gallagher, M., Fagg, J., Bestwick, C., Paul, T., Beale, M., et al. (1996). The hypersensitive reaction, membrane damage and accummulation of autofluorescent phenolics in lettuce cells challenged by *Bremia lactucae*. *The Plant Journal, 9*, 851–865.

Bolwell, G. P. (1993). Dynamic aspects of the plant extracellular matrix. *InternationalReview of Cytology, 146*, 261–324.

Carzaniga, R., Bowyer, P., & O'Connell, R. J. (2001). Production of extracellular matrices during development of infection structures by the downy mildew *Peronospora parasitica*. *New Phytologist, 149*, 83–93.

Channon, A. G. (1981). Downy mildew of Brassicas. In D. M. Spencer (Ed.), *The downy mildews* (pp. 321–336). London: Academic.

Constantinescu, O., & Fatehi, J. (2002). *Peronospora*-like fungi (Chromista, Peronosporales) parasitic on Brassicaceae and related hosts. *Nova Hedwigia, 74*, 291–338.

Dai, G. H., Andary, C., Mondolot-Cosson, L., & Boubals, D. (1995). Histochemical studies on the interaction between three species of grapevine, *Vitis vinifera*, *V. rupestris* and *V. rotundifolia* and the downy mildew fungus, *Plasmopara viticola*. *Physiological and Molecular Plant Pathology, 46*, 177–188.

Dickinson, C. H., & Greenhalgh, J. R. (1977). Host range and taxonomy of *Peronospora* on crucifers. *Transactions of the British Mycological Society, 69*, 111–116.

Dickman, M. B., Ha, Y. S., Yang, Z., Adams, B., & Huang, C. (2003). A protein kinase from *Colletotrichum trifolii* is induced by plant cutin and is required for appressorium formation. *Molecular Plant-Microbe Interactions, 16*, 411–421.

Diez-Navajas, A. M., Wiedemann-Merdinoglu, S., Greif, C., & Merdinoglu, D. (2008). Nonhost versus host resistance to the grapevine downy mildew, *Plasmopara viticola*, studied at the tissue level. *Phytopathology, 98*, 776–780.

Donofrio, N. M., & Delaney, T. P. (2001). Abnormal callose response phenotype and hypersensitivity to *Peronospora parasitica* in defense-compromised Arabidopsis *nim1-1* and salicylate hydroxylase-expressing plants. *Molecular Plant-Microbe Interactions, 14*, 439–450.

Fry, S. C. (1986). Cross-linking of matrix polymers in the growing cell walls of angiosperms. *Annual Review of Plant Physiology, 37*, 165–186.

Ge, X. T., Li, H., Han, S., Sivasithamparam, K., & Barbetti, M. J. (2008). Evaluation of Australian *Brassica napus* genotypes for resistance to the downy mildew pathogen, *Hyaloperonospora parasitica*. *Australian Journal of Agricultural Research, 59*, 1030–1034.

Gindro, K., Pezet, R., & Viret, O. (2003). Histological study of the responses of two *Vitis vinifera* cultivars (resistant and susceptible) to *Plasmopara viticola* infections. *Plant Physiology and Biochemistry, 41*, 846–853.

Goodman, R. N., & Novacky, A. J. (1994). *The hypersensitive reaction in plants to pathogens. A resistance phenomenon*. St. Paul: APS.

Graham, M. Y., & Graham, T. L. (1991). Rapid accumulation of anionic peroxidases and phenolic polymers in soybean cotyledon tissues following treatment with *Phytophthora megasperma* f. sp. *glycinea* wall glucan. *Plant Physiology, 97*, 1445–1455.

Heath, M. C. (1974). Light and electron microscope studies of the interactions of host and non-host plants with cowpea rust-*Uromyces phaseoli* var. *vignae*. *Physiological Plant Pathology, 4*, 403–414.

Heath, C. M. (1980). Reactions of nonsuscepts to fungal pathogens. *Annual Review of Phytopathology, 18*, 211–236.

Heath, M. C. (2000). Nonhost resistance and nonspecific plant defences. *Current Opinion in Plant Biology, 3*, 315–319.

Heintz, C., & Blaich, R. (1990). Ultrastructural and histochemical studies on interactions between *Vitis vinifera* L. and *Uncinula necator* (Schw.) Burr. *New Phytologist, 115*, 107–117.

Holub, E., & Cooper, A. (2004). Matrix, reinvention in plants: how genetics is unveiling secrets of non-host disease resistance. *Trends in Plant Science, 9*, 211–214.

Howlett, B. J., Ballinger, D. J., & Barbetti, M. J. (1999). Diseases of canola in Australia. In P. A. Salisbury, T. D. Potter, G. McDonald, & A. G. Green (Eds.), *Canola in Australia: the first thirty years* (pp. 47–52). Canberra: Organising Committee of the 10th International Rapeseed Congress.

Jahnen, W., & Hahlbrock, K. (1988). Differential regulation and tissue-specific distribution of enzymes of phenylpropanoid pathways in developing parsley seedlings. *Planta, 173*, 197–204.

Kamoun, S., Huitema, E., & Vleeshouwers, V. G. A. A. (1999). Resistance to oomycetes: a general role for the hypersensitive response? *Trends in Plant Science, 4*, 196–200.

Koch, E., & Slusarenko, A. (1990). Arabidopsis is susceptible to infection by a downy mildew fungus. *The Plant Cell, 2*, 437–445.

Lebeda, A., Pink, D. A. C., & Mieslerová, B. (2001). Host–parasite specificity and defense variability in the *Lactuca* spp.–*Bremia lactucae* pathosystem. *Journal of Plant Pathology, 83*, 25–35.

Lebeda, A., Sedlářová, M., Lynn, J., & Pink, D. A. C. (2006). Phenotypic and histological expression of different genetic backgrounds in interactions between lettuce, wild *Lactuca* spp., *L. sativa* x *L. serriola* hybrids and *Bremia lactucae*. *European Journal of Plant Pathology, 115*, 431–441.

Lebeda, A., Sedlářová, M., Petřivalský, M., & Prokopová, J. (2008). Diversity of defence mechanisms in plant-oomycete interactions: A case study of *Lactuca* spp.-*Bremia lactucae*. *European Journal of Plant Pathology, 122*, 71–89.

Li, H., Sivasithamparam, K., Barbetti, M. J., & Kuo, J. (2004). Differences in the germination and invasion by ascospores and pycnidiospores of *Leptosphaeria maculans* on canola varieties with varying susceptibility to blackleg. *Journal of General Plant Pathology, 70*, 261–269.

Li, H., Kuo, J., Barbetti, M. J., & Sivasithamparam, K. (2007). Differences in the responses of stem tissues of spring-type *Brassica napus* cultivars with polygenic resistance and single dominant gene-based resistance, to inoculation with *Leptosphaeria maculans*. *Canadian Journal of Botany, 85*, 191–203.

Mauch-Mani, B., & Slusarenko, A. J. (1993). *Arabidopsis* as a model host for studying plant-pathogen interactions. *Trends in Microbiology, 1*, 265–270.

Mauch-Mani, B., & Slusarenko, A. J. (1996). Production of salicylic acid precursors is a major function of phenylalanine ammonia-lyase in the resistance of *Arabidopsis* to *Peronospora parasitica*. *The Plant Cell, 8*, 203–212.

McDowell, J., Cuzick, A., Can, C., Beynon, J., Dangl, J. L., & Holub, E. B. (2000). Downy mildew (*Peronospora parasitica*) resistance genes in *Arabidopsis* vary in functional requirements for *NDR1*, *EDS1*, *NPR1* and salicylic acid accumulation. *The Plant Journal, 22*, 523–529.

Mellersh, D. G., & Heath, M. C. (2003). An investigation into the involvement of defense signaling pathways in components of the nonhost resistance of *Arabidopsis thaliana* to rust fungi also reveals a model system for studying rust fungal compatibility. *Molecular Plant Microbe Interactions, 16*, 398–404.

Mendgen, K., & Hahn, M. (2002). Plant infection and the establishment of fungal biotrophy. *Trends in Plant Science, 7*, 352–356.

Nashaat, N. I., Heran, A., Awasthi, R. P., & Kolte, S. J. (2004). Differential response and genes for resistance to *Peronospora parasitica* (downy mildew) in *Brassica juncea* (mustard). *Plant Breeding, 123*, 512–515.

Paul, V. H., Klodt-Bussmann, E., Dapprich, P. D., Capelli, C., & Tewari, J. P. (1998). Results on preservation, epidemiology, and aggressiveness of *Peronospora parasitica* and results with regard to the disease resistance of the pathogen on *Brassica napus*. *Bulletin OILB/SROP, 21*, 49–56.

Satou, M., & Fukumoto, F. (1996). The host range of downy mildew, *Peronospora parasitica*, from *Brassica oleracea*, cabbage and broccoli crops. *Annals of the Phytopathological Society of Japan, 62*, 393–396.

Scalbert, A., Monties, B., Lallemand, J. Y., Guittet, E., & Rolando, C. (1985). Ether linkage between phenolic acids and lignin fractions from wheat straw. *Phytochemistry, 24*, 1359–1362.

Sedlářová, M., & Lebeda, A. (2001). Histochemical detection and role of phenolic compounds in defence response of *Lactuca* spp. to lettuce downy mildew (*Bremia lactucae*). *Journal of Phytopathology, 149*, 693–697.

Silue′, D., Nashaat, N. I., & Trilly, Y. (1996). Differential responses of *Brassica oleraceae* and *B. rapa* accessions to seven isolates of *Peronospora parasitica* at the cotyledon stage. *Plant Disease, 80*, 142–144.

Tucker, S. L., & Talbot, N. J. (2001). Surface attachment and pre-penetration stage development by plant pathogenic fungi. *Annual Review of Phytopathology, 39*, 385–417.

Unger, S., Büche, C., Boso, S., & Kassemeyer, H. H. (2007). The course of two different *Vitis* genotypes by *Plasmopara viticola* indicates compatible and incompatible host–pathogen interactions. *Phytopathology, 97*, 780–786.

Viennot-Bourgin, G. (1981). Downy mildew of Brassicas. In D. M. Spencer (Ed.), *The downy mildews* (pp. 1–15). London: Academic.

Williams, P. H. (1985). 'Downy mildew.' Crucifer Genetics Cooperative (CrGc) Resource, Madison WI, University of Wisconsin, USA.

Wynn, W. K. (1976). Appressorium formation over stomates by the bean rust fungus: response to a surface contact stimulus. *Phytopathology, 66*, 136–146.

Yun, B. W., Atkinson, H. A., Gaborit, C., Greenland, A., Read, N. D., Pallas, J. A., et al. (2003). Loss of actin cytoskeletal function and EDS1 activity, in combination, severely compromises non-host resistance in *Arabidopsis* against wheat powdery mildew. *The Plant Journal, 34*, 768–777.

Zhou, N., Tootle, T. L., & Glazebrook, J. (1999). *Arabidopsis* PAD3, a gene required for camalexin biosynthesis, encodes a putative cytochrome P450 monooxygenase. *Plant Cell, 11*, 2419–2428.

Eur J Plant Pathol (2011) 129:101–121
DOI 10.1007/s10658-010-9627-8

Distribution of race-specific resistance against *Bremia lactucae* in natural populations of *Lactuca serriola*

Irena Petrželová · Aleš Lebeda

Accepted: 26 April 2010 / Published online: 23 May 2010

Abstract A metapopulation approach was applied to population studies of a common weed, *Lactuca serriola* (prickly lettuce). Seedlings grown from seed samples collected from 752 individual *L. serriola* plants in 50 populations occurring along an east-to-west transect across four European countries (Czech Republic, Germany, Netherlands and United Kingdom) were screened for resistance to 10 common races of *Bremia lactucae*. Based on the recorded reaction patterns, host individuals were characterized into specific resistance (R-) phenotypes. Diversity of R-phenotypes, their variation and distribution among and within European populations, was evaluated at different spatial scales, i.e. from a metapopulation involving the entire European study area to individual plants occurring in local populations. Generally, European populations of *L. serriola* have been shown to be highly susceptible to *B. lactucae*. However, large variation in *L. serriola* resistance was found both among and within individual countries. There was a clear gradient of increasing uniformity of race-specificity moving from central to western Europe, as well as a slight decrease in the diversity of R-phenotypes. Populations in the United Kingdom were the most divergent in terms of resistance structure from other geographic regions, and also were the most homogeneous, most likely a consequence of the relatively greater degree of spatial isolation from other regions. Metapopulation, inter- and intra-population variation in host resistance is discussed from the viewpoint of occurrence of race-specific interactions in this wild plant pathosystem.

Keywords Resistance complexity · Gene flow · Diversity · Host-pathogen interactions · Intra-population variation · Lettuce downy mildew · Metapopulation · Prickly lettuce · Resistance structure

I. Petrželová · A. Lebeda (✉)
Faculty of Science, Department of Botany,
Palacký University in Olomouc,
Šlechtitelů 11,
783 71 Olomouc, Czech Republic
e-mail: ales.lebeda@upol.cz

Introduction

It is well known that pathogens have the potential to play a substantial role in shaping the structure, as well as the dynamics and evolution of natural plant communities (e.g. Burdon et al. 2006). Importantly, studies of wild plant pathosystems have been found to be attractive from an applied perspective, particularly with regard to quantifying factors that determine the evolution and maintenance of variation in host resistance and pathogen virulence. They can yield much information useful for both resistance breeding and the practical management of plant disease (Lenné and Wood 1991; Lebeda et al. 2007b, 2009a). However, the number of wild plant-pathogen associations that have been studied so far is still limited, and there are large differences in the level of available information for individual interactions.

The most intensively studied natural plant-pathogen interaction is that involving *Linum marginale-Melampsora lini*. Many papers have been written concerning various aspects of the host plant-pathogen relationship (coevolution, spatial and temporal variation, as well as population dynamics and epidemiology) in that pathosystem (e.g. Burdon and Jarosz 1991; Burdon 1994; Burdon and Thompson 1995; Burdon et al. 1999; Burdon and Thrall 1999, 2000; Thrall et al. 2001, 2002). Some additional population studies have been published for other wild plant pathosystems, e.g. *Filipendula ulmaria-Triphragmium ulmariae* (Ericson et al. 2002; Smith et al. 2003); *Plantago lanceolata-Podosphaera plantaginis* (Laine 2004, 2005; Laine and Hanski 2006; Laine 2007); *Senecio vulgaris-Erysiphe fischeri* (Bevan et al. 1993a, b, c); *Silene* spp.-*Ustilago violacea* (*Microbotryum violaceum*) (Thrall and Antonovics 1995; Alexander et al. 1996; Delmotte et al. 1999; Carlsson-Granér and Thrall 2002; Carlsson-Granér 2006); and *Valeriana salina-Uromyces valerianae* (Ericson et al. 1999). However, there is a large knowledge gap regarding interactions between weed host plants and oomycetes (Holub 2008; Lebeda et al. 2008), despite the fact that oomycete pathogens represent an important group of species which are common in nature and have a world-wide distribution (Lebeda and Schwinn 1994).

In most published studies focused on wild pathosystems, great emphasis is put on the importance of considering host and pathogen interactions within a metapopulation context, i.e. local populations connected by migration events and gene flow (Burdon 1993; McDermott and McDonald 1993). Regional differences in migration rates may result in large differences and variation among individual local populations but these represent only small pieces of a highly diverse mosaic of resistance and virulence phenotypes over a larger area. Most of the studies mentioned above also highlighted the necessity for parallel studies of the population structure of both participants in interactions between plants and obligate biotrophic pathogens (Clarke and Akhka 2002; Holub 2008; Lebeda et al. 2008), especially because their populations are, of necessity, closely interconnected in a coevolutionary sense. Knowledge of populations of one species (either host or pathogen) within these interactions may help to explain much of the behavior of the populations of coevolving species.

In our study we focused on the wild plant pathosystem *Lactuca serriola* L. (prickly lettuce)-*Bremia lactucae* Regel (lettuce downy mildew). *B. lactucae* is an oomycete pathogen of cultivated lettuce (*Lactuca sativa*), as well as many other species of Asteraceae that have worldwide distribution (Lebeda et al. 2002, 2007b). For the 17 *Lactuca* species described from Europe (Lebeda et al. 2001a), only seven are known as natural hosts of *B. lactucae* (Lebeda et al. 2002). Of these seven species, *L. serriola* is the most common wild host of *B. lactucae*, as has recently been published for the Czech Republic (Petrželová and Lebeda 2004a; Lebeda et al. 2007a, 2008) and the Netherlands (Hooftman et al. 2006). In other European countries only sporadic occurrences of natural infections by *B. lactucae* were recorded during collection of *Lactuca* spp. in Austria, France, Germany, Switzerland (Lebeda et al. 2001a) and Slovenia (Doležalová et al. 2001).

Currently, there are no detailed studies focused on the natural distribution of *B. lactucae* in Europe or the other continents. However, we recently acquired a large amount of information for the Czech pathogen populations of *B. lactucae* on the wild host *L. serriola* (Lebeda 2002; Lebeda and Petrželová 2004a; Petrželová and Lebeda 2004b; Lebeda et al. 2008). However, to date there are no corresponding detailed data for resistance variation in Czech host plant populations. The same is true for the host populations in the other European countries (Lebeda et al. 2007a).

The aim of our present research was to: a) screen *L. serriola* samples originating from four European countries for their resistance to 10 isolates of *B. lactucae*; b) determine the diversity of resistance phenotypes, and c) assess both their distribution and variation at the intra- and inter-population levels within the European metapopulation of *L. serriola*. Some partial results had been published previously (Lebeda and Petrželová 2004b, 2007a, b; Lebeda et al. 2008). In this paper we present a more detailed analysis of various aspects of resistance variation within populations of *L. serriola*. Here we summarize the results of this analysis, which includes data from a broad range of different *L. serriola* populations across different geographic regions.

Materials and methods

Collection of plant materials

Lactuca serriola seed samples used for our research originated mostly from collections made in 2001 as part of the Fifth Framework Programme of the

European Union called "Gene-Mine". Fifty populations of *L. serriola* occurring in four European countries (Czech Republic (CZ), $n=16$; Germany (D), $n=16$; Netherlands (NL), $n=8$; and United Kingdom (UK), $n=10$) were investigated and seeds collected (Lebeda et al. 2007a). Collection sites were distributed along transects throughout the countries studied in order to collect the most representative samples of the spatial variation of resistance in the European *L. serriola* metapopulation (Fig. 1). In order to capture variation within individual local populations of *L. serriola*, seeds were taken separately from individual plants; if available, 16 plants per population were examined. Summaries of data on the origin of the *L. serriola* samples, including site geographical and ecological characterizations, as well as information on the observed populations of *L. serriola* and the occurrence of natural infections caused by fungal pathogens, have been published elsewhere (Lebeda et al. 2007a). Examples of the most common habitats of *L. serriola* are depicted in Fig. 2. Plants with symptoms of natural infection are shown in Fig. 3.

Plant and pathogen materials used to evaluate resistance patterns

Naturally-growing *L. serriola* plants did not provide enough seed usable for our experiments. The collection samples, originating from individual plants, had to be regenerated and multiplied in a greenhouse first, and the seeds derived from a total of 752 samples were included in resistance screening (Table 1). Thus, in the present study "samples" are considered as the sets of *L. serriola* plants grown from seeds of regenerants derived from original collection samples, and represent original individuals ($n=752$) from the natural populations ($n=50$) of *L. serriola* originally examined (Lebeda et al. 2007a).

L. serriola samples were screened for their resistance to 10 isolates representing the officially denominated races of *B. lactucae* Bl:1, Bl:5, Bl:12, Bl:14, Bl:15, Bl:16, Bl:17, Bl:18, Bl:21, and Bl:24 with known virulence patterns (van Ettekoven and van der Arend 1999; van der Arend et al. 2006). The *B. lactucae* isolates used in this research originated from cultivated lettuce (*Lactuca sativa*) and are maintained in the Czech National Collection of Microorganisms (http://www.vurv.cz/collections/vurv.exe/) at Palacký University in Olomouc, Department of Botany (http://botany.upol.cz).

Inoculation studies and resistance screening

The tests were carried-out according to the methods described in Lebeda (2002) and Petrželová and Lebeda (2004b). Pathogen races were maintained

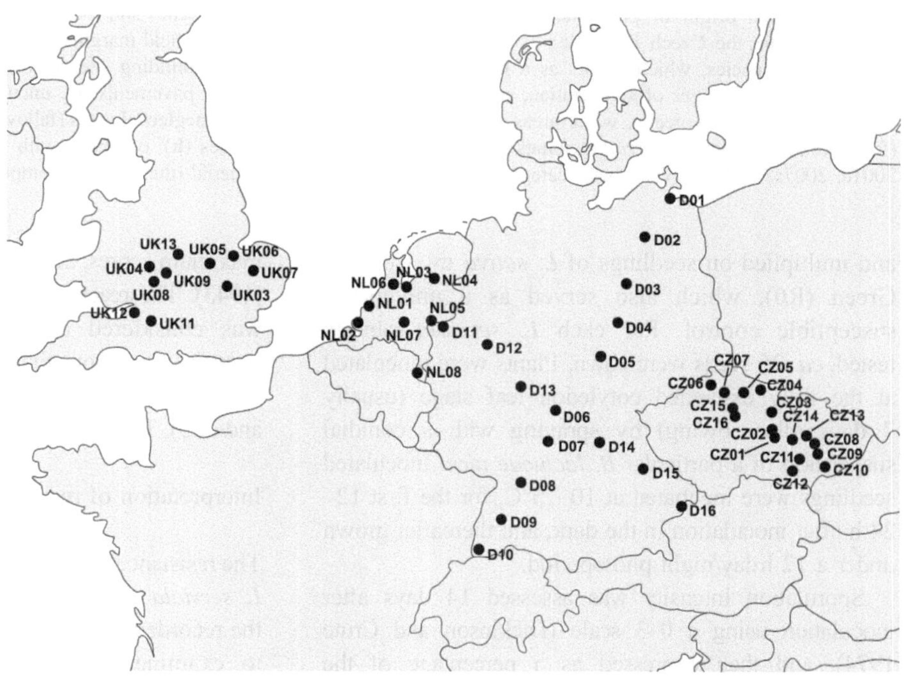

Fig. 1 Collection sites of *Lactuca serriola* populations in the Czech Republic, Germany, Netherlands and United Kingdom (Lebeda et al. 2007a)

Fig 2 Populations of *Lactuca serriola* in its common habitats, representing the origin of plant material used in this study (examples from the Czech Republic). *L. serriola* is a typical synanthropic species, which occurs as a pioneer plant in the early successional stages of a vegetation, and occupies various man-made habitats; typically, waste areas with disturbed soils (for more details about *L. serriola* habitats see e.g. Lebeda et al. 2001a, 2007a, 2008). Several basic categories (bold text) were distinguished: along the transport corridors such as railway (see **a**) or ditches and roadsides (**b**); agricultural areas, mostly field roads or field margins (**c**); ruderal areas such as dust heaps (**d**), debris, building sites or man-made piles of soil; urban areas, mostly pavements (**e**), uncultivated lawns (**f**) or parking sites; larger neglected areas (fallows) (**g**); moist habitats such as moist ditches (**h**), or places with a local accumulation of biological material (dunghills or compost-heaps) (**i**)

and multiplied on seedlings of *L. sativa* cv. Cobham Green (R0), which also served as a universally susceptible control. For each *L. serriola* sample tested, *ca.* 25 seeds were sown. Plants were inoculated at the fully expanded cotyledon leaf stage (usually 7 days after sowing) by spraying with a conidial suspension of a particular *B. lactucae* race. Inoculated seedlings were incubated at 10–15°C, for the first 12–24 h after inoculation in the dark, and thereafter grown under a 12 hday/night photoperiod.

Sporulation intensity was assessed 14 days after inoculation using a 0–3 scale (Dickinson and Crute 1974), and then expressed as a percentage of the maximum scores, according to Towsend and Heuberger (1943). The reaction of a particular *L. serriola* sample was considered as susceptible if the sporulation intensity was more than 30%, and at least half of the seedlings showed a high degree of infection (i.e. rated 2 and/or 3). Tests giving ambiguous results were repeated.

Interpretation of resistance data

The resistance phenotypes (R-phenotypes) of individual *L. serriola* samples were characterized on the basis of the recorded reaction patterns. This data was then used to examine patterns of variation in resistance to

Fig 3 *Lactuca serriola* plants with symptoms of natural infection by lettuce downy mildew. Weedy growing plants can be infected by *B. lactucae* in all developmental stages (Petrželová and Lebeda 2004a). Typically, *B. lactucae* causes light green, yellow or necrotic lesions surrounded by larger leaf veins, visible on the upper leaf surface (Lebeda et al. 2008). Figure shows the early stage of infection on rosette (**a**) or stem leaves (**b**) of the young plants. Sometimes, young plants may be completely covered with lesions (**c**). Under optimal microclimatic conditions, sporulation occurs on the lower surface of lesions (**d**). Figure (**e**) shows strong infection on stem leaves of adult *L. serriola* plants

B. lactucae within and between wild populations of *L. serriola*. Variation was assessed at several different spatial scales (European metapopulation, regional populations extended over larger areas of the European countries studied [in the sense of "spatially extended populations" suggested by Freckleton and Watkinson (2002)], local populations and individuals within populations). Our aim was to assess whether there were any geographic differences in the resistance of *L. serriola* populations, determine spatial patterns of

Table 1 Survey of *Lactuca serriola* populations and samples screened for resistance

Origin of seed samples	Number of populations	Number of seed samples
Czech Republic (CZ)	16	250
Germany (D)	16	241
Netherlands (NL)	8	120
United Kingdom (UK)	10	141
Totally	50	752

resistance variation, and to evaluate at which spatial scale the variation is most pronounced.

Results

Evaluation of the resistance of *L. serriola* populations in their natural habitats

As mentioned previously, collecting trips, designed to study weedy populations of *L. serriola*, were undertaken in 2001. Information about the occurrence of diseases on *L. serriola* plants was also recorded (Lebeda et al. 2007a). Surprisingly, the natural occurrence of *B. lactucae* infections in European populations of *L. serriola* was rarely observed within the host European metapopulation. In fact, in 2001, diseased populations of *L. serriola* were only found in the Czech Republic, where the disease incidence was very high. From a total of the sixteen populations that were examined for the presence of *B. lactucae*, fifteen were infected. However, disease prevalence (expressed for the whole local population as the degree of visually observable infection (DI) using the 0–3 scale (Lebeda et al. 2008)) was very low in most populations (two populations displayed a medium level of disease prevalence [DI2], and one a high level [DI3]).

Resistance of European metapopulation of *L. serriola* to *B. lactucae*—a general view

Resistance screening of *L. serriola* samples revealed broad variation in resistance within the European metapopulation, both at the level of populations and of individuals (Fig. 4). The relative resistance of populations of *L. serriola* was assessed, taking into consideration the prevalent reaction patterns of tested *L. serriola* samples to the set of 10 races of *B. lactucae* used. As shown in Fig. 4a, the overall resistance of the European metapopulation was quite low, which is in contrast with only slightly infected or infection-free populations observed during our field trips. Under laboratory conditions, only a few host populations were considered to be resistant; one third of the examined populations were highly susceptible; one third expressed a pronounced race-specific response; and the rest consisted of a mixture of resistant, susceptible, and clearly race-specific responding individuals.

When considering the resistance of the European metapopulation of *L. serriola*, from the viewpoint of the overall resistance of individuals, only 59 samples tested expressed complete resistance to all 10 races of *B. lactucae* used in the inoculation study. In total, 157 samples were susceptible to all races used, and 536 samples expressed more or less pronounced race-specific response (Fig. 4b).

Analysis of resistance variation in *L. serriola* originating from four European countries

Geographic variation in resistance of L. serriola *populations to individual races of* B. lactucae

Despite the overall low resistance of the European *L. serriola* metapopulation, clear differences in resistance to individual races of *B. lactucae* were observed

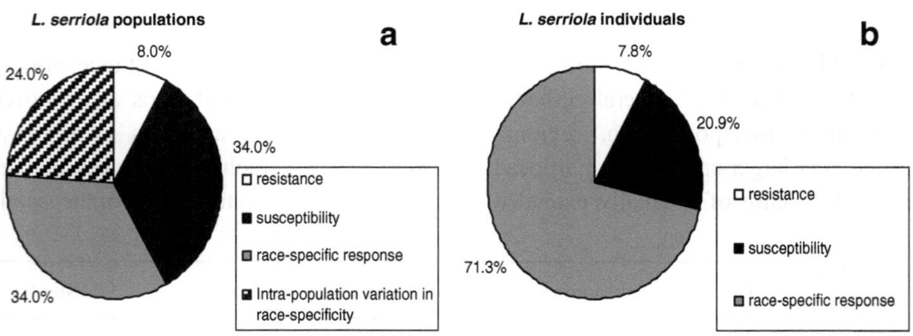

Fig. 4 Distribution of resistance to *Bremia lactucae* among *Lactuca serriola* populations ($n=50$) and individuals ($n=752$) within the entire European metapopulation of *L. serriola*. (**a**) Proportion of *L. serriola* populations with a high level of resistance, susceptibility, race-specificity (resistance to several races and susceptibility to other of 10 used races of *B. lactucae*) or intra-population variation in resistance (mixture of individuals with highly resistant and highly susceptible reaction patterns occurring in approximately equal ratio within one local population); (**b**) Proportion of *L. serriola* individuals which were completely resistant, completely susceptible or expressed race-specific response to the used set of *B. lactucae* races

among the four countries (Fig. 5). As shown in Fig. 5a, the Czech populations were found to be the most variable in overall resistance to individual pathogen races tested, with respect to both the prevalence of resistant and susceptible responses, and there was also a relatively high frequency of local populations characterized by variable responses of individuals. In Germany (Fig. 5b), overall the prevalence of resistant individuals was lower, and populations were generally either highly susceptible or showed variation in response to individual isolates of *B. lactucae*. Minimal resistance was only recorded for pathogen races Bl:1, Bl:5, Bl:12, Bl:14, Bl:15, and Bl:18. Dutch populations of *L. serriola* (Fig. 5c) were highly susceptible to most of the races used, and only showed high levels of resistance against race Bl:15, with lower levels of resistance to races Bl:1, Bl:5, Bl:12, and Bl:17. Populations in the United Kingdom (Fig. 5d) were characterized by very low variation in their responses. All populations were proven to be completely resistant to pathogen races Bl:1, Bl:5, and Bl:15, and highly susceptible to Bl:12, Bl:14, Bl:16, Bl:18, Bl:21, and Bl:24. Responses to race Bl:17 were the most variable.

Geographic variation in resistance to the set of B. lactucae races among L. serriola populations and individuals

A very useful characteristic for the evaluation of resistance of host populations is the complexity of resistance (i.e. the proportion of resistances recorded in a host individual from the total number of resistances examined, in our case from resistances to the 10 pathogen races used), representing the level of race-specificity of the host individuals. The results from the tested *L. serriola* samples from the Czech Republic (CZ), Germany (D), the Netherlands (NL) and United Kingdom (UK) (in our study representing individuals from *L. serriola* populations originally visited) revealed large differences among European countries, with a variety of resistance complexity patterns. Figure 6 shows that individuals with the most complex resistance patterns only occurred in the

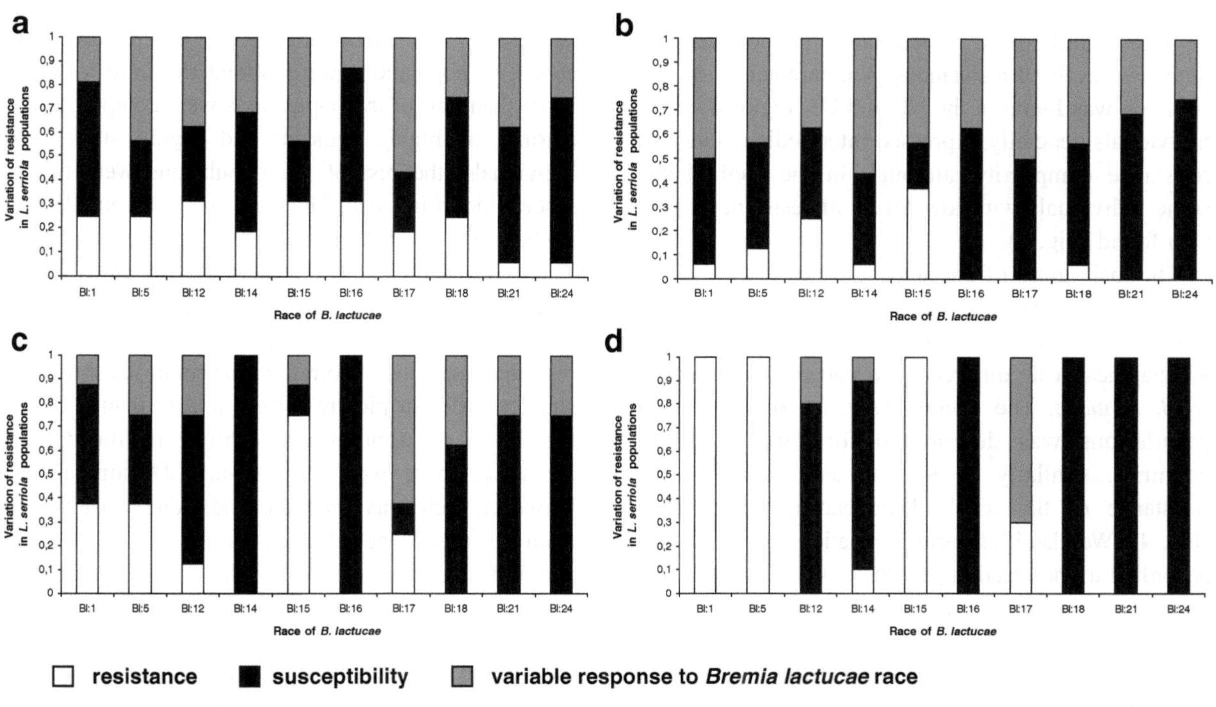

Fig. 5 Variation of resistance of *Lactuca serriola* populations to ten races of *Bremia lactucae* within and among four European countries (**a** Czech Republic, n=16 populations; **b** Germany, n=16; **c** Netherlands, n=8; **d** United Kingdom, n=10). In individual histograms each column shows a proportion of local populations of *L. serriola* within European country with a prevalence of resistant, susceptible and variable responses (mixture of resistant and susceptible individuals) to individual races of *B. lactucae*

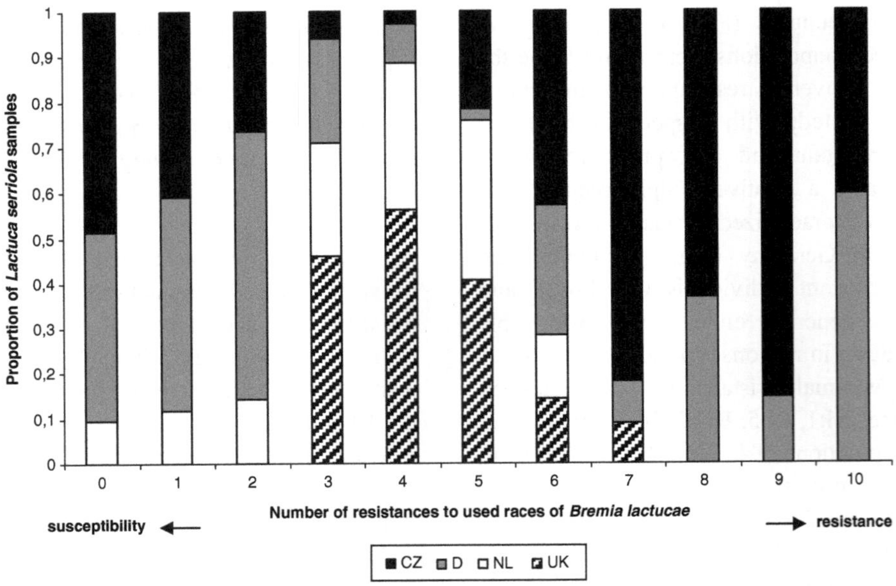

Fig. 6 Variation in complexity of resistance to the used set of *Bremia lactucae* races (n=10) among *Lactuca serriola* populations in four European countries. Complexity of resistance was evaluated for 752 individuals of *L. serriola* and was expressed as the number of recorded resistant responses of individuals: *0*=no resistance to used races of *B. lactucae*, *10*=resistance to all races. Each column of the histogram shows a relative proportion of CZ, D, NL and UK samples in the categories 0 to 10

CZ and D regions. However, there were also a large number of individuals found in these countries with a very low complexity of resistance phenotypes, i.e. with low levels of resistances. A completely different situation was found in the NL and UK regions, where individuals generally expressed intermediate levels of resistance complexity, although in the Netherlands some individuals with low levels of resistances were also found (Fig. 6).

The main goals of our study were to determine the distribution of race-specific resistance among populations of *L. serriola*, and to evaluate whether there are geographical-scale differences in patterns of resistance to *B. lactucae*. The overall resistance of *L. serriola* populations was determined for four European countries, similarly as for the assessment of the resistance of the whole European metapopulation (Fig. 4). We classified local populations of *L. serriola* according to the reaction patterns to the set of 10 races of *B. lactucae* used, (i.e. according to the prevalent level of resistance complexity) (Fig. 7). Our study revealed that race-specific resistance of *L. serriola* was unequally distributed among European regions. Populations with high overall resistance to the set of *B. lactucae* races used were found only in the Czech Republic. However, a large proportion of the CZ populations expressed high levels of susceptibility or strong intra-population variation in race-specific resistance (Fig. 7a). According to the criteria described above, D populations were among the most variable. More than half of the populations were composed of a mixture of highly resistant and highly susceptible individuals; the rest of the populations were highly susceptible (Fig. 7b). The NL and UK populations studied were highly uniform in their types of responses. The most common reaction pattern recorded was that of pronounced race-specificity (Figs. 7c, d).

At the level of the host individuals, comparison of resistance responses from the four countries showed a slightly different picture for CZ and D than it did at the level of populations. Unlike in the Czech Republic, there were no resistant D populations. However, there was a range of individuals originating from various D populations that expressed a completely resistant response (Figs. 8a, b). The number of resistant individuals was even higher in D populations (total of 35) than in the CZ populations (total of 24). In addition, in both the CZ and D populations there were large numbers of individuals with a relatively pronounced race-specific response.

The graphs showing proportions of resistant, susceptible, and race-specific responsive NL and UK

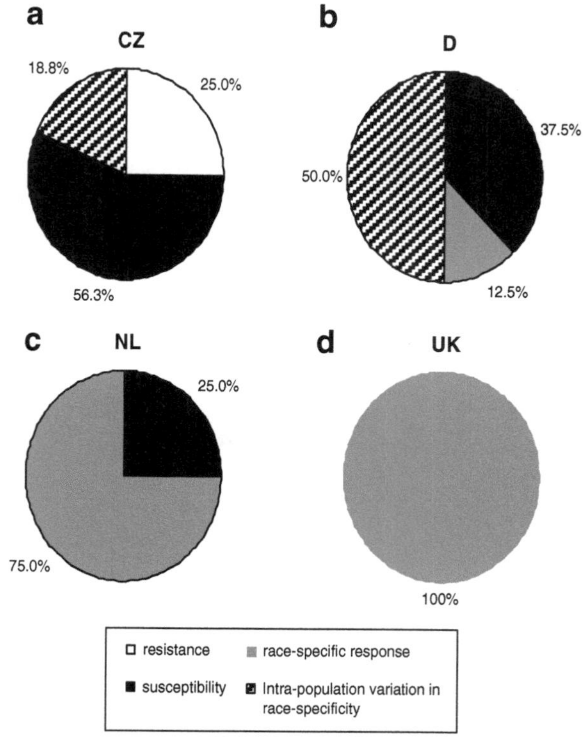

Fig. 7 Distribution of resistance to *Bremia lactucae* among *Lactuca serriola* populations in four European countries (adapted according to data published in Lebeda et al. 2008). Figure shows a proportion of local populations of *L. serriola* with a high level of resistance, susceptibility, race-specificity and intra-population variation in race-specificity to the used set of *B. lactucae* races in each of the studied European populations (**a-d**)

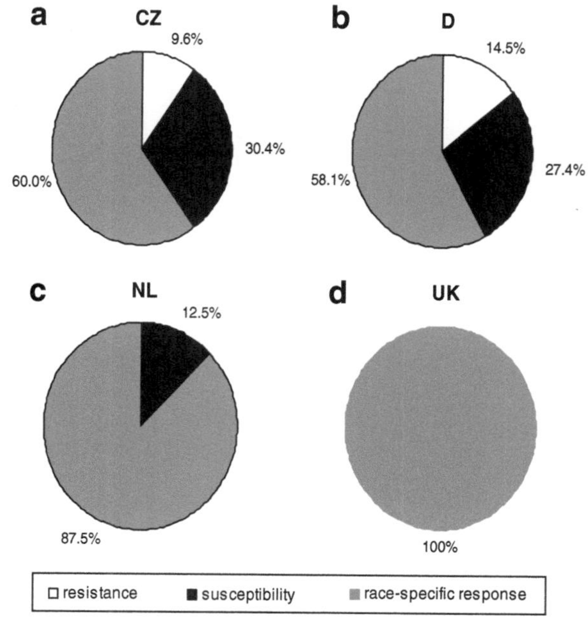

Fig. 8 Distribution of resistance to *Bremia lactucae* among *Lactuca serriola* individuals (n=752) in four European countries. Figure shows a proportion of *L. serriola* individuals which were completely resistant, completely susceptible or expressed race-specific response to the used set of *B. lactucae* races in each of the studied European populations (**a-d**)

individuals (Figs. 8c, d) are similar to those of the entire NL and UK populations.

Spatial distribution of race-specific resistance to B. lactucae across Europe

Of particular interest is the geographic distribution of race-specific resistance to *B. lactucae* within Europe. The spatial arrangement of local host populations showing resistant, susceptible and race-specific responses is depicted in Fig. 9. Within the Czech Republic, resistant populations of *L. serriola* were mainly located in the western parts. One resistant population was also found in the eastern part of the Czech Republic; however, susceptible populations (or those with mixtures of highly resistant and susceptible individuals) were more frequently found there. German populations of *L. serriola* were examined on two transects across the country (Fig. 1). Populations of *L. serriola* in the middle parts of Germany showed strong intra-population variation in their resistance to *B. lactucae*; while in the marginal parts of transects in eastern parts of Germany susceptible populations prevailed. Close to the border with the Netherlands one population with more pronounced race-specificity was found; another one was located in the south-western part of the country. As already mentioned, most Dutch populations displayed clear race-specificity. However, two highly susceptible populations were found in the border area near Germany. Interestingly, in the United Kingdom only populations with clear race-specific response were recorded (Fig. 9).

Analysis of within-population variation in resistance to *B. lactucae*

Variation in complexity of resistance of the local L. serriola populations to the used set of B. lactucae races

The other aim of our research was to determine the intra-population variation in resistance to *B. lactucae*,

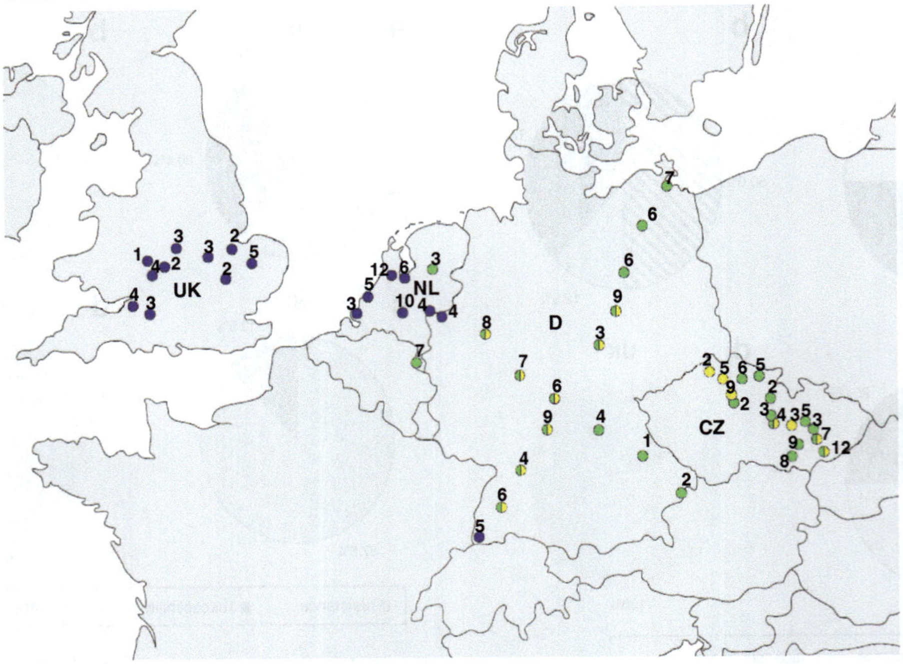

Fig. 9 Spatial distribution of race-specific resistance of *L. serriola* to *Bremia lactucae* across the European area. *Yellow spots* represent local populations of *L. serriola* with a high level of resistance; *green-yellow spots* mixture of highly resistant and highly susceptible individuals within a local population (i.e. intra-population variation in race-specificity); *violet spots* local populations with a high level of race-specificity (i.e. strong differential response to the used *B. lactucae* races); *green spots* local populations of *L. serriola* with a high level of susceptibility. Numbers close to the spots express the total amount of R-phenotypes described within a local population

and to compare particular local populations of *L. serriola* in the European countries studied. First, we quantified the mean number of resistances of individuals within 50 local populations of *L. serriola*. As shown in Fig. 10, the highest mean resistance (>8) was found for populations CZ6, CZ7, CZ14, CZ15, and D13, (which means that all individuals in these populations had highly complex resistance patterns to the races of *B. lactucae*). However, none of these populations as a whole displayed complete resistance to all 10 races used. A medium level complexity of resistance (the mean resistance of individuals to 5-6 races) was recorded for some other CZ and D populations (e.g. CZ1, CZ9, CZ10, D5, D6, D8, and D12); mostly due to a more pronounced intra-population variation in the resistance of individuals. However, in the Czech Republic and Germany there were many local populations (CZ2-CZ5, CZ8, CZ11-CZ13, CZ16, D1-D3, D10, D11, D14-D16) which were almost or completely susceptible, and thus observed complexity of resistance was low (mean resistance of individuals to only 1 or 2 races). Two almost completely susceptible populations (NL4 and NL8) were also found in the Netherlands. With that exception, all NL and UK local populations expressed medium levels of resistance complexity (typically mean resistance to 4 races; Fig. 10c, d).

To evaluate where to search for potential sources of new and efficient resistances we specified the overall resistance of individual local populations of *L. serriola* using the same criteria as shown in Figs. 4 and 7. Local populations were categorized according to the predominant responses of host individuals to 10 races of *B. lactucae* (Table 2). From the Table again it is evident that most of the resistance to *B. lactucae* recorded in the local populations of *L. serriola* studied was race-specific.

Heterogeneity of the local populations of L. serriola

Despite the highly uniform outward responses of many populations of *L. serriola* (resistant, susceptible, or race-specific) more or less pronounced intra-population variation in race-specificity was common in most local populations. Individuals with the same level of resistance complexity (either low or high) often

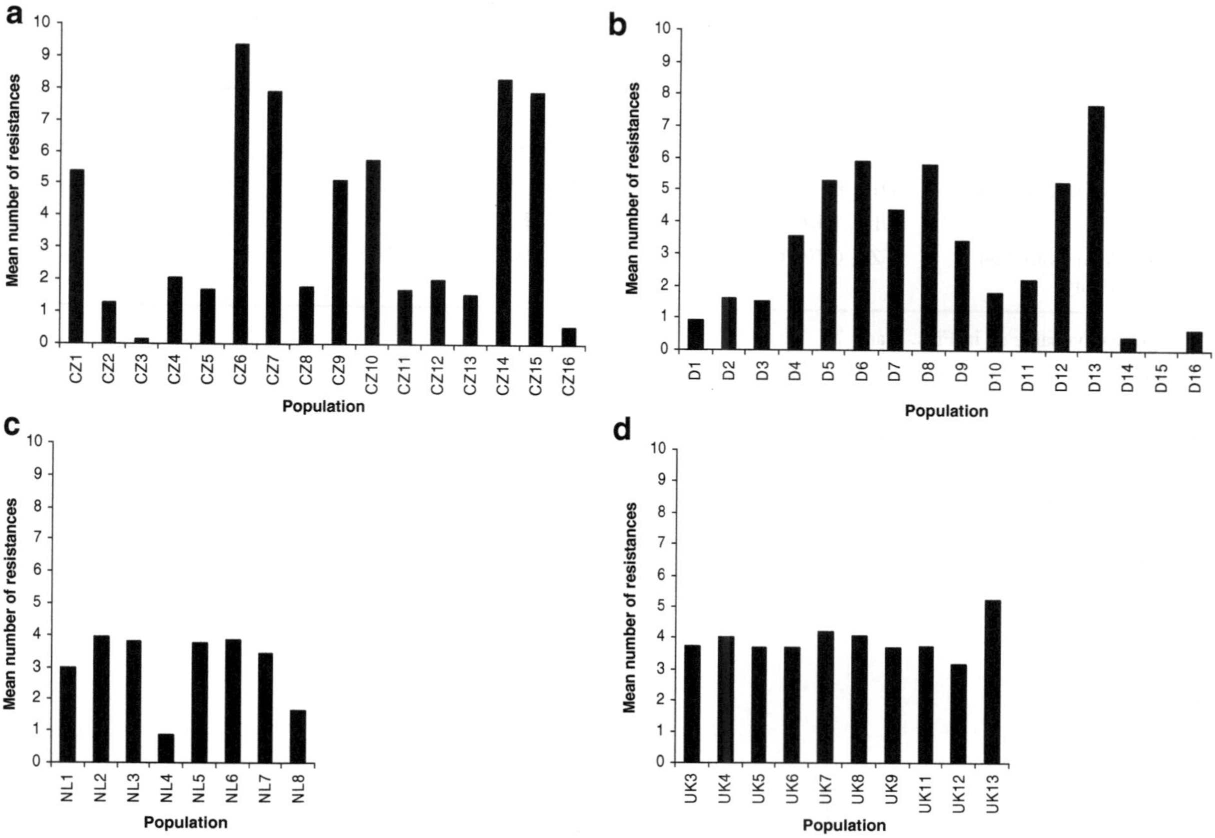

Fig. 10 Relative complexity of resistance of the local populations of *Lactuca serriola* to the used set of *Bremia lactucae* races (*n*= 10). Complexity of resistance was expressed as a mean number of races against which resistance of examined individuals was displayed within individual local populations of *L. serriola* in four European countries (**a-d**): *0*=no resistance to used races of *B. lactucae*, *10*=resistance to all used races

differed in the identity of those *B. lactucae* races to which they expressed resistance or susceptibility. Thus, local population resistance did not generally correlate with the total number of resistance (R-) phenotypes recognized. We expressed that discrepancy as heterogeneity or homogeneity of the local populations.

Figure 11 demonstrates that the degree of heterogeneity (proportion of individuals differing in their R-phenotype) varied broadly among the local populations, both within individual countries and within the entire European metapopulation. The D, NL and some CZ populations were among the most heterogeneous, while UK populations were the most homogeneous. The only exception to this was UK13; in this case the result is largely distorted by the small sample size (only 4 individuals examined). However, highly homogeneous local populations were also found in the Czech Republic (e.g. CZ1-CZ4, CZ6-CZ8, CZ13, CZ14, and CZ16), Germany (D5, D8, D14-D16), and the Netherlands (NL2, NL4, and NL5).

Phenotypic characterization of *L. serriola* individuals

Diversity of L. serriola resistance phenotypes in four European countries

R-phenotypes were described on the basis of reaction patterns of *L. serriola* seedlings derived from seed collections (as described above) to the set of *B. lactucae* races. On the whole, 89 different R-phenotypes were recorded in the studied European metapopulation, among the set of 752 *L. serriola* samples included in our study (Table 3). Regardless of the unequal numbers and sample sizes of the studied European populations, it is evident that there were differences in the diversity of R-phenotypes among countries with the highest

Table 2 Characterization of individual local populations of *Lactuca serriola* for resistance to *Bremia lactucae*

Response[a]	Local populations
High level of resistance[b]	CZ6, CZ7, CZ14, CZ15
High level of susceptibility[c]	CZ2, CZ3, CZ4, CZ5, CZ8, CZ11, CZ12, CZ13, CZ16, D1, D2, D3, D14, D15, D16, NL4, NL8
High level of race-specific response[d]	D10, D11, NL1, NL2, NL3, NL5, NL6[f], NL7, UK3, UK4, UK5, UK6, UK7, UK8, UK9, UK11, UK12, UK13
Intra-population variation in race-specificity[e]	CZ1, CZ9[f], CZ10[f], D4, D5[f], D6, D7[f], D8, D9[d], D12[d], D13[f]

[a] Categorization same as in Figs. 4 and 7

[b] Population composed of individuals with highly complex race-specific resistance patterns

[c] Population composed of individuals with lowly complex race-specific resistance patterns

[d] Whole populations or their big parts expressing high uniformity in their resistance resp. susceptibility to several races of *B. lactucae* (i.e. strong differential response to used races)

[e] Mixture of individuals with various race-specific reaction patterns characterized with high level of resistance and susceptibility occurring within one population

[f] Occurrence of individuals with intermediate susceptibility, i.e. reduced sporulation of a pathogen (ca 30–50%) on most plants included in a resistance test

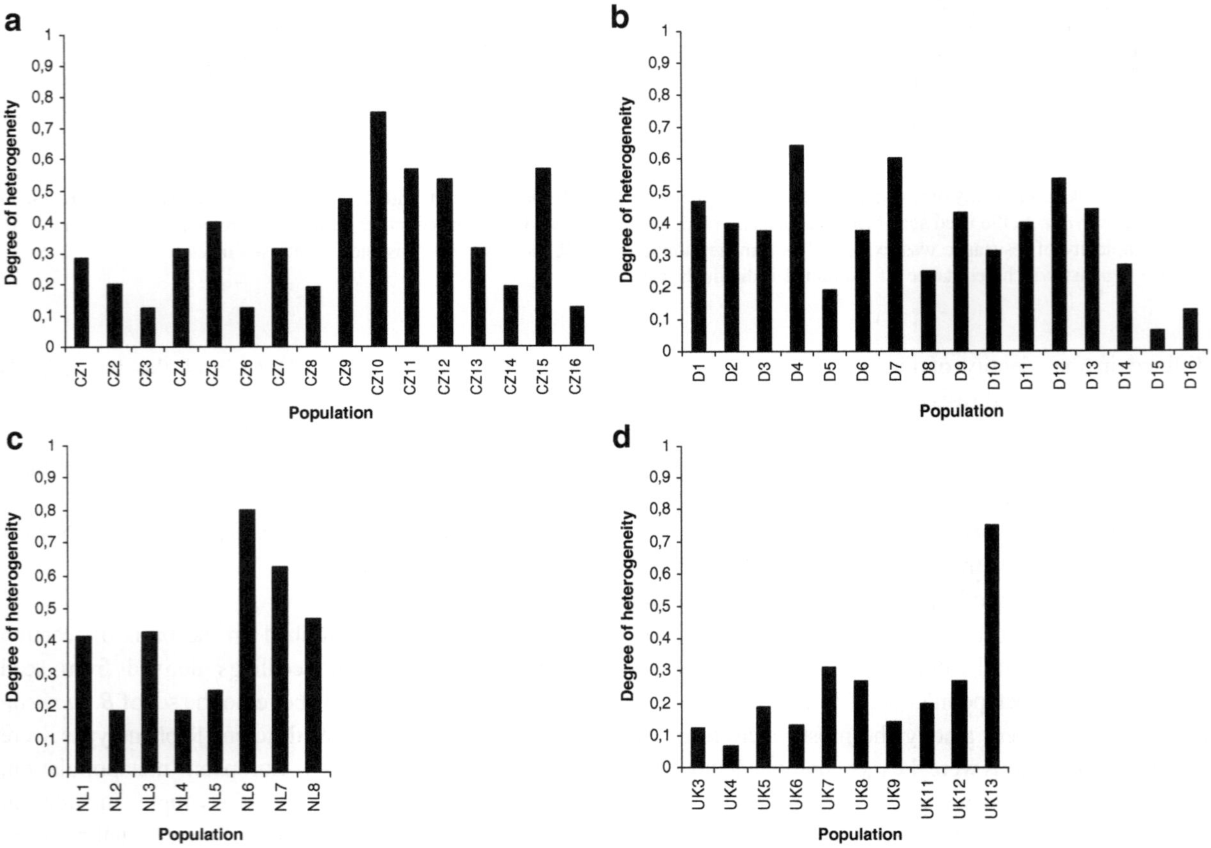

Fig. 11 Heterogeneity of the local populations of *Lactuca serriola* in four European countries (**a-d**). The degree of heterogeneity is expressed as a ratio between a total number of tested individuals in the local populations and a number of recognized resistance phenotypes: >*0*=high homogeneity of a local population, ≤*1* = high heterogeneity of a local population

Table 3 Diversity of resistance phenotypes of *Lactuca serriola* recognized in four European countries and within the entire European metapopulation

Population (n=number of samples)	Total number of recognized R-phenotypes[a]	Number of the most frequent R-phenotypes (% of samples)
CZ (n=250)	45	8 (78.8)
D (n=241)	34	7 (77.6)
NL (n=120)	27	5 (64.2)
UK (n=141)	10	2 (84.4)
European metapopulation (n=752)	89	14 (82.3)

[a] different reaction patterns of individual *L. serriola* samples are described as R-phenotypes (or resistance phenotypes) characterized by specific combinations of resistant (-) and susceptible (+) reactions to the used *B. lactucae* races (n=10)

relative diversity (in proportion to sample size) being recorded in the Netherlands (a total of 27 R-phenotypes). High numbers of *L. serriola* R-phenotypes were also recognized in the Czech Republic and Germany (a total of 45 and 34 phenotypes respectively). However, individual R-phenotypes were unevenly distributed among the local populations, both within and among the four countries studied; additionally, their prevalence were generally relatively low.

The lowest diversity of R-phenotypes was recorded in the United Kingdom. A total of 10 R-phenotypes were recognized, with only two prevalent among 84.4% of all of the UK samples tested (Table 3).

Resistance patterns and prevalence of recognized L. serriola resistance phenotypes

A list of the major R-phenotypes, with their prevalence across the entire European metapopulation of *L. serriola*, as well as within individual countries is given in Table 4. The most commonly distributed R-phenotype (20.9% of all samples) was susceptible to all ten races of *B. lactucae*, and it was recorded in all countries studied with the exception of the UK. It was the most dominant R-phenotype in CZ (distributed in 12 out of 16 populations, and represented by 30.4% of the CZ individuals) and D populations (in 14 out of 16 populations, with a prevalence of 27.4% of the D individuals). The second most frequent R-phenotype was resistant to pathogen lines Bl:1, Bl:5, Bl:15, and Bl:17 (18.1% of the samples tested), and was mostly distributed in the NL and UK populations, where it was the most prevalent R-phenotype. The third most frequent R-phenotype (7.8% of the samples) was resistant to all ten *B. lactucae* races and, as stated above, it was only present in some CZ and D populations.

When comparing the resistance patterns of R-phenotypes recognized in individual countries, large differences were found. Phenotypically, the CZ and D populations were very similar. They were characterized by four main groups of R-phenotypes (Table 4). As mentioned above, most frequent was the R-phenotype with susceptibility to all ten races of *B. lactucae*. A large number of CZ and D samples were described by high susceptibility to the majority of isolates used, with resistance only being expressed against Bl:12 and/or Bl:15, and/or Bl:17. The other most frequent R-phenotype was characterized by resistance to all ten isolates of *B. lactucae*. There also were quite a large number of highly resistant samples with susceptibility only to race Bl:21 and/or Bl:24. In the NL populations, the most frequent R-phenotypes were characterized by various combinations of resistance to Bl:1, Bl:5, Bl:12, Bl:15, and Bl:17, plus susceptibility to other races; however, the completely susceptible R-phenotype was also quite frequent. In the UK populations there were only two broadly distributed R-phenotypes (the first one with resistance only to Bl:1, Bl:5, Bl:15, and Bl:17; the second with resistance only to Bl:1, Bl:5, and Bl:15). Both of these were also recorded in the D and NL populations, but there the frequency of occurrence was much lower (Table 4).

Discussion

This study is probably the first detailed research focused on host resistance variation in plant-oomycete interactions in a wild plant pathosystem. When studying host-pathogen interactions it is difficult to draw authoritative conclusions from results based only on studies of local populations. Spatial structure is a critically important factor for host-pathogen dynamics in nature, and the metapopulation framework has proven to be useful for investigating the ecological and evolutionary dynamics of spatially structured host-pathogen associations in nature (Thrall and Burdon 1997). Most papers dealing with population biology and the ecology of hosts (as well as their

Table 4 Frequency and distribution of the most common resistance phenotypes in the European metapopulation of *Lactuca serriola*

Race of *B. lactucae*/resistance phenotype of *L. serriola*[a]										Total number of samples (%)	Number of samples/populations			
Bl:1	Bl:5	Bl:12	Bl:14	Bl:15	Bl:16	Bl:17	Bl:18	Bl:21	Bl:24		CZ	D	NL	UK
+	+	+	+	+	+	+	+	+	+	157 (20.9)	76/12	66/14	15/3	0/0
–	–	+	+	–	+	–	+	+	+	136 (18.1)	0/0	10/2	37/5	89/9
–	–	–	–	–	–	–	–	–	–	59 (7.8)	24/5	35/9	0/0	0/0
+	+	+	+	–	+	+	+	+	+	45 (6.0)	14/5	26/7	5/4	0/0
–	–	–	–	–	–	–	–	+	+	45 (6.0)	28/6	17/6	0/0	0/0
–	–	+	+	–	+	+	+	+	+	39 (5.2)	0/0	2/2	7/3	30/8
+	+	–	+	+	+	+	+	+	+	34 (4.5)	19/3	15/4	0/0	0/0
+	+	+	+	+	+	–	+	+	+	23(3.0)	11/4	3/2	9/1	0/0
+	+	–	+	–	+	+	+	+	+	22 (2.9)	1/1	18/7	3/1	0/0
–	–	–	–	–	–	–	–	+	–	17 (2.3)	17/4	0/0	0/0	0/0
–	–	–	+	–	+	–	+	+	+	14 (1.9)	0/0	0/0	9/6	5/3
–	–	–	–	–	–	–	–	–	+	9 (1.2)	8/4	1/1	0/0	0/0
+	+	+	+	–	+	–	+	+	+	8 (1.0)	6/3	1/1	1/1	0/0
–	–	+	–	–	+	–	+	+	+	8 (1.1)	0/0	0/0	0/0	8/3
Other (altogether 75 different host R-phenotypes, each represented by less than 8 samples)										136 (18.1)	46/11	47/12	34/4	9/4
Totally										752 (100)	250/16	241/16	120/8	141/10

[a] Symbols used in the table

+ susceptible response

– resistant response

pathogens) in wild pathosystems have shown that in a highly fragmented landscape the best approach to the study of population structure and its variation is to consider populations as semi-independent units which are inter-connected to varying degrees by migration and gene flow (Thrall et al. 2001). There may be regional differences in local selection pressures, exerted by species interacting with one another, and influenced by environmental factors, which result in variation among local populations within a metapopulation (Laine 2005). Gene flow causes changes in gene frequencies through the reduction of differentiation among populations, and an increase of within-population variation by the mixing of individuals from populations subjected to different selection pressures (Papa and Gepts 2004). Thus, patterns of local adaptation are intensely affected by dynamics and migration rates of both host and pathogen, as well as the scale of the gene flow (Gandon et al. 1996; Thrall and Burdon 1997; Laine 2005).

The metapopulation concept is of particular relevance to host plant-obligate parasite interactions where the pathogen lacks free-living stages. In such systems, where pathogen persistence is tightly linked to its host, the effect of fragmentation and the degree of population isolation may be exacerbated (Burdon et al. 2006). This is especially true in pathosystems, such as *Lactuca* spp.-*B. lactucae*, which adhere to the principles of the gene-for-gene theory (Lebeda et al. 2001b, 2002; Kuang et al. 2008). In this paper we only focused on the resistance structure of the host. Within the framework of the EU "Gene Mine" project the main task of our working group was to compare the spatial structure of resistance to *B. lactucae* in different parts of the European *L. serriola* metapopulation (Lebeda et al. 2007a) following an east-to-west transect of distribution including four countries (Czech Republic, Germany, Netherlands, and United Kingdom). The aim was to quantify variation in resistance within the entire European *L. serriola* metapopulation and its distribution among the four main regions, as well as among individual local populations (Lebeda et al. 2007a). At the same time, we searched for possible novel sources of resistance for lettuce breeding (Lebeda et al. 2007a, b, 2009a).

In general, the European metapopulation of *L. serriola* showed limited resistance to the races of *B. lactucae* used in the inoculation studies (Fig. 4); however broad variation in the complexity of resistance patterns was found at both population (Fig. 6) and individual (Fig. 10) levels. In total, 34% of the populations examined and about 21% of the samples tested displayed high susceptibility. On the other hand, completely resistant individuals only represented 7.8% of the samples of *L. serriola* tested; these were only found in CZ and D populations. In addition, there was considerable within and among-population diversity with regard to recorded R-phenotypes (Table 2). Our results showed that *L. serriola* plants with completely different R-phenotypes may grow in close proximity, within the local population. Similar patterns have also been observed in other wild pathosystems like *Linum marginale-Melampsora lini* (Burdon and Thompson 1995) or *Senecio vulgaris-Erysiphe fischeri* (Bevan et al. 1993b). Conversely, we observed the same phenomenon (i.e. variation in virulence at very fine spatial scales) within local populations of the pathogen, *B. lactucae* (Lebeda and Petrželová 2004a).

Because of the broad variation in resistance of *L. serriola* recorded within Europe, we were curious to know how resistance to *B. lactucae* was distributed across the transect area. The resistance data indicate that race-specific resistance is the most common type of resistance in the European *L. serriola* metapopulation, but resistance to some races is unequally distributed among individual geographical regions. Further, there were clear geographical differences in the genetic diversity of the resistance in individual areas. These results confirm the importance of viewing population studies over larger geographic areas in the metapopulation context (Thrall and Burdon 2002).

The Czech populations were the most variable of all those evaluated. They expressed high variability in their responses to individual races of *B. lactucae.* A large part of the CZ populations were highly susceptible to the races of *B. lactucae*; however, on the other hand, the Czech Republic was the only geographic region within the distribution of *L. serriola* where local populations with high levels of resistance were found (Lebeda et al. 2008). Within some of the CZ populations, high intra-population variation in resistance to the isolates of *B. lactucae* was recorded. At the level of individuals, high variation was found in both the structure (Table 4) and complexity (Fig. 10) of the recognized resistance phenotypes; and numerous individuals were characterized by race-specific responses.

With regard to interactions between *L. serriola* and *B. lactucae* in the Czech region, we have extensive knowledge of pathogen virulence patterns, as well as of their spatial and temporal variation, obtained during long-term research on this system (Lebeda et al. 2008; Lebeda and Petrželová 2004a; Petrželová and Lebeda 2004b). This provides an opportunity to compare spatially-structured patterns of host resistance and pathogen virulence. For example, over multiple years in the Czech population of *B. lactucae* there were one or a few v-phenotypes prevalent in the pathogen population, and a broad range of other phenotypes that were much less abundant, with number and identity of the prevalent ones varying among years (Lebeda et al. 2008). For the Czech population of *L. serriola* we only have comprehensive data for 2001, but it is evident that the basic patterns of the population structure are similar to that of *B. lactucae*. Eight R-phenotypes were plentiful in CZ populations, representing 78.8% of the individuals examined (the total number of recognized R-phenotypes in the Czech Republic is 45). Similar population structures (i.e. high heterogeneity among host populations and dominance by a few R-phenotypes) have been found in the *Senecio vulgaris-Erysiphe fischeri* (Bevan et al. 1993b; Clarke and Akhka 2002) and *Linum marginale-Melampsora lini* interactions (Burdon and Thrall 2000; Thrall et al. 2001). Theory suggests that in nature, the persistence of polymorphisms in host-pathogen associations are likely to be maintained by continual cycles of coevolution within local populations, combined with occasional immigration of new resistance and virulence genes (Frank 1992; Barrett et al. 2008).

Because we currently lack data on the temporal and spatial dynamics of resistance in populations of *L. serriola* over time we cannot assess the extent to which changes over time and space in host resistance are linked to shifts in pathogen virulence. However, several papers dealing with other wild pathosystems, e.g. *Linum marginale-Melampsora lini* (Burdon and Thrall 2000; Thrall and Burdon 2002) or *Microbotryum (=Ustilago) violaceum* (Delmotte et al. 1999), have shown that the within- and among-population spatial structure and dynamics are principally different for host and pathogen populations. For example in the *L. marginale-M. lini* interaction, within-population variation and diversity is much greater for host plants than for the pathogen (Thrall et al. 2001). Spatial models of gene-for-gene interactions suggest that this may be a general pattern (Thrall and Burdon 2002). At the among-population level, high overall differentiation has been recorded for both species in interactions between *M. violaceum* and its host plants, but it was much greater for the pathogen, which may indicate gene flow is lower among pathogen populations relative to the host (Delmotte et al. 1999).

The resistance structure of German and Czech populations of *L. serriola* was similar with regard to the number of highly susceptible individuals, but in Germany there were a much larger number of populations that had shown within-population differentiation in the resistance of individuals. In the German populations, high diversity of R-phenotypes was found and the identities of most dominant ones were the same as in the Czech Republic. Unlike in the Czech Republic, completely resistant individuals were widely distributed among populations. Interestingly, when looking at the distribution of resistance across Europe, one can see that resistant Czech populations are mainly located in the western part of the country. Westwards, towards central Germany, there are populations containing mixtures of susceptible and resistant individuals, while most of the other German populations are composed of only highly susceptible individuals. One study says that in pathogen metapopulations, where local populations are at least partially isolated but connected by gene flow, those populations are likely to contain genotypes that have coevolved locally with host populations as well as recently arrived genotypes from other host populations (Laine 2005). Similarly, for distribution patterns of resistant and susceptible individuals recorded within German populations of *L. serriola*, we hypothesize that resistant individuals may represent migrants from the resistant Czech populations to an area with otherwise susceptible populations.

There was a radical change in *L. serriola* resistance structure in the Netherlands. Here, most host individuals exhibited clear race-specific responses (i.e. medium complexity R-phenotypes). However, we found an overlap between German and Dutch populations of *L. serriola*. One local population, with clearly pronounced race-specificity, was found in Germany near to the Netherlands border; conversely, two local populations with high susceptibility were found in the Netherlands close to the border with Germany. As for the phenotypic composition, relatively high variation at the level of R-phenotypes was found in the Netherlands. Some· overlap between adjacent German and Dutch populations was also found in the occurrence of particular R-phenotypes (data not shown).

In the United Kingdom all populations were characterized by clear race-specific responses (Fig. 8) which agreed with previously reported results (Crute 1987). Interestingly, the diversity of R-phenotypes in the UK populations was quite low, with two types representing 84% of all UK individuals (Table 4). In addition, there were several other R-phenotypes with clear race-specificity, which were unique to the UK populations. Despite the overall low diversity, the phenotypic composition was very similar to that of the NL as UK populations shared the most dominant R-phenotypes with the NL populations. Thrall et al. (2001) also observed that populations of *Linum marginale* growing close together were more similar in phenotypic patterns of resistance to *Melampsora lini* isolates.

To summarize, large within- and among-population diversity of R-phenotypes of *L. serriola* recorded in this study was approximately evenly distributed across continental Europe (Fig. 9). In contrast, overall within- and among-population diversity of R-phenotypes was extremely low in the United Kingdom. However, there was a clear gradient of increasing race-specificity of *L. serriola* populations moving from central Europe towards the Atlantic coast, as seen in a preliminary analysis (Lebeda et al. 2008). There was a shift to balanced resistance towards western Europe, which was expressed in different ways among the studied regions. Within the Czech Republic we observed a mosaic of resistant, mixed and susceptible populations of *L. serriola*, although the latter prevailed. Towards the west, a more pronounced balance between resistant and susceptible individuals was recorded at the within-population level. In the Netherlands *L. serriola* individuals did not differ much in resistance levels (i.e. complexity), but they expressed a balance between resistance and susceptibility to a spectrum of pathogen races at the level of their phenotypes, which was described as a strong race-specific response. Populations were most balanced in the United Kingdom, where two R-phenotypes providing resistances to particular combinations of pathogen races were almost only distributed across the whole area studied; i.e. individuals did not differ largely either in their resistance complexity or in recognized reaction patterns in UK. Modern agriculture now frequently employs breeding strategies simulating the more diverse situations represented in nature to increase the effectiveness of crop plant resistance to diseases. Commonly used breeding approaches such as cultivar mixtures or gene pyramiding (Mundt 2002; Mundt et al. 2002; Pink 2002) may be considered as such an analogy to spatial structuring of resistance recently found in real natural populations of *L. serriola* across Europe. From our recent results it seems that, at least under natural conditions, combination of both strategies (i.e. mixture of different plant genotypes and combination of resistance to various pathogen races within individual genotypes) is the most efficient, and provides the highest overall degree of resistance at the population level, which was most pronounced in NL and UK populations.

Kuang et al. (2008) reported a shift towards lower R-gene diversity of *L. serriola* moving westwards from the center of its origin and genetic diversity (presumed to be situated in the eastern Mediterranean). They supposed that during dispersion of this species to Europe it may have lost a large part of the original populations' genetic diversity. Recently, D'Andrea et al. (2006) studied genetic diversity of a wide range of *L. serriola* populations from western and central Europe using chloroplast (PCR-RFLP) and nuclear (microsatellites, SSR) markers. They revealed high genetic differentiation among individual populations. However, spatial pattern in gene diversity variation was significant only for populations isolated by greater distances. This is supported also by isozyme data (Dziechciarkova et al. 2004) and recent AFLP studies of European populations of *L. serriola* (Lebeda et al. 2009b). Further, D'Andrea et al. (2006) made a prediction map of the distribution of *L. serriola* gene diversity across Europe, which showed a decrease of diversity westwards and northwards from the areas with the highest values, which supports data published by Kuang et al. (2008). Among European areas with high gene diversities were eastern and central Europe, South of France and North of Italy, while lower gene diversities were recorded for populations from west Mediterranean (Portugal and Spain), South of Italy, United Kingdom, the Alps and Southern-Scandinavia (D'Andrea et al. 2006). Our resistance data correlate well with the predictions of D'Andrea (2006). Most CZ and D populations included in our resistance study are situated in regions with higher gene diversity, which decreases towards the Netherlands and is the lowest in the United Kingdom.

Geographical-scale diversification of plant-pathogen systems has been shown in several other wild pathosystems, e.g. *Senecio vulgaris-Erysiphe fischeri* (Bevan et al. 1993b; Clarke and Akhka 2002), *Linum marginale-Melampsora lini* (Burdon and Thrall 2000), or *Plantago lanceolata-Podosphaera plantaginis* (Laine 2005). However, the first evidence for existence of a gradient of resistance in natural plant pathosystems was given by Springer (2007). He studied populations of California dwarf flax (*Hesperolinon californicum*) occurring along a south-north transect across California and observed a clear latitudinal cline in resistance structure to *M. lini*. Resistance levels were lowest in the north and increased gradually and continuously towards the south, and mirrored almost identically a cline in infection prevalence. For the wild *L. serriola-B. lactucae* pathosystem, we have no relevant data on pathogen occurrence across Europe except for the Czech Republic. Thus, we are not able to examine the extent to which geographic patterns of resistance correlate with disease prevalence in this interaction.

Significant role in shaping resistance structure of a host population play both the level of patchiness and connectivity of the host populations within the larger metapopulation (Carlsson-Granér and Thrall 2002). Theoretically, spatial differences in resistance of *L. serriola* populations may be based on a range of factors. In gene-for-gene interactions, the spatial structure of populations may result from the availability of compatible host-pathogen pairs within the spatial mosaic of individuals, at any specific location and time. Furthermore, ephemerality of this annual weed host, ecological circumstances of local host populations, as well as agronomic management and weed control practices that may vary among countries, may indirectly participate in shaping the spatial structure of populations by influencing survivorship of specific R-genotypes in the landscape, regardless of their resistance or susceptibility. Possibly, unequal distribution of the two morphological forms (*L. serriola* f. *serriola* and *L. serriola* f. *integrifolia*) within the European metapopulation of *L. serriola* (Lebeda et al. 2001a, 2007a) may also contribute to observed differences in spatial distribution of resistance among the studied regions (Lebeda et al. 2008).

L. serriola, which is native to Mediterranean area, has undergone a large expansion over the past

250 years and has colonized most of Europe (D'Andrea et al. 2009). The main factors responsible for such a large spread of *L. serriola* into the new regions are especially global climate warming, increasing anthropogenic disturbance and ruderalization of environment, and development of transport networks (Hooftman et al. 2006; D'Andrea et al. 2009). The result is greater availability and better connectivity of disturbed and ruderal habitats, which may be colonized by *L. serriola* (D'Andrea et al. 2009). This is probably true for continental Europe. Our studies, focused on the ecogeography of *L. serriola* (Doležalová et al. 2001; Lebeda et al. 2001a, 2004, 2007a), show that in most European countries the highest densities of *L. serriola* populations were along transport corridors. However, when comparing the four countries included in this study, *L. serriola* occupied the broadest range of different habitats in the Czech Republic and there is an apparent decrease in the range of habitats of *L. serriola* towards the western parts of Europe (Lebeda et al. 2001a, 2007a). We hypothesize, therefore, that the level of connectedness of individual populations also decreases, which may in turn greatly influence the extent of among-population differences in resistance patterns.

In contrast, increasing isolation leads to lower frequency of population mixing and higher uniformity of populations, which may explain the situation in United Kingdom. We hypothesize that isolation from continental Europe by sea was of crucial importance in shaping the resistance structure of the UK *L. serriola* population (Lebeda et al. 2008, 2009b), with two possible explanations of the origin of such specific spatial structure. It may either result from extremely strong adaptation to the local pathogen populations, or it may originate from several founder events from the continental Europe (mainland to island migration). However, we may only speculate on this, as for the United Kingdom no data are available on the natural distribution of *B. lactucae* on weedy growing plants of *L. serriola* (Crute 1987; Lebeda et al. 2007a).

Of course, results on distribution and variation of resistance in European metapopulation of *L. serriola* may be largely influenced by the spectrum of *B. lactucae* races used for our resistance screening. However, the data summarized in this paper clearly showed that host-pathogen interaction in the wild plant pathosystem *L. serriola-B. lactucae* is predominantly based on race-specificity. Patterns of race-specificity in host populations of *L. serriola* were geographically structured with decreasing diversity from continental (central) Europe to the Atlantic coast (UK). Host populations in central Europe (Czech Republic, Germany) were the most heterogeneous with regard to resistance diversity in clear contrast to the relatively homogeneous UK populations. These results support our previous suggestion that this pathosystem is suitable as a model system for the research of host-pathogen interactions in natural plant communities (Lebeda et al. 2008), as well as for further exploitation of resistance variation in lettuce breeding (Lebeda 1998; Lebeda et al. 2002; Lebeda and Zinkernagel 2003; Beharav et al. 2006; Lebeda et al. 2007b, 2009a).

Acknowledgements The authors are very grateful to Dr. Peter Thrall for critical reading of the first draft of the manuscript. Thanks also to the staff of Laboratory of Plant Pathology (Palacký University in Olomouc, Faculty of Science, Department of Botany) for their valuable help with preparation of plants and multiplication of seeds for the experiments. The research was supported by EU-funded project Gene-Mine "Improved use of germplasm collections with the aid of novel methodologies for integration, analysis and presentation of genetic data sets" (Contract no. QLK5-CT-2000-00722), project of Czech Ministry of Education (MSM 6198959215).

References

Alexander, H. M., Thrall, P. H., Antonovics, J., Jarosz, A. M., & Oudeman, P. V. (1996). Population dynamics and genetics of plant disease: a case study of anther-smut disease. *Ecology, 77*, 990–996.

Barrett, L. G., Thrall, P. H., Burdon, J. J., & Linde, C. C. (2008). Life history determines genetic structure and evolutionary potential of host-parasite interactions. *Trends in Ecology and Evolution, 23*, 678–685.

Beharav, A., Lewinsohn, D., Lebeda, A., & Nevo, E. (2006). New wild *Lactuca* genetic resources with resistance against *Bremia lactucae*. *Genetic Resources and Crop Evolution, 53*, 467–474.

Bevan, J. R., Clarke, D. D., & Crute, I. R. (1993a). Resistance to *Erysiphe fischeri* in two populations of *Senecio vulgaris*. *Plant Pathology, 42*, 636–646.

Bevan, J. R., Crute, I. R., & Clarke, D. D. (1993b). Diversity and variation in expression of resistance to *Erysiphe fischeri* in *Senecio vulgaris*. *Plant Pathology, 42*, 647–653.

Bevan, J. R., Crute, I. R., & Clarke, D. D. (1993c). Variation for virulence in *Erysiphe fischeri* from *Senecio vulgaris*. *Plant Pathology, 42*, 622–635.

Burdon, J. J. (1993). The structure of pathogen populations in natural plant communities. *Annual Review of Phytopathology, 31*, 305–323.

Burdon, J. J. (1994). The distribution and origin of genes for race-specific resistance to *Melampsora lini* in *Linum marginale*. *Evolution, 48*, 1564–1575.

Burdon, J. J., & Jarosz, A. M. (1991). Host-pathogen interactions in natural populations of *Linum marginale* and *Melampsora lini*: I. Patterns of resistance and racial variation in a large host population. *Evolution, 45*, 205–217.

Burdon, J. J., & Thompson, J. N. (1995). Changed patterns of resistance in a population of *Linum marginale* attacked by the rust pathogen *Melampsora lini*. *Journal of Ecology, 83*, 199–206.

Burdon, J. J., & Thrall, P. H. (1999). Spatial and temporal patterns in coevolving plant and pathogen associations. *The American Naturalist, 153*(supplement), S15–S33.

Burdon, J. J., & Thrall, P. H. (2000). Coevolution at multiple spatial scales: *Linum marginale-Melampsora lini*—from the individual to the species. *Evolutionary Ecology, 14*, 261–281.

Burdon, J. J., Thrall, P. H., & Brown, H. D. (1999). Resistance and virulence structure in two *Linum marginale-Melampsora lini* host-pathogen metapopulations with different mating systems. *Evolution, 53*, 704–716.

Burdon, J. J., Thrall, P. H., & Ericson, L. (2006). The current and future dynamics of disease in plant communities. *Annual Review of Phytopathology, 44*, 19–39.

Carlsson-Granér, U. (2006). Disease dynamics, host specificity and pathogen persistence in isolated host populations. *Oikos, 112*, 174–184.

Carlsson-Granér, U., & Thrall, P. H. (2002). The spatial distribution of plant populations, disease dynamics and evolution of resistance. *Oikos, 97*, 97–110.

Clarke, D. D., & Akhka, A. (2002). Population genetics of powdery mildew—natural plant pathosystems. In R. R. Bélanger, W. R. Bushnell, A. J. Dik, & T. L. W. Carver (Eds.), *The powdery mildews. A comprehensive treatise* (pp. 200–218). St. Paul: APS.

Crute, I. R. (1987). The geographical distribution and frequency of virulence determinants in *Bremia lactucae*: relationships between genetic control and host selection. In M. S. Wolfe & C. E. Caten (Eds.), *Populations of plant pathogens: Their dynamics and genetics* (pp. 193–212). Oxford: Blackwell.

D'Andrea, L., van de Wiel, C., Meirmans, P., Guadagnuolo, R., Kozlowski, G., den Nijs, H., & Felber, F. (2006). Chapter 2: Molecular biogeography of prickly lettuce (*L. serriola* L.) in Europe based on nuclear microsatellites and chloroplast DNA polymorphism. In L. D'Andrea *Historical and Molecular biogeography of prickly lettuce (L. serriola L.) and evaluation of hybridisation with the crop (L. sativa L.), Ph.D. Thesis* (pp. 41–68). http://doc.rero.ch/lm.php?url=1000,40,4,20060823144601-IC/these_DAndreaL.pdf.

D'Andrea, L., Broennimann, O., Kozlowski, G., Guisan, A., Morin, X., Keller-Senften, J., et al. (2009). Climate change, anthropogenic disturbance and the northward range expansion of *Lactuca serriola* (Asteraceae). *Journal of Biogeography, 36*, 1–15.

Delmotte, F., Bucheli, E., & Shykoff, J. A. (1999). Host and parasite population structure in a natural plant-pathogen system. *Heredity, 82*, 300–308.

Dickinson, C. H., & Crute, I. R. (1974). The influence of seedling age and development on the infection of lettuce by *Bremia lactucae*. *Annals of Applied Biology, 76*, 49–61.

Doležalová, I., Lebeda, A., & Křístková, E. (2001). Prickly lettuce (*Lactuca serriola* L.) germplasm collecting and distribution study in Slovenia and Sweden. *Plant Genetic Resources Newsletter, 128*, 41–44.

Dziechciarková, M., Lebeda, A., Doležalová, I., & Křístková, E. (2004). Isozyme variation in European *Lactuca serriola* germplasm. In J. Vollmann, H. Grausgruber, & P. Ruckenbauer (Eds.) *Genetic Variation for Plant Breeding* (pp. 103–107). Proceedings of the 17th Eucarpia General Congress, 8–11 September 2004, Tulln, Austria. Vienna. Vienna, Austria: BOKU-University of Natural Resources and Applied Life Sciences.

Ericson, L., Burdon, J. J., & Müller, W. J. (1999). Spatial and temporal dynamics of epidemics of the rust fungus *Uromyces valerianae* on populations of its host *Valeriana salina*. *Journal of Ecology, 87*, 649–658.

Ericson, L., Burdon, J. J., & Müller, W. J. (2002). The rust pathogen *Triphragmium ulmariae* as a selective force affecting its host, *Filipendula ulmaria*. *Journal of Ecology, 90*, 167–178.

Frank, S. A. (1992). Models of plant-pathogen coevolution. *Trends in Genetics, 8*, 213–219.

Freckleton, R. P., & Watkinson, A. R. (2002). Large-scale spatial dynamics of plants: metapopulations, regional ensembles and patchy populations. *Journal of Ecology, 90*, 419–434.

Gandon, S., Capowiez, Y., Dubois, Y., Michalakis, Y., & Olivieri, I. (1996). Local adaptation and gene-for-gene coevolution in a metapopulation model. *Proceedings of the Royal Society of London, Series B, 263*, 1003–1009.

Holub, E. B. (2008). Natural history of *Arabidopsis thaliana* and oomycete symbioses. *European Journal of Plant Pathology, 122*, 91–109.

Hooftman, D. A. P., Ostermeijer, J. G. B., & den Nijs, J. C. M. (2006). Invasive behaviour of *Lactuca serriola* (Asteraceae) in the Netherlands: Spatial distribution and ecological amplitude. *Basic and Applied Ecology, 7*, 507–519.

Kuang, H., Van Eck, H. J., Sicard, D., Michelmore, R., & Nevo, E. (2008). Evolution and genetic population structure of prickly lettuce (*Lactuca serriola*) and its RGC2 resistance gene cluster. *Genetics, 178*, 1547–1558.

Laine, A.-L. (2004). Resistance variation within and among host populations in a plant-pathogen metapopulation: implications for regional pathogen dynamics. *Journal of Ecology, 92*, 990–1000.

Laine, A.-L. (2005). Spatial scale of local adaptation in a plant-pathogen metapopulation. *Journal of Evolutionary Biology, 18*, 930–938.

Laine, A.-L. (2007). Detecting local adaptation in a natural plant-pathogen metapopulation: a laboratory vs. field transplant approach. *Journal of Evolutionary Biology, 20*, 1665–1673.

Laine, A.-L., & Hanski, I. (2006). Large-scale spatial dynamics of a specialist plant pathogen in a fragmented landscape. *Journal of Ecology, 94*, 217–226.

Lebeda, A. (1998). *Biodiversity of the interactions between germplasms of wild Lactuca spp. and related genera and lettuce downy mildew (Bremia lactucae)* (p. 70). Wellesbourne: Report on research programme of OECD "Biological Resource Management for Sustainable Agricultural Systems", Horticulture Research International.

Lebeda, A. (2002). Occurrence and variation in virulence of *Bremia lactucae* in natural populations of *Lactuca serriola*. In P. T. N. Spencer-Phillips, U. Gisi, & A. Lebeda (Eds.), *Advances in Downy Mildew research* (pp. 179–183). Dordrecht: Kluwer.

Lebeda, A., & Schwinn, F. J. (1994). The downy mildews—an overview of recent research progress. *Journal of Plant Diseases and Protection, 101*, 225–254.

Lebeda, A., & Zinkernagel, V. (2003). Characterization of new highly virulent German isolates of *Bremia lactucae* and efficiency of resistance in wild *Lactuca* spp. germplasm. *Journal of Phytopathology, 151*, 274–282.

Lebeda, A., Doležalová, I., Křístková, E., & Mieslerová, B. (2001a). Biodiversity and ecogeography of wild *Lactuca* spp. in some European countries. *Genetic Resources and Crop Evolution, 48*, 153–164.

Lebeda, A., Pink, D. A. C., & Mieslerová, B. (2001b). Host-parasite specificity and defense variability in the *Lactuca* spp.-*Bremia lactucae* pathosystem. *Journal of Plant Pathology, 83*, 25–35.

Lebeda, A., & Petrželová, I. (2004a). Variation and distribution of virulence phenotypes of *Bremia lactucae* in natural populations of *Lactuca serriola*. *Plant Pathology, 53*, 316–324.

Lebeda, A., & Petrželová, I. (2004b). Occurrence of race-specific resistance to *Bremia lactucae* in *Lactuca serriola* germplasm originating from four European countries. In J. Vollmann, H. Grausgruber, & P. Ruckenbauer (Eds.), *Genetic variation for plant breeding* (pp. 113–116). Vienna: EUCARPIA & BOKU—University of Natural Resources and Applied Life Sciences.

Lebeda, A., & Petrželová, I. (2007a). Comparison of resistance to *Bremia lactucae* in populations of *Lactuca serriola* occurring in Central Europe (Czech Republic) and the British Isles (England, U.K.). In S. Bullitta (Ed.) *Plant Genetic Resources of Geographical and "other" Islands (Conservation, evaluation and use for plant breeding). Proceedings of the XVII Eucarpia Genetic Resources Section Meeting* (pp. 59–62)". Castelsardo (Italy), 30 March-2 April 2005; CNR-ISPAAM, Italy: sezione Sassari Publisher.

Lebeda, A., & Petrželová, I. (2007b). Race-specific resistance to *Bremia lactucae* in European populations of *Lactuca serriola*. In *EUCARPIA Leafy Vegetables 2007*, Conference Abstracts (p. 11). 18–20 April 2007; University of Warwick, Warwick HRI, (U.K.); Oral Presentations.

Lebeda, A., Pink, D. A. C., & Astley, D. (2002). Aspects of the interactions between wild *Lactuca* spp. And related genera and lettuce downy mildew (*Bremia lactucae*). In P. T. N. Spencer-Phillips, U. Gisi, & A. Lebeda (Eds.), *Advances in Downy Mildew research* (pp. 85–117). Dordrecht: Kluwer.

Lebeda, A., Doležalová, I., Feráková, V., & Astley, D. (2004). Geographical distribution of wild *Lactuca* species (Asteraceae, Lactuceae). *The Botanical Review, 70*, 328–356.

Lebeda, A., Doležalová, I., Křístková, E., Dehmer, K. J., Astley, D., van de Wiel, C. C. M., et al. (2007a). Acquisition and ecological characterization of *Lactuca serriola* L. germplasm collected in the Czech Republic, Germany, the Netherlands and United Kingdom. *Genetic Resources and Crop Evolution, 54*, 555–562.

Lebeda, A., Ryder, E. J., Grube, R., Doležalová, I., & Křístková, E. (2007b). Lettuce (Asteraceae; *Lactuca* spp.), Chapter 9. In R. Singh (Ed.), *Genetic resources, chromosome engineering, and crop improvement Series, Volume 3—vegetable crops* (pp. 377–472). Boca Raton: CRC.

Lebeda, A., Petrželová, I., & Maryška, Z. (2008). Structure and variation in the wild-plant pathosystem: *Lactuca serriola—Bremia lactucae*. *European Journal of Plant Pathology, 122*, 127–146. Reprinted in: A. Lebeda, P. T. N. Spencer-Phillips, & B. M. Cooke (Eds.) The Downy Mildews—Genetics, Molecular Biology and Control. Dordrecht, the Netherlands: Springer.

Lebeda, A., Doležalová, I., Křístková, E., Kitner, M., Petrželová, I., Mieslerová, B., et al. (2009a). Wild *Lactuca* germplasm for lettuce breeding: current status, gaps and challenges. *Euphytica, 170*, 15–34. doi:10.1007/s10681-009-9914-7.

Lebeda, A., Kitner, M., Dziechciarková, M., Doležalová, I., Křístková, E., & Lindhout, P. (2009b). An insight into the genetic polymorphism among European populations of *Lactuca serriola* assessed by AFLP. *Biochemical Systematics and Ecology, 37*, 597-608. doi:10.1016/j.bse.2009.10.010.

Lenné, J. M., & Wood, D. (1991). Plant disease and the use of wild germplasm. *Annual Review of Phytopathology, 29*, 35–63.

McDermott, J. M., & McDonald, B. A. (1993). Gene flow in plant pathosystems. *Annual Review of Phytopathology, 31*, 353–373.

Mundt, C. C. (2002). Use of multiline cultivars and cultivar mixtures for disease management. *Annual Review of Phytopathology, 40*, 381–410.

Mundt, C. C., Cowger, C., & Garrett, K. A. (2002). Relevance of integrated disease management to resistance durability. *Euphytica, 124*, 245–252.

Papa, R., & Gepts, P. (2004). Gene flow between crops and their wild progenitors. In M. Dekker (Ed.), *Encyclopedia of plant and crop science* (pp. 488–491). New York: Dekker.

Petrželová, I., & Lebeda, A. (2004a). Occurrence of *Bremia lactucae* in natural populations of *Lactuca serriola*. *Journal of Phytopathology, 152*, 391–398.

Petrželová, I., & Lebeda, A. (2004b). Temporal and spatial variation in virulence of natural populations of *Bremia lactucae* occurring on *Lactuca serriola*. In P. T. N. Spencer-Phillips & M. Jeger (Eds.), *Advances in Downy Mildew research* (Vol. 2, pp. 141–163). Dordrecht: Kluwer.

Pink, D. A. C. (2002). Strategies using genes for non-durable disease resistance. *Euphytica, 124*, 227–236.

Smith, D. L., Ericson, L., & Burdon, J. J. (2003). Epidemiological patterns at multiple spatial scales: an 11-year study of a *Triphragmium ulmariae—Filipendula ulmaria* metapopulation. *Journal of Ecology, 91*, 890–903.

Springer, Y. P. (2007). Clinal resistance structure and pathogen local adaptation in a serpentine flax-flax rust interaction. *Evolution, 61*, 1812–1822.

Thrall, P. H., & Antonovics, J. (1995). Theoretical and empirical studies of metapopulations: population and genetic dynamics of the *Silene-Ustilago* system. *Canadian Journal of Botany, 73*(Suppl. 1), 1249–1258.

Thrall, P. H., & Burdon, J. J. (1997). Host-pathogen dynamics in a metapopulation context: the ecological and evolutionary consequences of being spatial. *Journal of Ecology, 85*, 743–753.

Thrall, P. H., & Burdon, J. J. (2002). Evolution of gene-for-gene systems in metapopulations: the effect of spatial scale of host and pathogen dispersal. *Plant Pathology, 51*, 169–184.

Thrall, P. H., Burdon, J. J., & Young, A. (2001). Variation in resistance and virulence among demes of a plant host-pathogen metapopulation. *Journal of Ecology, 89*, 736–748.

Thrall, P. H., Burdon, J. J., & Bever, J. D. (2002). Local adaptation in the *Linum marginale-Melampsora lini* host-pathogen interaction. *Evolution, 56*, 1340–1351.

Towsend, G. R., & Heuberger, W. (1943). Methods for estimating losses caused by diseases in fungicide experiments. *Plant Disease Report, 27*, 340–343.

Van der Arend, A. J. M., Gautier, J., Grimault, V., Kraan, P., van der Laan, R., Mazet, J., Michel, H., Schut, J. W., Smilde, D., & de Witte, I. (2006). Identification and denomination of "new" races of *Bremia lactucae* in Europe by IBEB until 2006. http://www.plantum.nl/pdf/IBEB_identification_and_nomination_2006.pdf.

Van Ettekoven, K., & van der Arend, A. J. M. (1999). Identification and denomination of "new" races of *Bremia lactucae.* In A. Lebeda & E. Křístková (Eds.), *Eucarpia leafy vegetables '99* (pp. 171–175). Olomouc: Palacký University.

Eur J Plant Pathol (2011) 129:123–133
DOI 10.1007/s10658-010-9712-z

Site-specific field resistance of grapevine to *Plasmopara viticola* correlates to altered gene expression and was not modulated by the application of organic amendments

Barbara Thuerig · Ana Slaughter · Elaheh Marouf · Matthias Held · Brigitte Mauch-Mani · Lucius Tamm

Accepted: 20 October 2010 / Published online: 18 November 2010

Abstract The influence of site on resistance of grapevine (*cv.* Chasselas) to *Plasmopara viticola* was evaluated. Grapevine leaves from three vineyards in the region of Lake Neuchâtel (Switzerland) were tested for their susceptibility to *P. viticola* in the lab in five successive years (2004–2008), and the expression levels of four selected defence-related genes (*Glucanase*, *Lipoxygenase 9*, *9-cis epoxycarotenoid dioxygenase*, *Stilbene synthase*) were studied in 1 year. In all 5 years of examination, differences between sites were substantial. In four out of 5 years, plants from site Hauvernier were much less susceptible to *P. viticola* than plants from site Auvernier. In another year, differences were less pronounced but still significant for one leaf age. Susceptibility of plants from a third site (Concise) varied from year to year. Differences in the genetic background were excluded by microsatellite analysis. Differences in susceptibility were mirrored in the constitutive expression pattern of four defence-related genes, with samples from the Hauterive site clearly separated from samples of the other two sites in redundancy analysis. Furthermore, it was evaluated whether site-specific resistance can be modulated by agronomic practices such as the application of organic amendments. In two commercial vineyards (*cv.* Pinot noir), soils had either not (control) or yearly (compost) been amended with a compost for the last 9 years. Leaves from plants grown in any of the two treatments did not differ in their susceptibility to *P. viticola* in both years of examination. Additionally, under controlled conditions, none of 19 different composts amended to the substrate of grapevine seedlings or cuttings affected their susceptibility to *P. viticola*, but 8 out of 19 composts reduced severity in the control bioassay *Arabidopsis thaliana*—*Hyaloperonospora arabidopsidis*, indicating that a modulation of site-specific susceptibility of grapevine plants by organic amendments is at the very least, difficult.

Keywords Compost · Induced systemic resistance · Soil · Terroir · *Vitis vinifera*

B. Thuerig (✉) · L. Tamm
Research Institute of Organic Agriculture,
Ackerstrasse,
CH-5070 Frick, Switzerland
e-mail: barbara.thuerig@fibl.org

A. Slaughter · E. Marouf · M. Held · B. Mauch-Mani
University of Neuchâtel, Institute of Biology,
Rue Emile Argand 11,
CH-2009 Neuchâtel, Switzerland

Introduction

Site has long been recognized as being an important factor in influencing plant well-being. Macro- and micro-climate, soil structure, soil nutrient content and the soil microbial population influence the well-being of plants directly but also indirectly by affecting occurrence and infectious potential of plant pathogens. Soils in which plants develop no or reduced disease symptoms, even if a pathogen is present or artificially added to the soil, are known as suppressive soils (Weller et al. 2002). Certain soils even influence the susceptibility of plants to foliar diseases (Tamm et al. 2010). Living soil microorganisms have been identified as a key factor to reduce the occurrence and severity of soil-borne diseases (Knudsen et al. 2002; Menzies 1959; Shipton et al. 1973; Stutz et al. 1986; Wiseman et al. 1996), as well as the resistance of plants to foliar diseases (Thuerig et al. 2009). Soil microorganisms can control soil-borne pathogens through competition, antibiosis, hyperparasitism and the induction of plant disease resistance (Haas and Défago 2005). In the case of foliar diseases, beneficial soil microorganisms and pathogens are spatially separated and the soil microorganism-induced defence-state of the plant is referred to as induced systemic resistance (ISR) (van Loon et al. 1998). Various root associated bacteria, mainly fluorescent pseudomonads, that enhance resistance of various plants to several foliar diseases have been identified (van Loon et al. 1998). In general, ISR is not associated with major changes in the constitutive expression of defence genes, but with a faster and/or stronger activation of basal defence mechanisms such as the expression of defence-related genes upon contact with a pathogen, also referred to as priming (Conrath et al. 2006). The occurrence of ISR against foliar diseases has been demonstrated in experiments under controlled conditions in annual plants mainly, while under field conditions and in woody perennial plants such as grapevine, only little is known about occurrence and relevance of this phenomenon (Kloepper et al. 1999). Yet, it has been shown that *in vitro* cultured grapevine plantlets can be colonised by beneficial pseudomonads, promoting growth of the plants and their resistance to *Botrytis cinerea* (Ait Barka et al. 2000), and that colonization of grapevine roots with a mycorrhizal fungus induced defence-responses against root-knot nematodes, including transcription of a chitinase (Li et al. 2006). Furthermore, the concept of 'terroir' is well known in viticulture and implies that there is a strong relationship between the composition of the grape, the characteristics of the wine and the territory of production (Morlat and Bodin 2006; Seguin 1986). Besides climate, plant material and anthropogenic influence, soil has been shown to be a main determinant of a defined 'terroir' and influences grape quality parameters such as sugar, anthocyanin, and phenolic contents (Downey et al. 2006; Jackson and Lombard 1993; Morlat and Bodin 2006). Many geological and pedological factors have been shown to have an important impact on vine and grape quality. Recently, Morlat and Bodin (2006) suggested that the type and weathering of parental rock are key factors for the grape quality within a certain climatic area, by affecting parameters such as mineral nutrient content, soil depth, clay content and water storing capacity.

For commercial agriculture, it is of crucial importance whether site-specific suppressiveness can be influenced by appropriate management techniques. Several studies have demonstrated that suppressiveness of natural soils to soil-borne diseases can be modulated by the application of organic soil amendments such as manure and compost (Fuchs et al. 2004; Litterick et al. 2004), and a few reports have shown the same for suppressiveness of soils to air-borne diseases (Tamm et al. 2010; Vallad and Goodman 2004). Yet, differences between management strategies were much smaller than differences between soil types (Tamm et al. 2010), and little is known about the effect of soil management practices on the resistance of woody perennial plants to foliar air-borne diseases.

The main objective of this study was to investigate the impact of site and organic soil amendments on the resistance of grapevine to *Plasmopara viticola*. We investigated whether i) grapevine plants of the same variety growing at different sites in the same region differ in their resistance to *P. viticola*, ii) differences in resistance correlate to altered expression levels of defence-related genes and iii) site-specific resistance can be modulated by soil management practices.

Methods

Vineyards The susceptibility of grapevine *cv* Chasselas grown in three vineyards in the area of Lake Neuchâtel was compared from 2004 to 2008. Vineyards are situated at Hauterive (HAU), Auvernier (AUV) and Concise (CON) (Table 1) and are managed and certified according to standards of organic agriculture. At all three sites, plants were sprayed with the same plant protection products that are allowed in certified organic farming in Switzerland according to similar

Table 1 Coordinates and soil properties of the three vineyards in the region of Lake Neuchâtel

		Auvernier (AUV)	Concise (CON)	Hauterive (HAU)
Coordinates		46°58′35″N, 6°52′45″E	46°51′0″, 6°43′12″E	47°0′55″N, 658′15″E
Altitude		433 m	440 m	501 m
Soil type		loam	highly loamy sand	sandy loam
Sand/silt/clay (%)		47/27/26	61/24/15	58/24/18
Mould (%)		2.6	2.3	2.1
pH		7.4	7.2	7.7
Nutrients (available/reserve)[1]	P	13/153	15/170	11/97
	K	71/354	72/212	64/262
	Mg	9/456	9/395	7/489
	Ca	154/55′164	144/56′262	153/78′982
	Cu	–/207	–/258	–/172
	Fe	–/177	–/235	–/180
Dehydrogenase activity (DHA)[2]		7.0	5.4	5.6

[1] $mg\,kg^{-1}$; [2] $mg\,g^{-1}\,h^{-1}$

spray schedules and dosages. The products are non-systemic contact fungicides (copper, sulphur) that loose efficacy after rainfall of more than 15–20 mm. Management of the vineyard at CON was changed from organic to integrated production in 2008 and was thus excluded from the analysis in 2008. The effect of long-term amendment of composts was assessed in two field-experiments, which had been established in 1996 in two organically managed vineyards (*cv.* Pinot noir) at Malans (MAL) and Walenstadt (WAL) in the eastern part of Switzerland. Soils were either not (control) or yearly amended with 50 (Walenstadt) or 40 (Malans) $m^3 ha^{-1}$ of a compost (10–30% farm yard manure, 40–60% biodegradable waste including wood chips, 10% top soil, 10% screen overflow of compost, 1.5% rock meal) (Maschinenring Zuger Berggebiet, Edlibach, Switzerland). The identity of the cultivar Chasselas was assessed with six microsatellite markers (VVMD5, VVMD7, VVMD27, VVS2, VrZAG62 and VrZAG79), the markers of choice for grape cultivar identification (This et al. 2004). The allele lengths were entered into the Swiss Vitis Microsatellite Database (SVMD) which contains data of about 170 *Vitis* spp. accessions cultivated in Switzerland (Vouillamoz et al. 2009).

Soil samples and soil analysis Soils were randomly sampled at AUV, CON and HAU in July 2005 with a 3 cm soil corer from the 0 to 20 cm soil layer in each field plot after removing organic matter at the soil surface. Soils were sieved to 1 cm, homogenised, watered to field capacity and stored at 2°C under aerobic conditions until further analysis. Soil dehydrogenase activity (DHA) was measured according to Tabatabai (1982) in 5 g of soil that were incubated at 30°C for 24 h in the presence of an alternative electron acceptor (triphenyltetrazoliumchloride). The red coloured product (triphenylformazan) was extracted with acetone and measured in a spectrophotometer at 546 nm. Other soil parameters including pH, sand/silt/clay content, mould content, P, K, Mg, Ca, Fe and Cu content (available and reserve) were determined according to the standards of the Swiss Federal Agronomic Research Stations Agroscope (Anonymous 1996).

Leaf disc bioassays Twenty plants were sampled at each of the AUV, CON and HAU sites between May 30 and June 9 in the years 2004–2008. At sites MAL and WAL, twelve plants per site and treatment were harvested in early June 2005 and 2006. From each plant, five (AUV, CON, HAU) or three to four (MAL; WAL) leaves from a main shoot were sub-sampled, starting with the youngest leaf of approximately 2/3 of the size of a fully grown leaf (leaf age 1), and the next older leaves. To remove residues of plant protection products, leaves were thoroughly cleaned by first soaking in water containing 0.01% Tween for

3 min, washing with a shower (30 s for each side of the leaf), and drying with clean paper towels. Two leaf discs were cut per leaf (14 mm diameter), placed on 1% water agar and inoculated with 10 μl droplets of *P. viticola* sporangia suspension (5×10^4 sporangia ml^{-1}). Within one leaf age, the spread of the droplets was homogeneous and did not vary between sites (data not shown). Droplets were dried with a paper towel 24 h after inoculation to avoid necrosis. Lesion diameters were measured 7 d post inoculation. In 2007, in order to analyze gene expression, in addition to the two leaf discs for susceptibility bioassays, 9 leaf discs (10 mm diameter) were cut per leaf and placed on 1% water agar. Leaf discs were sampled 0, 8, 12, 24 and 48 h after inoculation with *P. viticola* or mock-inoculation with H_2O, separately for each site, leaf age and type of inoculation (*P. viticola*-/mock-inoculation) (i.e. 20 leaf discs from 20 plants pooled per time point and inoculation type).

Grafting experiments Scions from the three sites were harvested in winter 2006 and grafted onto rootstock Kober 5BB by a commercial grapevine nursery (Rebschulen Andreas Meier, CH-Würenlingen). Young grafted vines were either grown in the greenhouse or planted in the vineyards at sites HAU or AUV in August 2007 to compare (i) the susceptibility of scions of different origin under identical conditions and (ii) the susceptibility of scions of identical origin grown at different sites. Leaves for bioassays were harvested on June 9, 2008 and processed as described above.

Compost bioassays In order to assess the potential of composts to reduce susceptibility of grapevine to *P. viticola*, 19 composts from nine commercial compost producers were tested under controlled conditions. Composts of each of the four quality groups (digestates for agricultural use, composts for agricultural use, composts for horticultural use, compost for covered cultures and private gardening) as defined by the guidelines of the Swiss Compost Plants were used (Fuchs et al. 2001). Feedstock and processing of each of the composts was known, and over 50 parameters including chemical and physical parameters, microbial activities, plant compatibility and suppressiveness to soil-borne diseases had been determined according to Kupper and Fuchs (2007) (unpublished data). Twenty percent (v/v) of compost was thoroughly mixed with a standard substrate ('Einheitserde Typ 0', Gebr. Patzer GmbH & Co. KG, D-Sinntal-Jossa) previously amended with 3 gl^{-1} of a mineral fertilizer (Tardit 3 M, Hauert Günther Düngerwerke GmbH, D-Erlangen). Plants were fertilized every second week with an ammonium nitrate solution (NH_4NO_3, 0.024 gNl^{-1} substrate), starting at the date of planting. Seedlings were used for greenhouse assays. Small seedlings of grapevine *cv.* 'Chasselas' (Syngenta AG, Stein, Switzerland) were transplanted to the substrate and fertilized as described above. Plants were grown in the greenhouse at a temperature of 18–28°C under natural light. In winter time, photoperiod was extended with mercury lamps to 16 h. Plants were drop inoculated with a sporangia suspension of *P. viticola* when they had 6 fully expanded leaves (two 5 μl drops on each fully expanded leaf, 5×10^4 sporangia ml^{-1}). After inoculation, plants were incubated at 100% relative humidity at 20°C for 24 h, then transferred to growth chambers (60% RH, 14 h light [80 $\mu E\ m^{-2}s^{-1}$] 20°C during day, 18°C during night) for 5 days and returned to the humidity chamber the evening before scoring in order to initiate sporangia production.

The same 19 composts were tested in parallel in the bioassay *Arabidopsis thaliana* L.—*Hyaloperonospora arabidopsidis*. *A. thaliana* is known to be able to respond to appropriate organic soil amendments or beneficial soil microorganisms with altered disease susceptibility (Vallad et al. 2003; van Loon et al. 1998). *A. thaliana* seeds accession Columbia *(Col-0)* were stratified in 0.1% water agar at 4°C for 4 d and planted directly on the substrate (pots ∅ 9 cm, 275 ml), with six replicates per treatment. Seven to 10 days after planting, plants were thinned to 5 per pot. Plants were grown in a growth chamber with a 10 hday (100–120 $\mu E\ m^{-2}s^{-1}$ at 23°C) and 14 h night (18°C) cycle at 70% relative air humidity for 18 days before inoculating with *H. arabidopsidis* strain NOCO (2×10^4 conidia ml^{-1}) with a hand sprayer. Plants were then incubated at 19°C, 100% relative air humidity and 10 h light during 24 h. Subsequently, relative air humidity was reduced to 80% for 5 d, then increased again to 100% to induce sporulation before assessing disease severity and harvesting. Disease severity (percentage leaf area covered by conidiophores) was assessed for each plant separately by assessing upper and lower sides of the leaves. In both bioassays, six replicates per treatment were used. The

19 composts were assessed in two distinct experimental sets, each including an internal control (standard substrate amended with mineral fertilizer as described above). The experiment was repeated with grapevine *cv.* Chasselas cuttings (Agroscope Changins-Wädenswil, Changins, Switzerland) instead of seedlings as described above with 10 composts, five of which had shown a disease suppressive effect in the *A. thaliana—H. arabidopsidis* bioassay in previous experiments (data not shown).

Preparation of cDNAs and real-time PCR for expression analysis of VvGluc, VvLox9, VvNCED and VvSTS Total RNA was isolated from frozen leaf tissue using a modified CTAB extraction and lithium chloride precipitation method according to Iandolino et al. (2004). The quantity of total RNA was determined with a Nano-Drop ND-1000 spectrophotometer (NanoDrop Technologies, Wilmingon, DE, USA). For quantitative real-time PCR analysis, RNA was treated with Turbo-DNase I (Ambion) according to the manufacturer's instructions. For cDNA synthesis, 1 μg of RNA was reverse transcribed using oligo$(dT)_{18}$ and Superscript III reverse transcriptase (Invitrogen Life Technologies) following the instructions of the manufacturer.

Expression analysis of the genes *VvGluc, VvLox9, VvNCED* and *VvSTS* was done by real-time PCR, using SYBR green method on an iCycler (Bio-Rad) real-time cycler. Each PCR reaction (20 μl) contained 0.25 mM of each primer, cDNA and 1 x SYBR Green Master Mix Reagent (Bio-Rad). The thermal cycling conditions were 95°C for 3 min followed by 95°C for 30 s, 56°C for 30 s, and 72°C for 35 s for 40 cycles, followed by a melt cycle from 60°C to 95°C. The primers used were as follows: VvGluc-F (5′-ACCAC CAGCATCACAAGTGGTA-3′) and VvGluc-R (5′-TGTTAAGTCGATTGCGGTGGAG-3′) for *VvGluc*, VvLox9-F (5′-TCAGTTCTGGTGGAAGGAAGT-3′) and VvLox9-R (5′-AGGCATGAATCTGCGGCT TATC-3′) for *VvLox9*, VvNCED-F (5′-TGATG TGGTCCAGAAGCCATAC-3′) for *VvNCED* and VvStSy-F (5′-CTCGAACCATCCGTCAGAAGAG-3′) and VvStSy-R (5′-CCTACGATTACAGCTGCA GACC-3′) for *VvSTS*. With all cDNAs used, the above primer sets gave single PCR products, which were verified by determining the melt curves for the products at the end of each run and by analysis of the products using gel electrophoresis. The efficiency of the primers was tested in preliminary experiments with serial dilutions of cDNA samples and maintained an E value of between 0.97 and 0.98. The expression of the four genes was normalised relative to Elongation Factor 1-α (*VvEF1-α*). All samples were measured in triplicate, every run included the *VvEF1-α* control for each sample. The Gene-X software (Bio-Rad) was used to calculate the mean normalised expression of the genes.

Statistical analysis Susceptibility of grapevine seedlings and *A. thaliana* plants grown in different composts or in the control were compared by an ANOVA ($\alpha=0.05$) followed by a Tukey-B test at $\alpha=0.05$ in case of significant results. Variation in the gene expression pattern of the four defense-related genes was evaluated separately for each timepoint after inoculation by a constrained linear ordination using Redundancy Analysis (RDA) with the environmental factors site, leaf age, and type of inoculation (mock-inoculation/inoculation *with P. viticola*), and results were visualized in an ordination diagram (Software CANOCO 4.5).

Results

Effect of site on the susceptibility of grapevine to *P. viticola*

The susceptibility of grapevine *cv.* Chasselas to *P. viticola* grown at three sites in the region of Lake Neuchâtel was studied in the years 2004 to 2008. The three vineyards are located a maximum of 26 km apart in the same geographic region (south-east facing slopes at the north-west shore of Lake Neuchâtel) (Table 1). Soils at the three sites were similar in their contents of available and reserve P, K, Mg, Ca and Fe, but differed in their sand-silt-clay contents, the percentage of stones and in soil depth (Table 1). Grapevine grown at the three sites differed significantly in their susceptibility to *P. viticola* in all 5 years of examination (Fig. 1). In 4 out of 5 years, plants from the HAU site were much less susceptible than plants from site AUV. In another year (2005), differences between HAU and AUV were less pronounced but still significant for one leaf age (Fig. 1). Susceptibility of plants from the CON site varied from year to year, being either as resistant as plants from the HAU site (2004), nearly as (2007), or

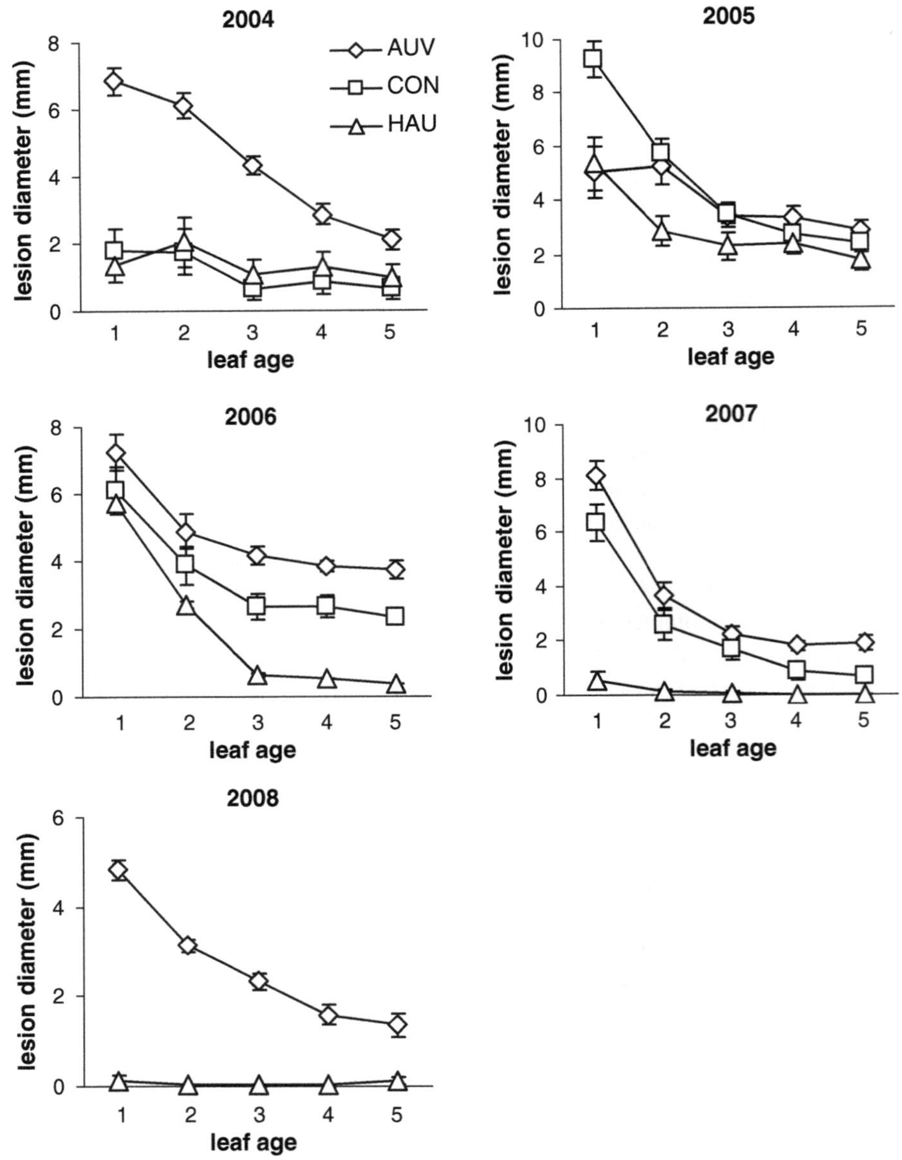

Fig. 1 Downy mildew susceptibility of grapevine leaves from plants (*cv.* Chasselas) grown in three commercial organic vineyards in the area of Lake Neuchâtel (AUV, CON and HAU) in five successive years. At each location, leaves of twenty plants were sampled, starting with a leaf of approximately 2/3 the size of a fully grown leaf (leaf age 1), and the next four older leaves. After washing thoroughly, two leaf discs were cut from each leaf, and inoculated with *P. viticola*. Lesion diameters were measured 7 d post inoculation. The figures show means±SE

even more (2005) susceptible than plants from the AUV site (2007) or intermediate (2006). Growth status of grapevine was comparable in all three sites (data not shown). Microsatellite analysis revealed that plants from all three sites have the same genetic background. Furthermore, grafted vines produced from scions of plants of each of the three sites did not differ in their susceptibility to *P. viticola* if grown either in the vineyard at the HAU site or in a standard substrate in the greenhouse (Fig. 2a). In contrast, grafted plants from the three sites grown at the AUV site tended to be more susceptible to *P. viticola* than plants grown at the HAU site in the first growing season after planting (Fig. 2b).

Effect of site on the expression pattern of four selected defence-related genes

In 2007, the expression patterns of four selected defence-related genes (β-1,3-Glucanase (*Glu*), stilbene synthase (*STS*), lipoxygenase 9 (*Lox9*) and 9-cis-epoxycarotenoid dioxygenase (*NCED*)) in leaves from the three sites and in five leaf ages were studied. Redundancy analysis on the constitutive expression

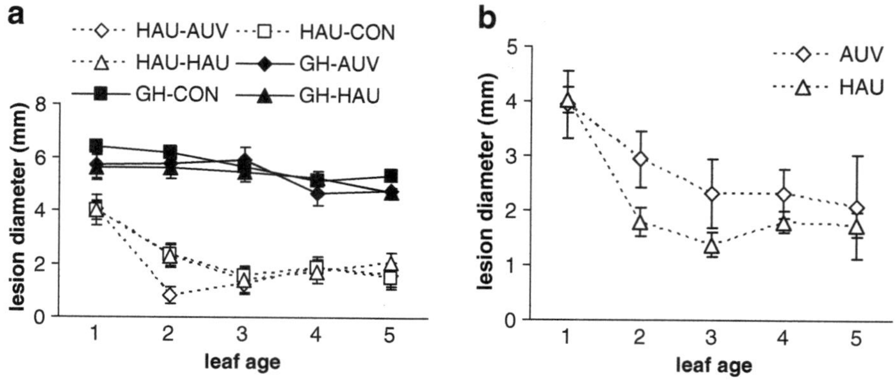

Fig. 2 Downy mildew susceptibility of young grafted vines (*cv.* Chasselas) produced from scions of three vineyards (AUV, CON and HAU) (grafted onto rootstock 5BB). Figures show (**a**) susceptibility of grafted vines produced from scions of the three origins grown for 9 months either in the greenhouse (GH) (filled symbols, solid lines) or in the field at site HAU (empty symbols, dashed lines) or (**b**) susceptibility of grafted vines produced from scions from the three vineyards and planted at sites AUV or HAU, in their first season after planting (data pooled for each site). From each plant, two leaf discs per leaf were cut from 5 leaf ages, starting with a leaf of approximately 2/3 the size of a fully grown leaf (leaf age 1), and the next four older leaves. Leaf discs were inoculated with *P. viticola* and lesion diameters measured 7 d post inoculation. The figure shows means±SE, n=25 (HAU), 10 (GH-AUV, GH-CON, GH-HAU), 9 (HAU-AUV, HAU-CON), 7 (HAU-HAU) or 4 (AUV). Legends indicate (**a**) site of growth—origin of scions or (**b**) site of growth

pattern revealed that 41% of the total variance was explained by the factor site (p=0.002), and 24% by the factor leaf age (p=0.014) (in total 65% explained variance) (Tab. 2). Visualisation of the results showed a clear separation of the HAU site from the other two sites (Fig. 3). High expression levels of *Glu* and *Lox9* were associated with samples from the HAU site and positively correlated with resistance to *P. viticola*. The expression level of *NCED* was mainly correlated with leaf age, with increased levels in older leaves. *STS* had only a minor impact on the ordination (Fig. 3). In addition to constitutive expression levels, the four genes were monitored for 48 h after cutting leaf discs and inoculation. Redundancy analysis revealed that (i) total variance explained by the factors site, leaf age and type of inoculation decreased from 65% to 25% over time (ii) percentage variance explained by the factor site decreased over time from 41% to 12%, but remained significant, and (iii) after 12 h and later on, inoculated and non-inoculated leaf discs did not differ in their expression patterns (Table 2).

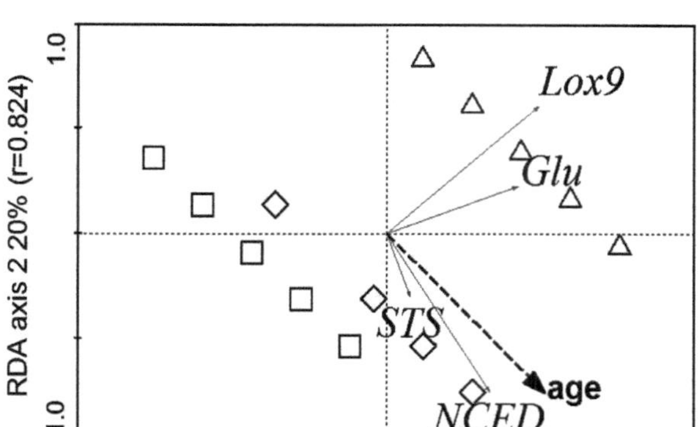

Fig. 3 Distribution of gene expression levels of grapevine before inoculation with *P. viticola*, in the ordination biplot of a redundancy analysis (RDA). Samples were labelled by site (triangles AUV, diamonds CON, circles HAU). The vectors represent the individual genes (solid vectors) and the environmental factor leaf age (dashed vector). Expression levels are relative to expression level of *VvEF1-α*, and the data were log-transformed

Modulation of site-specific susceptibility of grapevine to *P. viticola* by the application of composts

The effect of the long-term application of compost on resistance of grapevine *cv.* Pinot noir against *P. viticola* was examined in two field experiments, which had been established in 1996. After 10 years of continuous compost amendment, no significant differences between grapevine plants grown in compost-treated or control plots were detected in two consecutive years of sampling (Fig. 4b). Yet,

Table 2 Variance in the expression level of four defence-related genes before inoculation (constitutive expression level, $t=0$) and 8, 12, 24 and 48 h post (mock-) inoculation (p.i.) as explained by the factors site, leaf age, and inoculation type (mock- or pathogen-inoculation) determined by redundancy analysis

Factors	Explained variance				
	Constitutive expression ($t=0$)	$t=8$ h p.i.	$t=12$ h p.i.	$t=24$ h p.i.	$t=48$ h p.i.
All factors	64.6%	53%	44%	25%	39%
Sites	41% *	31% *	23% *	12% *	14% *
Leaf age	23.6% *	11% *	18% *	7% *	20% *
Inoculation	–	10% *	1%	3%	3%

* significant for all canonical axes, Monte Carlo Permutation Test, $p<0.05$

grapevines grown at the WAL site were less susceptible to *P. viticola* than plants grown at the MAL site in both years of examination (Fig. 4). To further study the effect of composts on the resistance of grapevine to *P. viticola*, 19 different composts were evaluated in the greenhouse on grapevine seedlings *cv.* Chasselas (Fig. 5a). As a control, the same 19 composts were evaluated in the system *A. thaliana-H. arabidopsidis*. None of the 19 composts had a significant effect on the resistance of grapevine seedlings. Interestingly, 8 of the 19 composts significantly reduced disease severity in the control system *A. thaliana- H. arabidopsidis* (Fig. 5b), and these results were repeatable (data not shown).Testing ten selected composts, eight of which had shown significant effects in the *A.thaliana-H.arabidopsidis* control system in previous experiments, on grapevine cuttings gave similar results as on grapevine seedlings (data not shown).

Discussion

Our research points to a substantial influence of site-specific properties on field resistance of *cv.* Chasselas and *cv.* Pinot noir. In the case of Chasselas, differences in resistance between the different investigated sites were substantial and relevant for commercial wine production, with leaves grown at the HAU site hardly developing disease symptoms upon inoculation with *P. viticola* in 3 of the 5 years examined. Such large site-specific differences in suppressiveness have been reported before, mainly for suppressiveness to soil-borne diseases (e.g. reviewed by Weller et al. (2002)). In addition, some studies showed that site-specific factors can have a similarly high impact on the resistance of plants to air-borne diseases. For instance, *A. thaliana* plants grown on different soil types differed in their susceptibility to downy mildew caused by *H. arabidopsidis* by a factor of six (i.e.

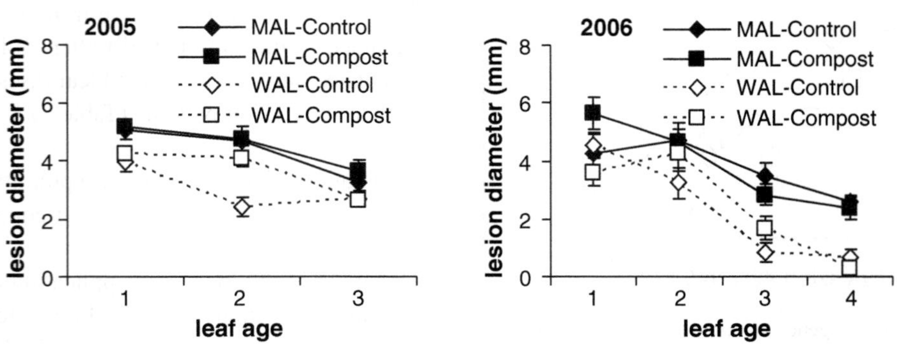

Fig. 4 Downy mildew susceptibility of grapevine *cv.* Pinot noir grown in two vineyards at Malans (MAL) or Walenstadt (WAL) amended or not-amended (control) with a compost after nine (2005) (**a**) or 10 years (2006) (**b**) of compost application. At each location, and in each treatment, leaves of twelve plants were sampled, starting with a leaf of approximately 2/3 the size of a fully grown leaf (leaf age 1), and the next two or three older leaves respectively. After washing thoroughly, ten leaf discs were cut from each leaf and inoculated with *P. viticola*. Lesion diameters were measured 7 d post inoculation. The figures show means±SE

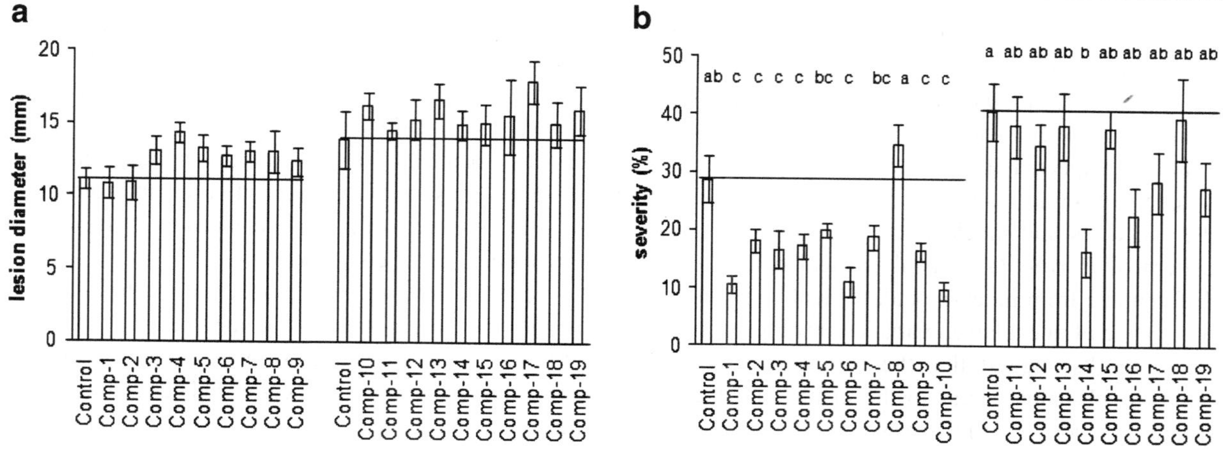

Fig. 5 Susceptibility of a) seedlings of grapevine *cv.* Chasselas to *P. viticola* and b) *A. thaliana* to *H. arabidopsidis* grown in a substrate amended with one out of 19 different composts (comp-1 to -19) or in the standard substrate only amended with a mineral fertilizer (control). The 19 composts were tested in two distinct experimental sets, each with the same internal control. *A. thaliana* plants were sprayed with *H. arabidopsidis* at the size of 5 true leaves. The area covered with conidiophores was assessed 7 d post inoculation. Each grapevine leaf was inoculated with 2 drops of *P. viticola,* and lesion diameters were measured 7 d post inoculation. The figures show means± SE, $n=6$. The treatments were compared by an ANOVA ($\alpha=0.05$) followed by a Tukey-B test ($\alpha=0.05$) in case of significant differences

plants grown on the most suppressive soil were six times less infected than plants grown on the least suppressive soil), whereas tomatoes grown on the same soils only differed by a factor of 0.2 (Tamm et al. 2010). To our knowledge, no such relevant differences between sites have been reported before for resistance of woody perennial plants such as grapevine to air-borne diseases. Although the site, i.e. the terroir, has been shown before to have a great impact on grape berry quality including the contents of many plant metabolites and on the resulting vine (Jackson and Lombard 1993; Morlat and Bodin 2006; Seguin 1986), the relative contribution of site-specific factors such as various soil parameters, climate, plant material, and anthropogenic impacts to the differentiation between sites is difficult to evaluate, mainly because of interactions between the different parameters (Seguin 1986). Nevertheless, soil has been identified as a key determinant by many studies, affecting various parameters relevant for grape berry and vine quality, such as mineral nutrient availability, water holding capacity and rooting depth (i.e. water uptake conditions), or the temperature in the root zone (Downey et al. 2006; Morlat and Bodin 2006; Seguin 1986).

Here we show that differences in susceptibility to *P. viticola* were correlated to altered constitutive gene expression patterns, even though only four defence-related genes were investigated. Studying the expression pattern of more defence-related genes might even lead to a further differentiation, e.g. between sites AUV and CON. The four genes were selected as representatives of different signalling pathways known to be involved in constitutive or induced resistance of different plants. Plants grown at the HAU site showed increased constitutive expression levels of *Lox9* and *Glu*. *Lox9* catalyses the rate-limiting step in jasmonic acid (JA) biosynthesis. JA has been shown to be involved in resistance of *P. viticola*-resistant varieties such as Solaris, but also in BABA-induced resistance (Hamiduzzaman et al. 2005). *Glu* is a marker gene of the salicylic acid (SA) pathway. As *Lox9*, its expression can be induced or potentiated by BABA or laminarin (Aziz et al. 2003; Hamiduzzaman et al. 2005). Yet, activation of the SA-pathway has been reported not to be sufficient in enhancing resistance of grapevine to *P. viticola*. Increased leaf age was not only correlated to increased levels of resistance to *P. viticola*, but also to increased expression levels of *NCED*, which is involved in the synthesis of abscisic acid (ABA). ABA was not effective in potentiating resistance of grapevine *cv.* Chasselas in leaf disc experiments (Hamiduzzaman et al. 2005), but can induce resistance against necrotrophic pathogens in *A. thaliana* (Ton and Mauch-Mani 2004). In contrast to differences in the constitutive expression pattern, there was no evidence that the four defence-related genes are activated faster and/or stronger upon contact with *P. viticola* in plants from the HAU site than in plants from any of the two other sites, as could be expected in classical soil microorganism-induced systemic resistance. Yet, 12 h after cutting the leaf discs and mock inoculation, the expression pattern of *P. viticola*—inoculated leaf discs did not differentiate from mock-inoculated leaf discs and the percentage of unexplained variance steadily increased. These results suggest that the cutting of leaf discs imposes substantial stress to the plants, and that the plant's reactions to this stress might overlay other reactions such as a putative priming effect. Thus, to monitor the reaction of the plants over a prolonged time-period, using whole leaves or whole plants instead of leaf-discs might give more detailed results. However, to compare plants grown in the field, using whole plants is not a feasible approach.

For an application in commercial agriculture, it is highly relevant whether site-specific suppressiveness of soils can be modulated by management techniques such as the application of organic amendments. Several studies have demonstrated that suppressiveness of natural soils to soil-borne, but also to air-borne diseases can be affected by the application of organic soil amendments such as manure and compost (Fuchs et al. 2004; Litterick et al. 2004; Tamm et al. 2010; Vallad and Goodman 2004). However, the effect of soil-amendments on resistance of various plant species to soil- and air-borne diseases was much smaller than site-specific effects (Tamm et al. 2010). In the present study, we found no evidence that the resistance of grapevine to *P. viticola* can be influenced by the application of composts. In two field experiments, no difference in the resistance of grapevine plants to *P. viticola* grown in control- or compost-treated plots were found, neither in the field (data not shown), nor when leaves were tested under controlled conditions in the lab, even though compost had been applied before annually for 12 years. Composts are known to differ in their quality and their ability to suppress soil-borne diseases depending on their composition, the production process, and maturity (Fuchs et al. 2004). Therefore, we have also evaluated the potential of 19 composts with different chemical, physical and microbial properties to reduce the susceptibility of grapevine seedlings and cuttings under controlled conditions. None of them had any effect in the system grapevine-*P. viticola*, even though eight of them significantly reduced disease of *A. thaliana* with *H. arabidopsidis*, our positive control known to be inducible via the root system. These results indicate that, in contrast to other plant-pathogen systems, it is unlikely to positively modulate site-specific resistance of grapevine plants to *P. viticola* by the application of composts.

In conclusion, we have shown that site has a major impact on the resistance of grapevine to *P. viticola*. Differences in resistance were correlated to altered expression patterns of defence-related genes, indicating that induced resistance might be the underlying mechanism for the observed differences in disease resistance. However, more research is needed to identify the site-specific factors involved in inducing resistance. In this study, site-specific suppressiveness to *P. viticola* could not be modulated by the application of composts, making an exploitation of the phenomenon in commercial agriculture unlikely.

Acknowledgements The authors are grateful to Nicole Mahlberg, Nicole Berger, Noemy Kraus, Thomas Amsler and many students for excellent technical assistance, Alfred Berner for maintenance of the long-term compost trial, Jacques Fuchs for supplying composts, Claire Arnold for microsatellite analysis and the Swiss National Foundation NCCR Plant Survival for funding.

References

Ait Barka, E., Belarbi, A., Hachet, C., Nowak, J., & Audran, J.-C. (2000). Enhancement of *in vitro* growth and resistance to gray mould of *Vitis vinifera* co-cultured with plant growth-promoting rhizobacteria. *FEMS Microbiology Letters, 186*, 91–95.

Anonymous. (1996). *Referenzmethoden der Schweizerischen Forschungsanstalten Agroscope—1. Bodenuntersuchungen zur Düngeberatung; Änderungen 1997–2009*. Zürich-Reckenholz: Agroscope.

Aziz, A., Poinssot, B., Daire, X., Adrian, M., Bézier, A., Lambert, B., et al. (2003). Laminarin elicits defense responses in grapevine and induces protection against *Botrytis cinerea* and *Plasmopara viticola*. *Molecular Plant-Microbe Interactions, 16*, 1118–1128.

Conrath, U., Beckers, G. J. M., Flors, V., Garcia-Agustin, P., Jakab, G., Mauch, F., et al. (2006). Priming: getting ready for the battle. *Molecular Plant Microbe Interactions, 19*, 1062–1071.

Downey, M. O., Dokoozlian, N. K., & Krstic, M. P. (2006). Cultural practice and environmental impacts on the flavonoid composition of grapes and wine: a review of recent research. *American Journal of Enology and Viticulture, 57*, 257–268.

Fuchs, J. G., Galli, U., Schleiss, K., & Wellinger, A. (2001). *ASCP guidelines 2001. Quality criteria for composts and digestates from biodegradable waste management.* Schönbühl: ASCP Association of Swiss Compost Plants in collaboration with the Swiss Biogas Forum.

Fuchs, J. G., Bieri, M., & Chardonnes, M. (2004). *Auswirkungen von Komposten und von Gärgut auf die Umwelt, die Bodenfruchtbarkeit, sowie die Pflanzengesundheit.* Frick: Research Institute of Organic Agriculture.

Haas, D., & Défago, G. (2005). Biological control of soil-borne pathogens by fluorescent pseudomonads. *Nature Reviews Microbiology, 3*, 307–319.

Hamiduzzaman, M. M., Jakab, G., Barnavon, L., Neuhaus, J.-M., & Mauch-Mani, B. (2005). β-Aminobutyric acid-induced resistance against downy mildew in grapevine acts through the potentiation of callose formation and jasmonic acid signaling. *Molecular Plant Microbe Interactions, 18*, 819–829.

Iandolino, A. B., Goes Da Silva, F., Lim, H., Choi, H., Williams, L. E., & Cook, D. R. (2004). High-quality RNA, cDNA, and derived EST libraries from grapevine (*Vitis vinifera* L.). *Plant Molecular Biology Reporter, 22*, 269–278.

Jackson, D. I., & Lombard, P. B. (1993). Environmental and management practices affecting grape composition and wine quality—a review. *American Journal of Enology and Viticulture, 44*, 409–430.

Kloepper, J. E., Rodriguez-Ubana, R., Zehnder, G. W., Murphy, J. F., Sikora, E., & Fernandez, C. (1999). Plant root-bacterial interactions in biological control of soilborne diseases and potential extension to systemic and foliar diseases. *Australasian Plant Pathology, 28*, 21–26.

Knudsen, I. M. B., Larsen, K. M., Jensen, D. F., & Hockenhull, J. (2002). Potential suppressiveness of different field soils to *Pythium* damping-off of sugar beat. *Applied Soil Ecology, 21*, 119–129.

Kupper, T., & Fuchs, J. G. (2007). *Kompost und Gärgut in der Schweiz.* Berne: Bundesamt für Umwelt.

Li, H.-Y., Yang, G.-D., Shu, H.-R., Yang, Y.-T., Ye, B.-X., Nishida, I., et al. (2006). Colonization by the arbuscular mycorrhizal fungus *Glomus versiforme* induces a defense response against the root-knot nematode *Meloidogyne incognita* in the grapevine (*Vitis amurensis* Rupr.), which includes transcriptional activation of the class III chitinase gene *VCH3*. *Plant and Cell Physiology, 47*, 154–163.

Litterick, A. M., Harrier, L., Wallace, C., Watson, C. A., & Wood, M. (2004). The role of uncomposted materials, composts, manures, and compost extracts in reducing pest and disease incidence and severity in sustainable temperate agricultural and horticultural crop production—a review. *Critical Reviews in Plant Sciences, 23*, 453–479.

Menzies, J. G. (1959). Occurrence and transfer of a biological factor in soil that suppresses potato scab. *Phytopathology, 49*, 648–652.

Morlat, R., & Bodin, F. (2006). Characterization of viticultural terroirs using a simple field model based on soil depth—II. Validation of the grape yield and berry quality in the Anjou vineyard (France). *Plant and Soil, 281*, 55–69.

Seguin, G. (1986). 'Terroirs' and pedology of wine growing. *Experientia, 42*, 861–873.

Shipton, P. J., Cook, R. J., & Sitton, J. W. (1973). Occurrence and transfer of a biological factor in soil that suppresses take-all of wheat in Eastern Washington. *Phytopathology, 63*, 511–517.

Stutz, E. W., Défago, G., & Kern, H. (1986). Naturally occurring fluorescent *Pseudomonads* involved in suppression of black root rot of tobacco. *Phytopathology, 76*, 181–185.

Tabatabai, M. A. (1982). Soil Enzymes. In A. L. Page, R. H. Miller, & D. R. Keeney (Eds.), *Methods of soil analysis, part 2 chemical and microbiological properties* (pp. 903–947). Madison: American Society of Agronomy & Soil Science Society of America 9.

Tamm, L., Thuerig, B., Bruns, C., Fuchs, J. G., Köpke, U., Laustela, M., et al. (2010). Soil type, management history, and soil amendments influence the development of soil-borne (*Rhizoctonia solani, Pythium ultimum*) and air-borne (*Phytophthora infestans, Hyaloperonospora parasitica*) diseases. *European Journal of Plant Pathology, 127*, 465–481.

This, P., Jung, A., Boccacci, P., Borrego, J., Botta, R., Costantini, L., et al. (2004). Development of a standard set of microsatellite reference alleles for identification of grape cultivars. *Theoretical and Applied Genetics, 109*, 1448–1458.

Thuerig, B., Fliessbach, A., Berger, N., Fuchs, J., Kraus, N., Mahlberg, N., et al. (2009). Re-establishment of suppressiveness to soil- and air-borne diseases by re-inoculation of soil microbial communities. *Soil Biology & Biochemistry, 41*, 2153–2161.

Ton, J., & Mauch-Mani, B. (2004). β-amino-butyric acid-induced resistance against necrotrophic pathogens is based on ABA-dependent priming for callose. *The Plant Journal, 38*, 119–130.

Vallad, G. E., Cooperband, L., & Goodman, R. M. (2003). Plant foliar disease suppression mediated by composted forms of paper mill residuals exhibits molecular features of induced resistance. *Physiological and Molecular Plant Pathology, 63*, 65–77.

Vallad, G. E., & Goodman, R. M. (2004). Systemic acquired resistance and induced systemic resistance in conventional agriculture. *Crop Science, 44*, 1920–1934.

Van Loon, L. C., Bakker, P. A. H. M., & Pieterse, C. M. J. (1998). Systemic resistance induced by rhizosphere bacteria. *Annual Review of Phytopathology, 36*, 453–483.

Vouillamoz, J. F., Arnold, C., & Frei, A. (2009). Swiss vitis microsatellite database (SVMD). In E. Peterlunger, G. Di Gaspero, & G. Cipriani (Eds.), *Proceedings of the IXth International conference on grape genetics and breeding* (pp. 477–480). Udine: Acta Horticulturae (ISHS) 827.

Weller, D. M., Raaijmakers, J. M., Mcspadden Gardener, B. B., & Thomashow, L. S. (2002). Microbial populations responsible for specific soil suppressiveness to plant pathogens. *Annual Review of Phytopathology, 40*, 309–348.

Wiseman, B. M., Neate, S. M., Keller, K. O., & Smith, S. E. (1996). Suppression of *Rhizoctonia solani* anastomosis group 8 in Australia and its biological nature. *Soil Biology & Biochemistry, 28*, 727–732.

Eur J Plant Pathol (2011) 129:135–148
DOI 10.1007/s10658-010-9626-9

Influence of nitric oxide and reactive oxygen species on development of lettuce downy mildew in *Lactuca* spp.

Michaela Sedlářová · Marek Petřivalský · Jana Piterková · Lenka Luhová · Jindra Kočířová · Aleš Lebeda

Accepted: 26 April 2010 / Published online: 16 May 2010

Abstract The role of nitric oxide and reactive oxygen species, molecules indispensable for plant-pathogen signalling, was studied in the *Lactuca* spp.-*Bremia lactucae* pathosystem. Using a leaf disc model the translaminar effect of various compounds affecting their metabolism was studied by light microscopy. Time course studies revealed a slowdown in the development of *B. lactucae* (race BL16) infection structures by rutin (scavenger of reactive nitrogen and oxygen species) and SNP (NO donor) within 48 h post inoculation, followed by a retardation of sporulation. Application of the specific NO scavenger, PTIO, accelerated penetration of *B. lactucae* but had no further effects on the plant-pathogen interaction. Inhibitors of NO synthase (L-NAME) and nitrate reductase (sodium tungstate) did not influence pathogen development. Our results suggest that drastic change in the NO: hydrogen peroxide ratio seems to determine the pathogen's fate. NO synthase-like activity significantly increased early after *B. lactucae* challenge in resistant *L. virosa*. Confocal laser scanning microscopy revealed the accumulation of nitric oxide in the penetrated cells, pointing to a role in the initiation of the hypersensitive reaction. The tips of germ tubes and appressoria of *B. lactucae* also accumulated NO, suggesting an essential role for this molecule in penetration of the biotrophic pathogen. Additionally, temporal changes in endogenous levels of rutin and quercetin in extracts from *Lactuca* spp. leaves will be discussed in connection to their role as part of the antioxidative machinery that influences the plants' susceptibility/resistance to lettuce downy mildew.

Keywords Antioxidants · *Bremia lactucae* · *Lactuca sativa* · *Lactuca saligna* · *Lactuca virosa* · Oomycete · Oxidative stress · Quercetin · Rutin · Sodium nitroprusside

Abbreviations

DAF-FM DA	4-amino-5-(N-methylamino)-2′,7′-difluorofluorescein diacetate
hpi	hours post inoculation
HR	hypersensitive response
L-NAME	N^G-nitro-L-arginine-methylester
NOS	nitric oxide synthase
NR	nitrate reductase
PTIO	2-phenyl-4,4,5,5-tetramethylimidazoline-1-oxyl-3-oxide

M. Sedlářová (✉) · A. Lebeda
Department of Botany, Faculty of Science,
Palacký University in Olomouc,
Šlechtitelů 11,
783 71 Olomouc, Czech Republic
e-mail: michaela.sedlarova@upol.cz

M. Petřivalský · J. Piterková · L. Luhová · J. Kočířová
Department of Biochemistry, Faculty of Science,
Palacký University in Olomouc,
Šlechtitelů 11,
783 71 Olomouc, Czech Republic

PV	primary vesicle
ROS	reactive oxygen species
RNS	reactive nitrogen species
SNP	sodium nitroprusside
SD	standard deviation
SV	secondary vesicle

Introduction

Throughout their co-evolution, plants and pathogens have continually confronted each other in a struggle for growth and survival. Numerous components involved in pathogen perception, signal generation and transmission, and activation of defence responses have already been identified whilst others wait to be understood. Reactive oxygen species (ROS) and reactive nitrogen species (RNS) rank among the crucial components of signalling pathways in stressed plants (Delledonne et al. 2001, 2002; De Gara et al. 2003; Wendehenne et al. 2004; Yoshioka et al. 2009; Zaninotto et al. 2006).

Nitric oxide (NO) is an essential regulatory molecule of plant immunity in synergy with reactive oxygen species (Arasimowicz and Floryszak-Wieczorek 2007). Induction of local biochemical defence responses in plant cells surrounding the infection site correlates with the accumulation of RNS. NO as a significant component of signalling networks is able to diffuse through membranes thanks to its hydrophobicity and small size (De Gara et al. 2003). The production of NO in plants was shown to be catalyzed by nitrate reductase (EC 1.7.1.1.) and nitric oxide synthase-like enzyme(s). However, the identity and structure of plant NO synthase has not been elucidated so far (Corpas et al. 2009; Wilson et al. 2008). NO can be also generated in plant cells by non-enzymatic processes like NO_2^- dismutation or reduction (Cooney et al. 1994; Stöhr and Ullrich 2002; Yamasaki 2000). In interactions with fungal pathogens *sensu lato* ROS and RNS are required for the initial signalling between both partners as well as for the activation of plant defence. Local generation of ROS is a typical feature of fungal pathogenesis, related to the active penetration of host cells (Sedlářová et al. 2007; Tománková et al. 2006). In resistant plants ROS and RNS modulate gene expression, induce structural defences or a local cell death, i.e. hypersensitive reaction (HR). Since the HR is accompanied by the generation of excess amounts of oxidative molecules, the fate of cells surrounding the infection site is determined by their ability to buffer this oxidative stress (Mur et al. 2008). Host cell collapse generally coincides with cessation of the attempted colonization, thus biotrophic pathogens must avoid triggering this process by, for example, elicitors released during cell wall penetration (Panstruga 2003). Additionally, ROS have direct antimicrobial effects (Shetty et al. 2008). The strategy of biotrophs may involve reduction of oxidative stress in the penetrated cell, e.g. by peroxidase production in the infection structures (Sedlářová et al. 2007). Antioxidant mechanisms involve specialized enzymes (superoxide dismutase (EC 1.15.1.1), peroxidase (EC 1.11.1.7) and catalase (EC 1.11.1.6)), as well as non-enzymatic systems (ascorbate, glutathion, α-tocoferol, β-caroten, phenolic compounds) (Blokhina et al. 2003; De Gara et al. 2003; Grace 2005).

Lettuce and its wild relative *Lactuca* spp. are threatened by *Bremia lactucae* infection that causes downy mildew with enormous economic losses in crop yield (Lebeda et al. 2002). Due to its high genotypic and phenotypic variability, this pathosystem is a widely used model to study plant-biotrophic oomycete interactions from the field to molecular level (Lebeda et al. 2008a, b). The role of ROS and antioxidative enzymes in these interactions has been already shown (Sedlářová et al. 2007). An intensive accumulation of H_2O_2 early after inoculation was linked to race-specific resistance by initiation of HR, although variations in timing and intensity were found among genotypes. The tissues surrounding the HR showed an induction in antioxidative capacity, mainly due to an increased POX activity. This allowed the confinement of the HR cells with their excess in ROS. H_2O_2 and POX were also observed in *B. lactucae* infection structures during initial stages of the infection process (Sedlářová et al. 2007). Recently, accumulation of NO in germ tubes, appressoria and penetration pegs of powdery mildew, was proposed as an important determinant for successful invasion into host tissues (Piterková et al. 2009; Prats et al. 2008).

Our recent work describes the role of NO in the development of infection structures of downy mildew, *Bremia lactucae*, as well as in interacting host cells of *Lactuca* spp. The effect of external application of compounds modulating NO levels was studied in a series of experiments involving four *Lactuca* spp.-*Bremia lactucae* interactions with differential phenotype of susceptibility or resistance.

Materials and methods

Pathogen isolate

The experiments were performed with *Bremia lactucae* Regel race BL 16 (sextet code EU-A63/31/02/00) from the collection of the Department of Botany, Palacký University in Olomouc, included in the Czech National Collection of Microorganisms (collection number UPOC-FUN-013). The isolate was maintained and multiplied on seedlings of susceptible lettuce, *L. sativa* (cv. Cobham Green or cv. British Hilde), grown as reported previously (Sedlářová et al. 2001, 2007).

Plant material and cultivation

Eight-week-old plants of *Lactuca* spp. genotypes with different phenotypes of response to *B. lactucae* (race BL16) were used: susceptible *L. sativa* L. (UCDM2) and resistant *L. sativa* L. (Mariska), *L. saligna* L. (CGN 05271) and *L. virosa* L. (NVRS 10.001 602). Histology of these plant genotype-pathogen race interactions has been described previously (Sedlářová et al. 2007). Plants were grown in a soil/peat (2:1, *v/v*) mixture in a growth chamber at 18/15°C (day/night), 12 h photoperiod, irradiance of 100 $\mu E.m^{-2}.s^{-1}$ provided by warm-white fluorescent lamps. Following inoculation, the cultivation temperature was decreased to 15/10°C and the plants were kept in dark during first 24 h to establish optimal conditions for pathogen development.

Inoculation, incubation and collection of samples

Whole plants/leaf discs were inoculated by spraying with a suspension of *B. lactucae* conidia in distilled water (approx. 5×10^5 ml^{-1}) and further incubation followed previously described procedures (Sedlářová et al. 2007). The 4th to 6th youngest leaves per plant were harvested 0, 4, 8, 24, 48, 72, 96, 168 or 216 h post inoculation (hpi), and either subjected to histochemical staining followed by confocal microscopy or frozen in liquid nitrogen and stored at −80°C until enzyme activity or concentration of antioxidants was studied in the leaf extract.

For light microscopy experiments, leaf discs (12 mm in diameter), derived from 4th–6th youngest leaf of a plant, were placed abaxial side up on wet filter paper in Petri dishes and subsequently inoculated.

Modulation of NO levels

Most of the following chemicals were purchased from Sigma-Aldrich, Czech Republic, except the glycerol (Tamda, Czech Republic) and NO-specific probe (Axxora, U.S.A.).

Different NO level modulators were applied to the filter papers under the leaf tissues in Petri dishes. The filters were saturated with 10 ml of the solution to be tested. The final concentrations of the studied substances were 0.1 mM SNP (NO donor), 0.1 mM PTIO (NO scavenger), 0.1 mM rutin (NO/ROS scavenger), 1 mM L-NAME (NOS inhibitor), and 5 mM Na_2WO_4 (inhibitor of nitrate reductase). In case of enzyme inhibitors, the tissues were pre-incubated 1 h in vacuum prior to the inoculation, other variants were inoculated directly. The control leaf discs were incubated with deionised water. Leaf discs were collected 4, 8, 24 and 48 hpi. Each Petri dish contained five leaf discs from one plant for each time interval and three Petri dishes per solution variant were used, i.e. each sample contained 15 leaf discs.

Light microscopy of *B. lactucae* development and *Lactuca* spp. tissue reactions

Leaf discs were fixed in 100% acetic acid for 2 days and then transferred and stored in 85% glycerol. Prior to observation the specimens were stained with 1% Aniline Blue (*w/v*). Pathogen development and host tissue reactions were examined by light microscopy (Olympus BX-60) and documented with a CCD camera (Olympus DP70; Olympus C&S, Prague, Czech Republic). In each variant five leaf discs and 15 spores/infection sites per leaf disc were randomly evaluated. The results were expressed as a mean frequency and standard deviation for each substance and time interval.

Statistical analysis

For each genotype the effect of the treatment on pathogen development and tissue response was statistically analyzed using NCSS 2001 software (Kaysville, UT, U.S.A.). The frequency of infection structures or hypersensitive cells was analyzed by ANOVA applying the Tukey-Kramer multiple-comparison test. Significantly different mean values (at 0.05 level) were marked by asterisks in Figs. 2, 3, 4 and 5.

Localisation of NO production by confocal microscopy

Small pieces of *Lactuca* spp. leaf tissues were cut off plants or leaf discs incubated with either water or modulators of NO production, incubated in 20 μM solution of DAF-FM DA (4-amino-5-(N-methylamino)-2′,7′-difluorofluorescein diacetate) (Axxora, U.S.A.) for 30 min, washed with buffer and mounted in 20 mM K-phosphate buffer (pH 7.0) on microscopic slides and subjected to confocal laser scanning microscopy (Olympus Fluorview 1000 attached to an inverted microscope IX81). Excitation was provided by the 488 nm line of an argon ion laser; a 505-nm dichroic filter and 519-nm longpass emission filter were applied to detect the signal. The base intensity of lasers was set according to samples from uninoculated plants at the beginning of each experiment. Negative controls were pre-treated with NO scavenger (0.1 mM c-PTIO) or NO synthase inhibitor (10 mM L-NAME) for 30 min prior to staining. The positive control was pre-incubated with a NO generator (0.1 mM SNP) (Piterková et al. 2009).

Preparation of plant extracts

Frozen leaf tissues were homogenized (1:2; *w*/*v*) with ice-cold 20 mM HEPES buffer, (pH 7.5), containing 2 mM ethylenediamine tetraacetic acid (EDTA), 0.5% polyvinylpolypyrrolidone (PVPP) and 10% glycerol in a chilled mortar. Prior to homogenisation, 5 mM dithiothreitol (DTT), 2 mM aprotonin and 2 mM phenylmethylsulphonyl fluoride (PMSF) were added to the extraction HEPES buffer. Leaf homogenates were centrifuged (16,000 × *g*, 15 min, 4°C) and supernatants immediately used for the determination of NO production.

Determination of NOS-like activity

NOS-like activity in plant extracts was determined spectrophotometrically by an oxyhemoglobin method based on the reaction of NO with oxyhemoglobin to yield methemoglobin and nitrate (Hevel and Marletta 1994) as previously described in tomato (Piterková et al. 2009). The reaction medium contained 10 μM oxyhemoglobin, 1 mM L-Arg, 1 mM $CaCl_2$, 1 mM Mg^{2+}, 1 μM calmodulin, 140 μM NADPH, 1 μM FAD, 1 μM FMN and 10 μM tetrahydrobiopterin in 100 mM HEPES buffer, pH 7.5. Norvaline as a competitive inhibitor of potentially interfering arginase was added to give a final 1 mM concentration. Superoxide dismutase (50 U) and catalase (100 U) were added to prevent interference by $O_2^{\cdot -}$ and/or H_2O_2. All measurements were performed at 37°C and the absorbance was determined at λ=401 nm on a microplate reader Synergy (Biotek, U.S.A.). The blank values were subtracted from all measured values (leaf extracts replaced by the same volume of buffer in the reaction mixture). For each variant, three measurements for each of the above mentioned three independent sets of samples were conducted, i.e. nine values were obtained. Data are presented as mean ± standard deviation (SD).

Rutin and quercetin content in leaves

Leaf samples were frozen in liquid nitrogen and lyophilised. Dried leaf samples were pulverized in a chilled mortar and transferred to plastic microtubes. Samples were extracted with 2 ml of 70% methanol by sonication (5 min) and incubation for 1h, followed by centrifugation (13,000 × *g*, 10 min, RT). Supernatants of methanol extracts were transferred to clean microtubes and evaporated at 40°C under a nitrogen stream. Samples were analysed by HPLC (system Gold; Beckman, U.S.A.) on HS PEG column (150× 4 mm, 5 μm, Supelco, U.S.A.), using isocratic elution with the mobile phase 50% methanol in 0.1 M formic acid at 30°C. The retention time of rutin and quercetin was 2.1 and 6.8 min, respectively. The amount of rutin and quercetin in leaf extracts was quantified using a calibration set of commercial rutin and quercetin standards. Results were expressed as molar concentration per dry weight and represent mean of 9 measurements and standard deviation.

Results

Development of *B. lactucae* infection structures and reaction of *Lactuca* spp. tissues

Germination and development of *Bremia lactucae* (race BL16) primary infection structures was studied on leaf discs incubated with solutions of compounds that influenced NO concentration within host tissues (Fig. 1). Pathogen spores usually germinate and form appressoria few hours after inoculation (Fig. 1a, b).

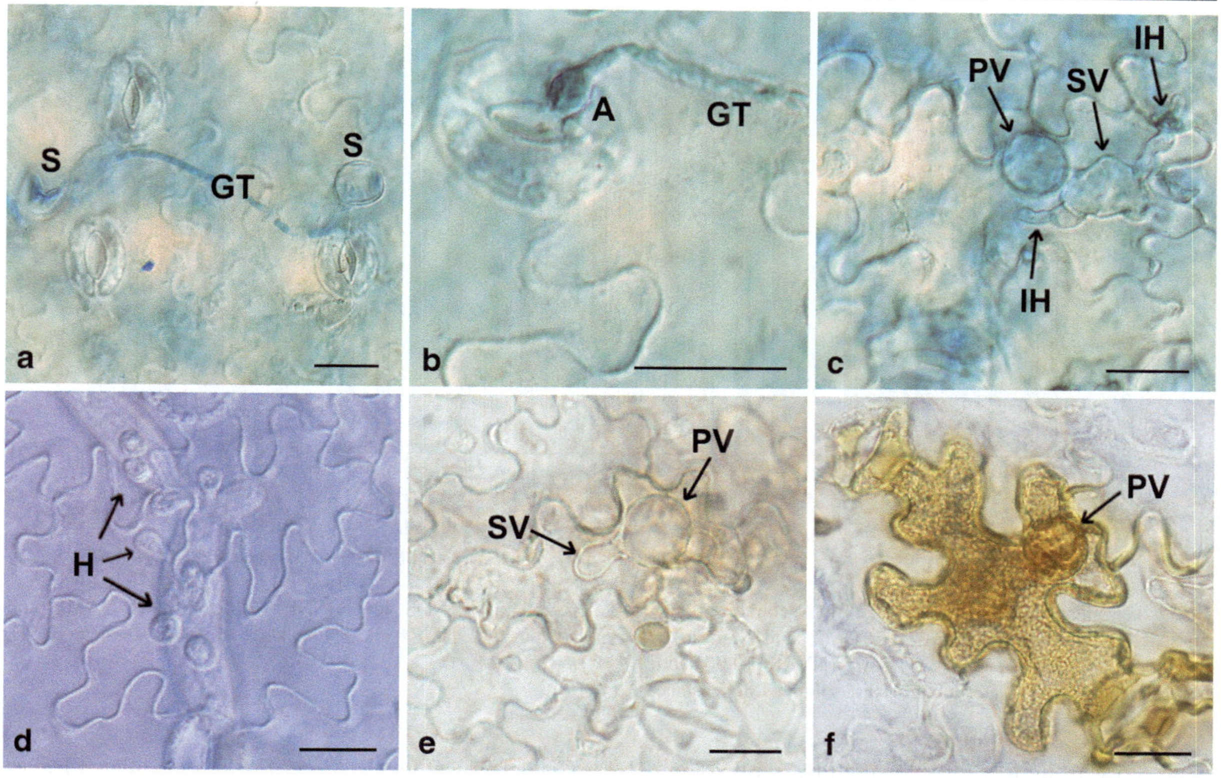

Fig. 1 Infection structures of *Bremia lactucae* (race BL16) and response of *Lactuca* spp. tissues. **a** Germination. **b** Detail of germ tube with appressorium on *L. virosa* (NVRS 10.001 602) 24 hpi. **c, d** Epidermal cells of susceptible *L. sativa* (UCDM2) penetrated by: **c,** primary infection structures 24 hpi and **d,** haustoria 168 hpi. **e** Onset of hypersensitive reaction in *L. saligna* (CGN 05271) 48 hpi. **f** Late stage of hypersensitive reaction in *L. virosa* (NVRS 10.001 602) 48 hpi. Infection structures: spore (*S*), germ tube (*GT*), appressorium (*A*), primary vesicle (*PV*), secondary vesicle (*SV*), intracellular hypha (*IH*), haustorium (*H*). The bar corresponds to 20 μm

Penetration of epidermal cell is followed by development of primary and secondary vesicles (Fig. 1c, e) and either 1/ further intercellular growth with abundant haustoria formation in susceptible host tissues (Fig. 1d); or 2/ cessation of pathogen growth in resistant plants often followed by hypersensitive reaction (Fig. 1f).

Pathogen germination on susceptible *L. sativa* (UCDM2) and resistant *L. sativa* (Mariska), *L. saligna* (CGN 05271) and *L. virosa* (NVRS 10.001 602) was recorded at 4 and 8 hpi (Fig. 2). Application of the NO donor SNP reduced conidial germination. Rutin had an even stronger inhibitory effect on germ tube formation in all genotypes, namely 4 hpi. Similarly, these compounds reduced the penetration rate in all genotypes from 8 hpi on (data not shown). The application of rutin had the opposite effect on *L. saligna*, where it intensified *B. lactucae* penetration 24 hpi (Fig. 3). The frequency of primary vesicles was increased in PTIO (a specific NO scavenger)-treated tissues, in all genotypes from 24 hpi (Fig. 3). The NOS inhibitor L-NAME and the NR inhibitor sodium tungstate had no significant effect on *B. lactucae* germination and primary vesicle development (Figs. 2 and 3).

Development of secondary vesicles was not affected by PTIO; all other compounds delayed growth and decreased frequency of secondary vesicles, especially in *L. sativa* genotypes (Fig. 4). The strongest effect was recorded for rutin and SNP both 24 and 48 hpi. Additionally, variation in expression of HR (Fig. 1e, f) was evaluated by light microscopy in all samples. Significant differences were recorded only for *L. virosa* (NVRS 10.001 602) 48 hpi, as presented in Fig. 5. The proportion of infection sites with hypersensitively responding cells

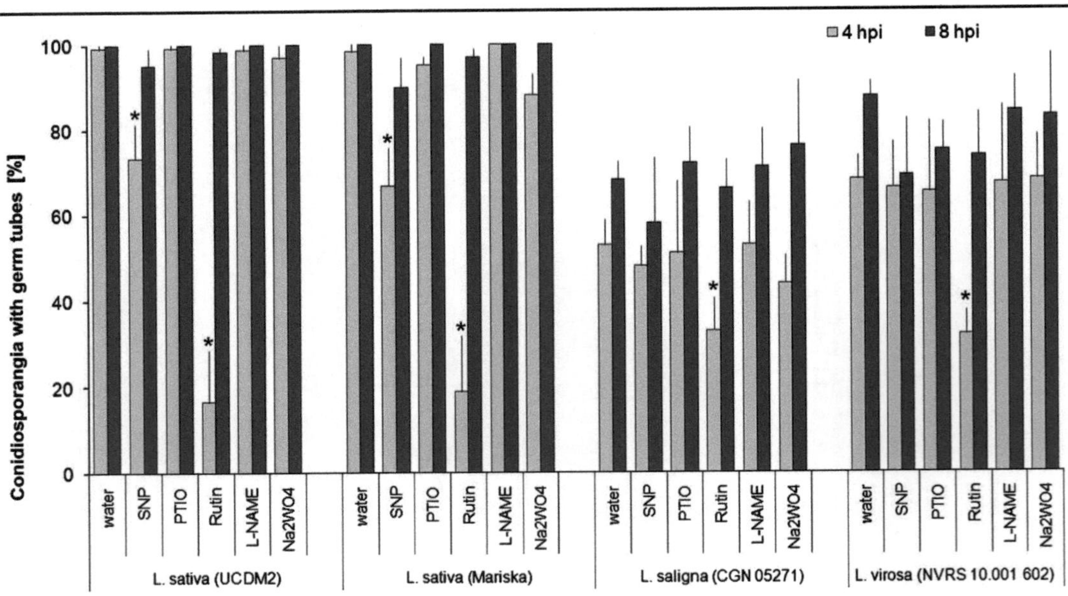

Fig. 2 Germination of *B. lactucae* (race BL16) conidiosporangia on leaf discs of *Lactuca sativa* (UCDM2, and Mariska), *L. saligna* (CGN 05271) and *L. virosa* (NVRS 10.001 602) incubated on distilled water (control) and solutions of substances modulating RNOS level (SNP, PTIO, rutin, L-NAME and sodium tungstate) 4 and 8 hpi. Data are given as mean ± SD, $n=15$

(Fig. 5a) was significantly lowered by application of SNP, PTIO, L-NAME and strongest by rutin. The average number of cells per HR (Fig. 5b) was increased by the enzyme inhibitors.

Histochemical localization of NO in inoculated leaves

Localization of nitric oxide within tissues of *Lactuca* spp. and infection structures of *Bremia lactucae* (race

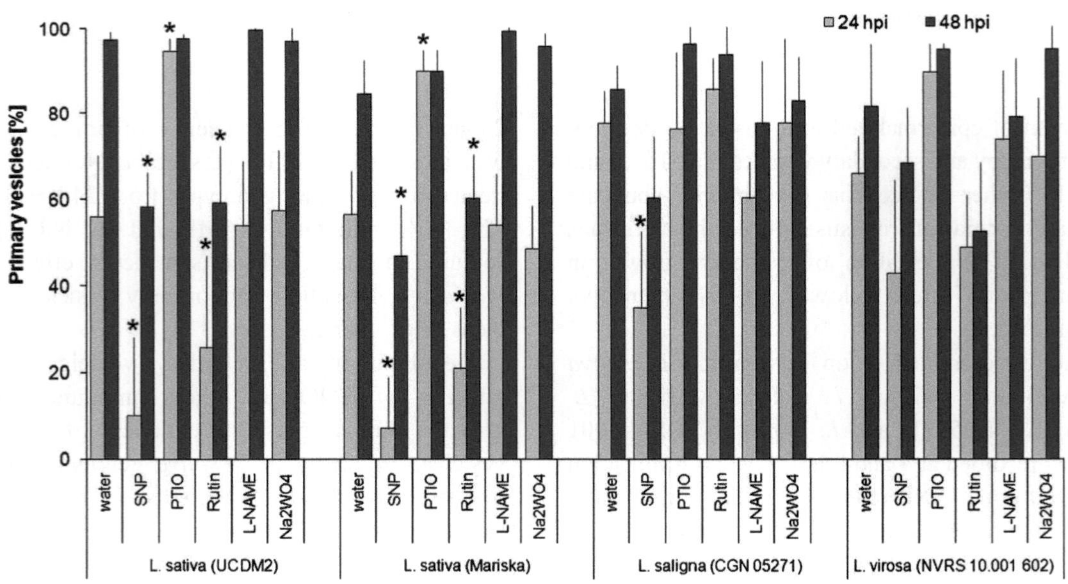

Fig. 3 Penetration rate of *B. lactucae* (race BL16), expressed as percentage of epidermal cells in contact with pathogen that contained primary vesicles. Leaf discs of *Lactuca sativa* (UCDM2, and Mariska), *L. saligna* (CGN 05271) and *L. virosa* (NVRS 10.001 602) incubated on distilled water (control) and solutions of substances modulating RNOS level (SNP, PTIO, rutin, L-NAME and sodium tungstate) are compared 24 and 48 hpi. Data are given as mean ± SD, $n=15$

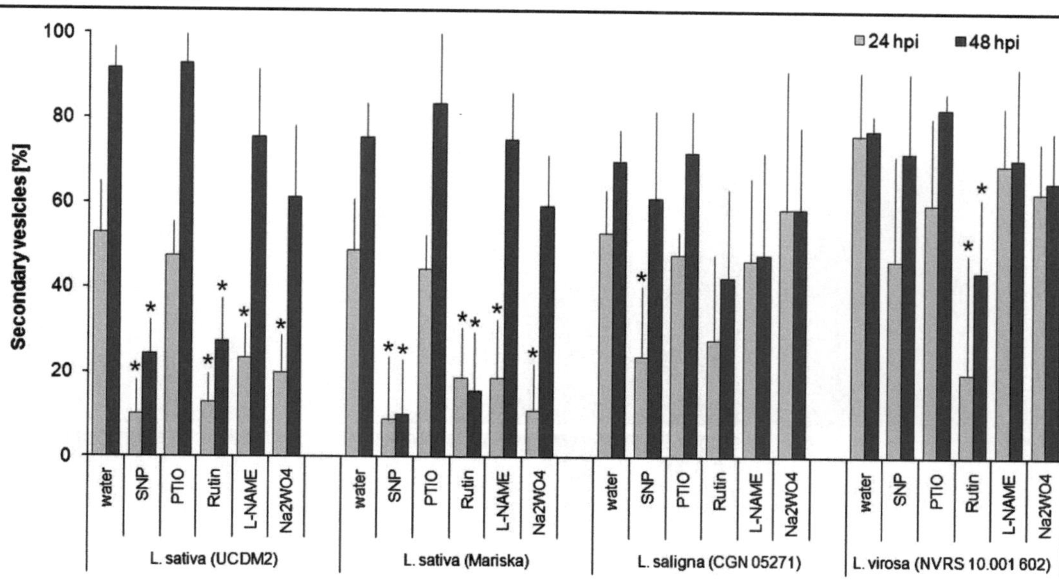

Fig. 4 Proportion of epidermal cells with developed secondary vesicle of *B. lactucae* (race BL16). Data recorded for leaf discs of *Lactuca sativa* (UCDM2, and Mariska), *L. saligna* (CGN 05271) and *L. virosa* (NVRS 10.001 602) incubated on distilled water (control) and solutions of substances modulating RNOS level (SNP, PTIO, rutin, L-NAME and sodium tungstate) are compared 24 and 48 hpi. Data are given as mean ± SD, n=15

BL16) was studied by confocal laser scanning microscopy following staining with the fluorescent probe DAF-FM DA (Fig. 6). Production of NO was observed in the infection structures of *B. lactucae* grown both on susceptible and resistant genotypes, although these differed in timing of the pathogen development. An intensive signal in the tip of the germ tube and appressorium (Fig. 6a) indicates an essential role of the molecule in the process of pathogen infection structures growth and penetration. A weaker NO signal was detected also in developing primary and secondary vesicles (Fig. 6f), intracellular hyphae and in haustoria on susceptible *L. sativa* (UCDM2) (Fig. 6d). In *L. virosa* (NVRS 10.001 602), a genotype with an abundant hypersensitive response, NO was frequently localized in penetrated cells undergoing HR before occurrence of detectable necrosis (Fig. 6b, c).

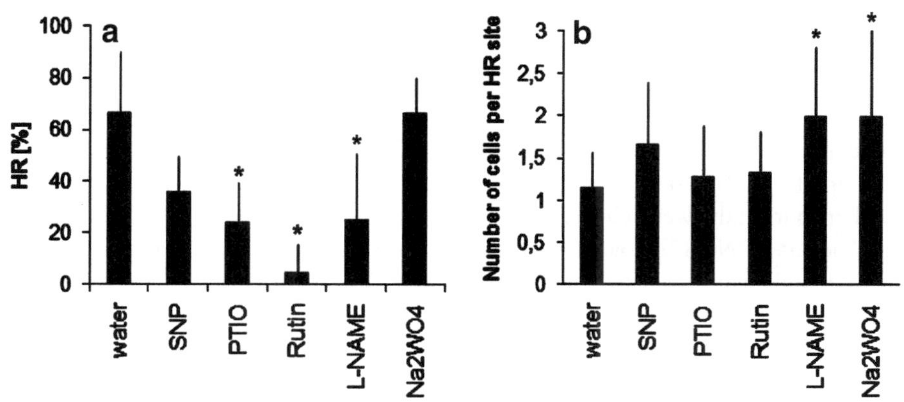

Fig. 5 *L. virosa* (NVRS 10.001 602)-*B. lactucae* (race BL16) interaction 48 hpi: **a** proportion of infection sites with hypersensitive response of epidermal cells; **b** average number of cells per HR. Leaf discs were incubated on distilled water (control) and solutions of substances modulating RNOS level (SNP, PTIO, rutin, L-NAME and sodium tungstate). Data are given as mean ± SD, n=45

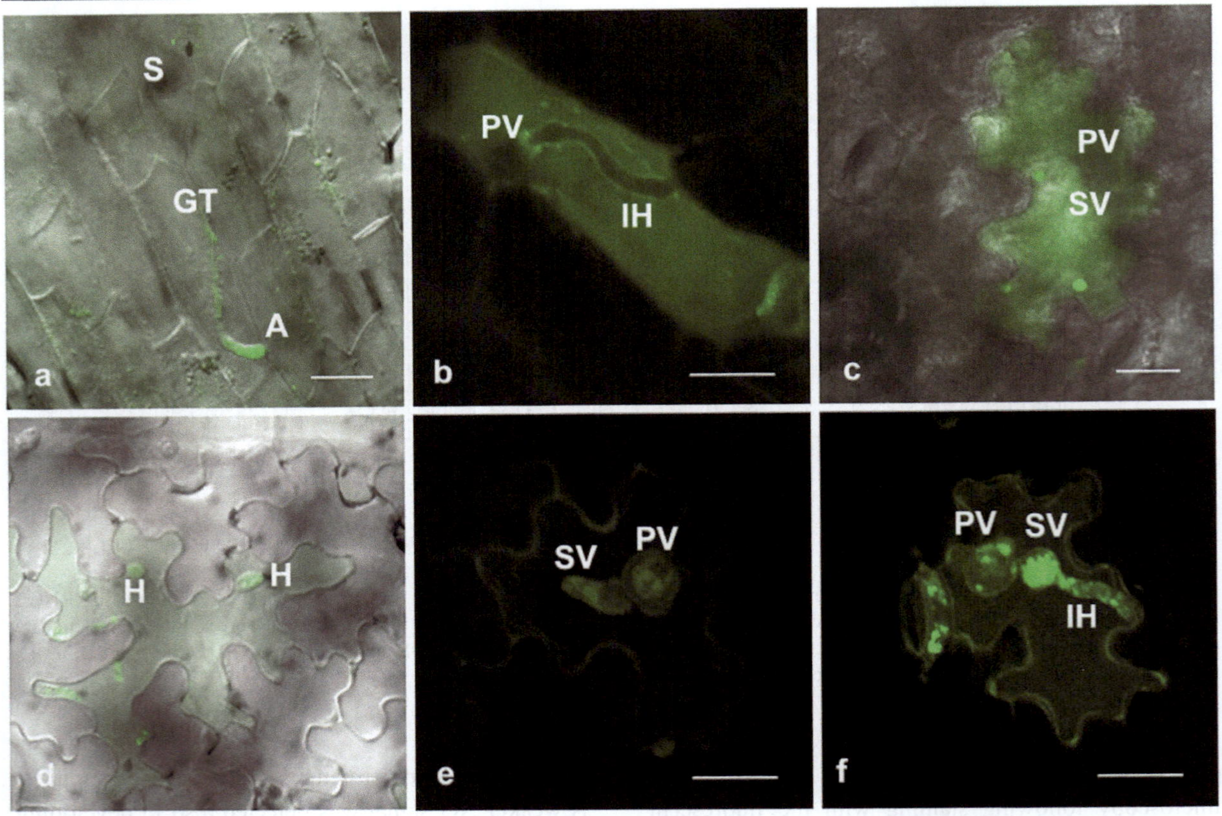

Fig. 6 Nitric oxide production in lettuce downy mildew and challenged host epidermal cells visualized by confocal laser scanning microscopy. **a** During germination the signal for NO was detected in *B. lactucae* germ tubes and appressoria, *L. virosa* (NVRS 10.001 602) 24 hpi. **b, c** In resistant genotypes NO accumulates in pre-necrotic cells 24 hpi: **b,** *L. saligna* (CGN 05271); **c,** *L. virosa* (NVRS 10.001 602). **d** Interestingly, signal for NO was found in epidermal cells of susceptible *L. sativa* (UCDM2) stressed by formation of pathogen haustoria 168 hpi. **e** At late stages of infection the autofluorescence of infection structures within HR co-localized with **f,** a signal of NO-sensitive fluorescent dye DAF-FM DA, *L. virosa* (NVRS 10.001 602) 168 hpi. Infection structures: spore (*S*), germ tube (*GT*), appressorium (*A*), primary vesicle (*PV*), secondary vesicle (*SV*), intracellular hypha (*IH*), haustorium (*H*). Bar corresponds to 20 μm

Changes of NOS-like activity in leaves inoculated by *B. lactucae*

NOS-like activity was determined as L-arginine and NADPH-dependent NO production by the oxyhemoglobine method in extracts of 4th–6th youngest plant leaves during 216 hpi (Fig. 7). Four *Lactuca* spp. genotypes involved in our study were characterized by differential basal levels of NOS-like activity. The highest activity was found in non-infected plants of the resistant genotypes of *L. virosa* (NVRS 10.001 602) and *L. saligna* (CGN 05271), whereas almost no activity was detected in *L. sativa* genotypes 0 hpi (Fig. 7). Following inoculation, a rapid increase in NOS-like activity was recorded in *L. virosa* (NVRS 10.001 602), to 300% of constitutive level at 4 and 8 hpi. The activity rapidly decreased from 24 hpi to the level of uninoculated plants. The resistant *L. sativa* (Mariska) displayed an increased activity of NOS-like enzyme upon inoculation with a maximum between 48 and 168 hpi. On the other hand, the NOS-like activity in resistant *L. saligna* (CGN 05271) with reduced hypersensitivity moderately increased at 4 hpi but was lower again at later stages of the infection process. Minor fluctuation of NOS-like activity characterized also susceptible *L. sativa* (UCDM2), with moderate increase between 4–24 hpi and 72–216 hpi.

Rutin and quercetin content in leaves

Contents of rutin and its aglycone form, flavonol quercetin, were determined by HPLC analysis in leaf extracts of *Lactuca* spp. plants during 216 h following

Fig. 7 Changes in NOS-like activity in leaves of *Lactuca* spp. plants infected with *B. lactucae* (BL 16) during 216 hpi. Data are given as mean ± SD, n=9

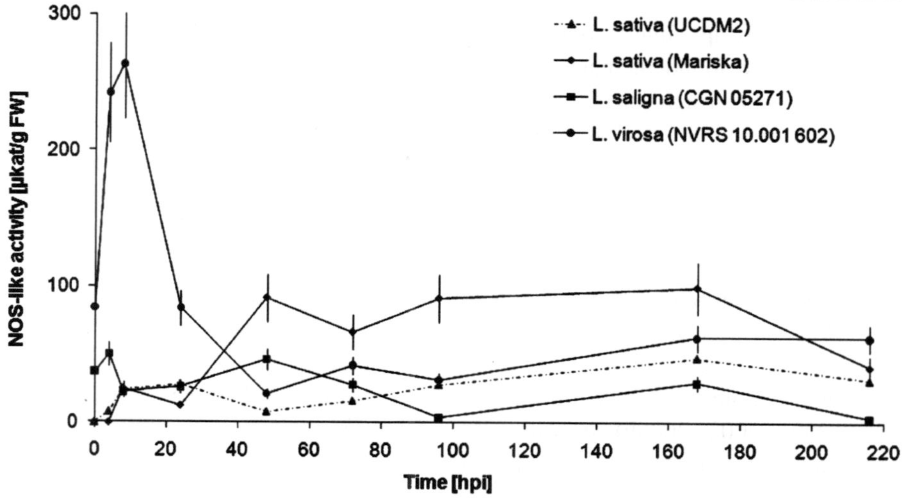

inoculation by *B. lactucae*. The highest level of rutin and quercetin, 2.4 and 7.5 mmol/g FW, respectively, was detected in leaf tissues of *L. virosa* (Fig. 8). Rutin content decreased 24 hpi, however, increased during 48–216 hpi. Susceptible *L. sativa* (UCDM2) was characterized by rather low constitutive level of both compounds which increased at late stages of infection, 216 hpi. During pathogenesis of *B. lactucae* on resistant *L. sativa* (Mariska) the content of both substances gradually decreased to approx. 50%

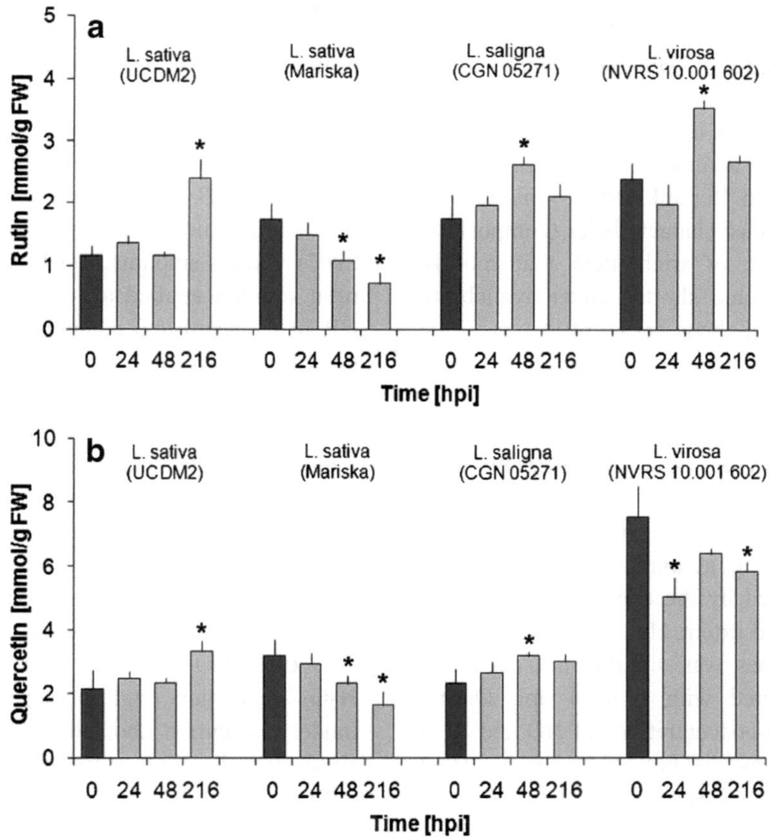

Fig. 8 Content of rutin (**a**) and quercetin (**b**) in leaf extracts from plants infected with *B. lactucae* (BL 16) (grey column) during 216 hpi. Controls are non-infected plants 0 hpi (black column). Data are given as mean ± SD, n=9

amount at 216 hpi. Intermediate levels of rutin and quercetin were found in intact tissues of *L. saligna* (CGN 05271) which increased in infected plants starting at 24 hpi and peaking at 48 hpi (Fig. 8).

Discussion

Histological and biochemical features of the *Lactuca* spp.-*Bremia lactucae* interaction, a model plant-biotrophic oomycete pathosystems, have been studied intensively in our laboratory for two decades (Lebeda et al. 2001, 2002, 2008b). It is a well accepted fact that spore germination of the pathogen isolates reaches the highest rate on genotypes of the original host plant species (e.g. Sedlářová et al. 2001). Thus, in the present study almost 100% of the conidiosporangia of *B. lactucae* (race BL16) germinated on lettuce leaf tissues (cvs. UCDM2 and Mariska), whereas on wild *Lactuca* spp. plants, *L. saligna* (CGN 05271) and *L. virosa* (NVRS 10.001 602), the number was reduced to approx. 60%. *B. lactucae* germ tubes also had a reduced length on the wild *Lactuca* spp. compared to *L. sativa* genotypes and the interspecific breeding lines (Lebeda and Pink 1998), and significant differences in the length of germ tubes between seedlings and adult plants of *L. sativa* were described (Lebeda and Reinink 1991). This can be explained by leaf surface characteristics (composition of waxes, abundance of trichomes) that modify attachment of spores and shorten water availability for their germination (Lebeda et al. 2001).

Although oxidative processes are considered to be vital for plant resistance, including *Lactuca* spp. (reviewed in Lebeda et al. 2008b), the importance of NO and ROS for oomycete pathogenesis has not been reported in detail yet. In previous work (Petřivalský et al. 2007) we found that the processes localized outside of the leaves, e.g. germination and appressoria formation, are affected by the compounds infiltrated into leaves to a limited extent. Here, we report that the number of germ tubes was significantly decreased only in tissues treated with rutin 4 hpi. During penetration, the local concentrations of NO and ROS change due to a release by both pathogen infection structures and host cells. Application of a NO donor or a strong antioxidant (rutin) reduced the development of primary as well as secondary vesicles. The only exception was an increase of the downy mildew penetration rate in *L. saligna* upon rutin treatment. On the other hand, lower NO concentrations after PTIO application led to an increase in frequency of primary vesicles.

The response of the individual lettuce genotypes to *B. lactucae* (race BL16) infection was reported in a previous study (Sedlářová et al. 2001), where initiation of the HR was shown to correlate with pathogen developmental stages rather than the time-scale of pathogenesis. Here we show that application of NO level modulators mainly altered the response of plant genotypes with a pronounced HR response, i.e. *L. virosa* (NVRS 10.001 602). NO level modulators lowered the proportion of infection sites with HR but had only a minor effect on the extent of the HR (Fig. 5). The synergistic action of NO and H_2O_2 is believed to orchestrate the HR and restrict pathogen invasion (Zaninotto et al. 2006), although the interactions between NO and ROS are still a matter of discussion (Delledonne et al. 2003; Wilson et al. 2008). Interaction of NO and H_2O_2 leads to the production of either highly reactive singlet oxygen or hydroxyl radicals, both of which can induce the processes of cell death. Reaction of NO with O_2^- leads to the formation of a strong oxidant compound, peroxynitrite (ONOO–; Saito et al. 2006). Accumulation of superoxide has not been recorded in the *Lactuca* spp.-*B. lactucae* interactions, however, H_2O_2 production colocalizes with NO (Sedlářová et al. 2007). Excess amounts of antioxidant, in the form of rutin solution applied to leaf tissues, diminished the HR in *L. virosa* and concurrently retarded pathogen growth. A spectrum of methodological approaches to study NO is available, although the specificity of some is not yet clear (Miller and Chang 2007; Vandelle and Delledonne 2008). We localized NO in the infected cells of resistant *Lactuca* spp. before expression of symptoms as visualized by DAF FM-DA staining (Fig. 6b–d) which is in agreement with previous experiments based on application of the most sensitive assays (reviewed by Mur et al. 2006). This early NO burst has been reported to precede H_2O_2 generation, and in a plant-powdery mildew model it occurred about 6 h before the first visible signs of cell death (Mur et al. 2006). There are indications that NO can affect the generation of ROS, which are closely associated with the HR in plants. Zeier et al. (2004) demonstrated that the pathogen-challenged nitric oxide dioxygenase transgenic lines of *Arabidopsis* exhibited not only reduced NO generation, but also a reduced oxidative burst (decreased H_2O_2 levels) and a blockage of phenylalanine ammonia lyase expression in incompatible interactions with *Pseudomonas syringae*. Interestingly, NO can influence apoptosis at many points and can be either pro- or anti-apoptotic. In plants, there have also been reports of NO suppressing cell death (Bethke et al. 2004). The relative concentrations of particular ROS and NO appear to be vital in the initiation or suppression in cell death. Our results confirm the need for NO in plant race-specific defence to oomycetes. However, no activation of host NO production has been found in interactions between animals and pathogenic oomycetes (Phillips et al. 2008).

Recently, a special emphasis has been put on approaches taking into account the life strategy of pathogens, because huge differences exist in pathogen biology, physiology, toxicology and the reactions of host plants (Glazebrook 2005). Nitric oxide was shown to participate not only in plant resistance to biotrophic pathogens but also to necrotrophs. NO and oxidative bursts were induced during compatible interaction of *Nicotiana benthamiana* with the necrotrophic fungal pathogen *Botrytis cinerea*. NO was shown to play a pivotal role also in basal defence against *B. cinerea* and *PR-1* gene expression in *N. benthamiana*. By contrast, ROS had a negative function in resistance, i.e. had a positive role in expansion of disease lesions during the *B. cinerea–N. benthamiana* interaction (Asai and Yoshioka 2009).

Both partners in plant-fungus interactions have been shown to produce NO (Mur et al. 2006; Prats et al. 2008), thus, we expected a similar phenomenon in oomycete pathogenesis. Our experiments showed a strong signal for NO in the apical parts of *B. lactucae* mycelium, tips of germ tubes and appressoria, which is prerequisite for penetration (Fig. 6a). In compatible interactions NO was detected also in haustoria (Fig. 6d) which indicates an involvement of NO in pathogen development. Sources of NO in oomycetes have not been studied but the presence of an enzyme similar to NOS is highly probable. Nevertheless, one methodological aspect of our work may be a matter of discussion. We are aware that we measured NOS-like activity in extracts of infected leaves covering the enzyme activity from both organisms. However, *B. lactucae* is a biotrophic endoparasite which makes its cultivation or detachment from infected tissues impossible (Lebeda et al. 2008b). Thus, we presume that application of constant inoculum concentration and standard cultivation conditions diminished the variation among genotypes. On the other hand, the influence of pathogen-derived enzyme activity might increase with increasing pathogen biomass during infection, especially in the susceptible genotype *L. sativa* (UCDM2) or the resistant *L. saligna* (CGN 05271), where *B. lactucae* (BL16) is arrested at later stages of development (Sedlářová et al. 2001). However, tissue and cell responses of non-host resistance of *L. saligna* accessions (Lebeda et al. 2001, 2002) are different from race-specific and field resistance of *L. sativa* (Lebeda and Reinink 1994). In *L. saligna* (CGN 05271) at 48 hpi significantly lower frequencies of hyphae and haustorium formation were observed, however, higher a frequency of HR in infection sites, in comparison with susceptible *L. sativa* (Lebeda and Reinink 1994). Nevertheless, there is also a large variation in formation and development of infection structures and tissue response in different accessions of *L. saligna* (Lebeda and Pink 1998; Lebeda and Reinink 1994; Lebeda et al. 2006) showing delicate differences in metabolic pathways (including ROS and RNS metabolism) in individual host genotype-pathogen race interactions (Sedlářová et al. 2007), primarily due to the genetic background of these interactions (Zhang et al. 2009).

Our results suggest that plants use multiple ways for the synthesis of NO, the critical molecule in plant-pathogen communication. Nitric oxide (NO) is a highly diffusible and short-lived physiological messenger. Despite its diffusible nature, NO modifies thiol groups of specific cysteine residues in target proteins and alters protein function via S-nitrosylation. Tyrosine nitration of proteins by NO was recently reported to be increased in susceptible but not in the resistant cultivar of sunflower following downy mildew (*Plasmopara halstedii*) infection. Therefore, post-translational protein modifications were proposed as a marker of nitrosative stress in the plant–pathogen interaction (Chaki et al. 2009).

External application of rutin had a striking impact on *B. lactucae* development so we analyzed the *in planta* concentrations among *Lactuca* spp. genotypes. Rutin, a plant flavonoid glycoside with antioxidant properties utilized in human medicine, is composed of flavonol quercetin and the disaccharide rutinose.

Rutin helps to maintain levels of the biological antioxidant reduced glutathione. Importantly, following nitrosation, both molecules may become pro-oxidants (Laughton et al. 1989). Only minor increases in levels of both rutin and quercetin in *L. saligna* tissues were recorded. Generally, there is no need to overstimulate the antioxidant machinery in this genotype characterized by reduction of downy mildew without large expression of HR (Lebeda and Pink 1998; Lebeda and Reinink 1994; Sedlářová et al. 2001). On the other hand, *L. virosa* is generally showing a high proportion of infection sites with HR (Lebeda and Pink 1998; Lebeda et al. 2006), expressed also phenotypically as tissue necrosis (Lebeda et al. 2002; Norwood et al. 1981). Tissues of this species are rich in both rutin and quercetin as there is a necessity to protect tissue surrounding HR cells. Higher levels of both antioxidants were found in the resistant cv. Mariska that decreased during the course of downy mildew infection. In the susceptible cv. UCDM2, an increase was found very late after inoculation (216 hpi), probably induced in order to buffer oxidative stress linked to *B. lactucae* intercellular growth. Our results show evidence for a dual (dose-dependent and balanced) role of NO and ROS in this host-pathogen interaction, that can be summarized as follows. In *L. virosa* (NVRS 10.001 602) high levels of rutin together with phenolic acids (Grúz and Sedlářová, unpublished), and peroxidase activity (Sedlářová et al. 2007) are important components of resistance mechanisms that buffer oxidative molecules released in relation to: 1) *B. lactucae* penetration of epidermal cells to limit formation of infection structures; 2) the onset of HR following *B. lactucae* recognition in the cytoplasm. Intensive hydrogen peroxide release in *L. virosa* leads to HR which often involves more than one cell from not only epidermal but also mesophyll tissue, in comparison to other resistant *Lactuca* spp. genotypes (Lebeda and Pink 1998; Lebeda et al. 2006; Sedlářová et al. 2001). On the other hand, still little is known about the physiology and biochemistry of resistance mechanisms of *L. saligna* to downy mildew, which is primarily characterized by limited HR (Lebeda and Reinink 1994; Lebeda et al. 2001, 2002) and minor changes in all of parameters investigated so far (Sedlářová et al. 2007; this paper). However, recent genetic investigations suggest that non-host resistance in *L. saligna* is the results of cumulative effects of many QTLs (at least 15) differentially operating at various developmental stages (Zhang et al. 2009). This is also supporting our recent biochemical data and their variation.

In many plants, resistance has been associated with a higher activity of antioxidant enzymes (peroxidase, catalase, superoxiddismutase) or antioxidant substances (Bolwell and Daudi 2009). However, there is no such direct link between any of the above-mentioned traits and susceptibility/resistance in *Lactuca* spp.-*B. lactucae*. Lettuce genotypes differ from other wild relative *Lactuca* spp. in many physiological features (Lebeda et al. 2008b). These differences demand further detailed study of specific host-pathogen interactions and various developmental stages. HR-based resistance in *L. virosa* as well as non-HR resistance in *L. saligna* are promising alternatives to be utilized in lettuce resistance breeding (Lebeda et al. 2009).

Acknowledgements Authors thank Mrs. Danuše Krátká and M.A. Iveta Hnízdová for proceeding of light microscopy samples, and Dr. Brigitte Mauch-Mani (Université de Neuchâtel, Switzerland) for critical reading of the manuscript. Provision of *Lactuca virosa* seeds by Prof. D. A. C. Pink (Warwick HRI, Warwick University, U.K.) as well as cooperation with Olympus Czech Group (Prague, Czech Republic) is gratefully acknowledged. The work was financed by the Czech Ministry of Education, Youth and Sports (grants No. MSM 6198959215 and 2E08018).

References

Arasimowicz, M., & Floryszak-Wieczorek, J. (2007). Nitric oxide as a bioactive signalling molecule in plant stress responses. *Plant Science, 172*, 876–887.

Asai, S., & Yoshioka, H. (2009). Nitric oxide as a partner of reactive oxygen species participates in disease resistance to necrotrophic pathogen *Botrytis cinerea* in *Nicotiana benthamiana*. *Molecular Plant-Microbe Interactions, 22*, 619–629.

Bethke, P. C., Badger, M. R., & Jones, R. L. (2004). Apoplastic synthesis of nitric oxide by plant tissues. *Plant Cell, 16*, 332–341.

Blokhina, O., Virolainen, E., & Fagerstedt, K. V. (2003). Antioxidants, oxidative damage and oxygen deprivation stress: a review. *Annals of Botany, 91*, 179–194.

Bolwell, G. P., & Daudi, A. (2009). Reactive oxygen species in plant–pathogen interactions. In L. A. del Río & A. Puppo (Eds.), *Reactive oxygen species in plant signalling* (pp. 113–133). Berlin: Springer-Verlag.

Chaki, M., Fernández-Ocaña, A. M., Valderrama, R., Carreras, A., Esteban, F. J., Luque, F., et al. (2009). Involvement of reactive nitrogen and oxygen species (RNS and ROS) in sunflower–mildew interaction. *Plant and Cell Physiology, 50*, 665–679.

Cooney, R. V., Harwood, P. J., Custer, L. J., & Franke, A. A. (1994). Light-mediated conversion of nitrogen dioxide to nitric oxide by carotenoids. *Environmental Health Perspectives, 102*, 460–462.

Corpas, F. J., Hayashi, M., Mano, S., Nishimura, M., & Barroso, J. B. (2009). Peroxisomes are required for *in vivo* nitric oxide (NO) accumulation in the cytosol following salinity stress of *Arabidopsis* plants. *Plant Physiology*. doi:10.1104/pp.109.146100.

De Gara, L., de Pinto, M. C., & Tommasi, F. (2003). The antioxidant systems vis-à-vis reactive oxygen species during plant–pathogen interaction. *Plant Physiology and Biochemistry, 41*, 863–870.

Delledonne, M., Zeier, J., Marocco, A., & Lamb, C. (2001). Signal interactions between nitric oxide and reactive oxygen intermediates in the plant hypersensitive disease resistance response. *Proceedings of the National Academy of Sciences of the United States of America, 98*, 13454–13459.

Delledonne, M., Murgia, I., Ederle, D., Sbicego, P., Biondani, A., Polverari, A., et al. (2002). Reactive oxygen intermediates modulate nitric oxide signalling in the plant hypersensitive disease-resistance response. *Plant Physiology and Biochemistry, 40*, 605–610.

Delledonne, M., Polverari, A., & Murgia, I. (2003). The functions of nitric oxide-mediated signalling and changes in gene expression during the hypersensitive response. *Antioxidants and Redox Signalling, 5*, 33–41.

Glazebrook, J. (2005). Contrasting mechanisms of defense against biotrophic and necrotrophic pathogens. *Annual Review of Phytopathology, 43*, 205–227.

Grace, S. C. (2005). Phenolics as antioxidants. In N. Smirnoff (Ed.), *Antioxidants and reactive oxygen species in plants* (pp. 141–168). Oxford: Blackwell Scientific.

Hevel, J. M., & Marletta, M. A. (1994). Nitric-oxide synthase assays. *Methods in Enzymology, 233*, 250–258.

Laughton, M. J., Halliwell, B., Evans, P. J., Robin, J., & Hoult, S. (1989). Antioxidant and pro-oxidant actions of the plant phenolics quercetin, gossypol and myricetin. Effects on lipid peroxidation, hydroxyl radical generation and bleomycin-dependent damage to DNA. *Biochemical Pharmacology, 38*, 2859–2865.

Lebeda, A., & Pink, D. A. C. (1998). Histological aspects of the response of wild *Lactuca* spp. and their hybrids, with *L. sativa* to lettuce downy mildew (*Bremia lactucae*). *Plant Pathology, 47*, 723–736.

Lebeda, A., & Reinink, K. (1991). Variation in the early development of *Bremia lactucae* on lettuce cultivars with different levels of field resistance. *Plant Pathology, 40*, 232–237.

Lebeda, A., & Reinink, K. (1994). Histological characterization of resistance in *Lactuca saligna* to lettuce downy mildew (*Bremia lactucae*). *Physiological and Molecular Plant Pathology, 44*, 125–139.

Lebeda, A., Pink, D. A. C., & Mieslerová, B. (2001). Host-parasite specificity and defense variability in the *Lactuca* spp.-*Bremia lactucae* pathosystem. *Journal of Plant Pathology, 83*, 25–35.

Lebeda, A., Pink, D. A. C., & Astley, D. (2002). Aspects of the interactions between wild *Lactuca* spp. and related genera and lettuce downy mildew (*Bremia lactucae*). In P. T. N. Spencer-Phillips, U. Gisi, & A. Lebeda (Eds.), *Advances in downy mildew research* (pp. 85–117). Dordrecht: Kluwer Academic.

Lebeda, A., Sedlářová, M., Lynn, J., & Pink, D. A. C. (2006). Phenotypic and histological expression of different genetic backgrounds in interactions between lettuce, wild *Lactuca* spp., *L. sativa* × *L. serriola* hybrids and *Bremia lactucae*. *European Journal of Plant Pathology, 115*, 431–441.

Lebeda, A., Petrželová, I., & Maryška, Z. (2008a). Structure and variation in the wild-plant pathosystem: *Lactuca serriola-Bremia lactucae*. *European Journal of Plant Pathology, 122*, 127–146.

Lebeda, A., Sedlářová, M., Petřivalský, M., & Prokopová, J. (2008b). Diversity of defence mechanisms in plant-oomycete interactions: a case study of *Lactuca* spp.-*Bremia lactucae*. *European Journal of Plant Pathology, 122*, 71–89.

Lebeda, A., Doležalová, I., Křístková, E., Kitner, M., Petrželová, I., Mieslerová, B., et al. (2009). Wild *Lactuca* germplasm for lettuce breeding: current status, gaps and challenges. *Euphytica, 170*, 15–34.

Miller, E. W., & Chang, C. J. (2007). Fluorescent probes for nitric oxide and hydrogen peroxide in cell signalling. *Current Opinion in Chemical Biology, 11*, 620–625.

Mur, L. A. J., Carver, T. L., & Prats, E. (2006). NO way to live; the various roles of nitric oxide in plant-pathogen interactions. *Journal of Experimental Botany, 57*, 489–505.

Mur, L. A. J., Kenton, P., Lloyd, A. J., Ougham, H., & Prats, E. (2008). The hypersensitive response; the centenary is upon us but how much do we know? *Journal of Experimental Botany, 59*, 501–520.

Norwood, J. M., Crute, I. R., & Lebeda, A. (1981). The location and characteristics of novel sources of resistance to *Bremia lactucae* Regel (Downy mildew) in wild *Lactuca* L. species. *Euphytica, 30*, 659–668.

Panstruga, R. (2003). Establishing compatibility between plants and obligate biotrophic pathogens. *Current Opinion in Plant Biology, 6*, 320–326.

Petřivalský, M., Kočířová, J., Sedlářová, M., Piterková, J., Luhová, L., & Lebeda, A. (2007). On the role of nitric oxide in *Bremia lactucae* pathogenesis on *Lactuca sativa*. In A. Lebeda & P. T. N. Spencer-Phillips (Eds.), *Advances in downy mildew research* (Vol. 3, pp. 175–184). Olomouc: Palacký University in Olomouc and JOLA.

Phillips, A. J., Anderson, V. L., Robertson, E. J., Secombes, C. J., & van West, P. (2008). New insights into animal pathogenic oomycetes. *Trends in Microbiology, 16*, 13–19.

Piterková, J., Petřivalský, M., Luhová, L., Mieslerová, B., Sedlářová, M., & Lebeda, A. (2009). Local and systemic production of nitric oxide in tomato responses to powdery mildew infection. *Molecular Plant Pathology, 10*, 501–513.

Prats, E., Carver, T. L. W., & Mur, L. A. J. (2008). Pathogen-derived nitric oxide influences formation of the appressorium infection structure in the phytopathogenic fungus *Blumeria graminis*. *Research in Microbiology, 159*, 476–480.

Saito, S., Yamamoto-Katou, A., Yoshioka, H., Doke, N., & Kawakita, K. (2006). Peroxynitrite generation and tyrosine nitration in defense responses in tobacco BY-2 cells. *Plant Cell Physiology, 47*, 689–697.

Sedlářová, M., Lebeda, A., & Pink, D. A. C. (2001). The early stages of interaction between effective and non-effective race-specific genes in *Lactuca sativa*, wild *Lactuca* spp. and *Bremia lactucae* (race NL16). *Journal of Plant Diseases and Protection, 108*, 477–489.

Sedlářová, M., Luhová, L., Petřivalský, M., & Lebeda, A. (2007). Localization and metabolism of reactive oxygen species during *Bremia lactucae* pathogenesis in *Lactuca sativa* and wild *Lactuca* spp. *Plant Physiology and Biochemistry, 45*, 607–616.

Shetty, N. P., Jørgensen, H. J. L., Jensen, J. D., Collinge, D. B., & Shetty, H. S. (2008). Roles of reactive oxygen species in interactions between plants and pathogens. *European Journal of Plant Pathology, 121*, 267–280.

Stöhr, C., & Ullrich, W. R. (2002). Generation and possible roles of NO in plant roots and their apoplastic space. *Journal of Experimental Botany, 53*, 2293–2303.

Tománková, K., Luhová, L., Petřivalský, M., Peč, P., & Lebeda, A. (2006). Biochemical aspects of reactive oxygen species formation in interaction *Lycopersicon* spp.-*Oidium neolycopersici*. *Physiological and Molecular Plant Pathology, 68*, 22–32.

Vandelle, E., & Delledonne, M. (2008). Methods for nitric oxide detection during plant–pathogen interactions. *Methods in Enzymology, 437*, 573–592.

Wendehenne, D., Durner, J., & Klessig, D. F. (2004). Nitric oxide: a new player in plant signalling and defence responses. *Current Opinion in Plant Biology, 7*, 449–455.

Wilson, I. D., Neill, S. J., & Hancock, J. T. (2008). Nitric oxide synthesis and signalling in plants. *Plant, Cell and Environment, 31*, 622–631.

Yamasaki, H. (2000). Nitrite-dependent nitric oxide production pathway: implications for involvement of active nitrogen species in photoinhibition *in vivo*. *Philosophical Transactions of the Royal Society B: Biological Sciences, 355*, 1477–1488.

Yoshioka, H., Asai, S., Yoshioka, M., & Kobayashi, M. (2009). Molecular mechanisms of generation for nitric oxide and reactive oxygen species, and role of the radical burst in plant immunity. *Molecules and Cells, 28*, 321–329.

Zaninotto, F., La Camera, S., Polverari, A., & Delledonne, M. (2006). Cross talk between reactive nitrogen and oxygen species during the hypersensitive disease resistance response. *Plant Physiology, 141*, 379–383.

Zeier, J., Delledonne, M., Mishina, T., Severi, E., Sonoda, M., & Lamb, C. (2004). Genetic elucidation of nitric oxide signalling in incompatible plant-pathogen interactions. *Plant Physiology, 136*, 2875–2888.

Zhang, N. W., Lindhout, P., Niks, R. E., & Jeuken, M. J. W. (2009). Genetic dissection of *Lactuca saligna* nonhost resistance to downy mildew at various lettuce developmental stages. *Plant Pathology, 58*, 923–932.

Eur J Plant Pathol (2011) 129:149–169
DOI 10.1007/s10658-010-9676-z

Identification of genes expressed during the compatible interaction of grapevine with *Plasmopara viticola* through suppression subtractive hybridization (SSH)

Guillaume Legay · Elaheh Marouf · Dave Berger · Jean-Marc Neuhaus · Brigitte Mauch-Mani · Ana Slaughter

Accepted: 23 August 2010 / Published online: 10 September 2010

Abstract Grapevine (*Vitis vinifera*) is the most widely cultivated and economically important fruit crop, but is susceptible to a large number of diseases. Downy mildew, caused by the obligate biotrophic oomycete pathogen *Plasmopara viticola,* is a common disease present in all regions where vines are cultivated. We used suppression subtractive hybridization (SSH) to generate two cDNA libraries enriched for transcripts induced and repressed, respectively, in the susceptible grapevine cultivar Chasselas 24 h after inoculation with *P. viticola*. Differential screening on glass slide microarrays yielded over 800 putative genes that were up-regulated in response to *P. viticola* infection and over 200 that were down-regulated. One hundred and ninety four of these, were sequenced, identified and functionally categorised. Transcript abundance of twelve genes over a 48 h time course was examined by reverse transcriptase quantitative real-time PCR (RT-qPCR). Ten of these genes were induced/enhanced by *P. viticola* challenge, confirming the results of the SSH. The vast majority of the genes identified are related to defence. Interestingly, many genes involved in photosynthesis were down-regulated.

Keywords Compatible interaction · Reverse transcriptase quantitative real-time PCR · *Plasmopara viticola* · Suppression subtractive hybridization · *Vitis vinifera*

G. Legay · E. Marouf · J.-M. Neuhaus ·
B. Mauch-Mani (✉) · A. Slaughter
Laboratory of Molecular and Cellular Biology,
University of Neuchâtel,
Rue Emile-Argand 11, Case Postal 158, 2009 Neuchâtel,
Switzerland
e-mail: Brigitte.mauch@unine.ch

D. Berger
Department of Plant Science, Forestry and Agricultural Biotechnology Institute (FABI), University of Pretoria,
Lunnon Road,
Pretoria 0002, South Africa

Introduction

Downy mildew caused by the obligate biotrophic pathogen *Plasmopara viticola*, is an economically important disease of grapevine. The pathogen attacks all green parts of the grapevine. Within a few hours after inoculation, the pathogen has penetrated the tissue and formed the first haustoria in a susceptible cultivar and after 3 days, the intercellular spaces are entirely filled with mycelium and sporulation is abundant under favourable environmental conditions (Unger et al. 2007). In a resistant cultivar, the first infection steps are the same, however, the invasive growth of *P. viticola* is delayed and further development ceases before the intercellular spaces colonized (Unger et al. 2007). Most of the widely grown grapevine cultivars are highly susceptible to *P.*

viticola and the control of downy mildew requires regular fungicide applications. Application of copper-containing fungicides to control downy mildew causes accumulation of this heavy metal in soil and groundwater, resulting in toxic effects to the environment. Constant use of fungicides is also a problem since it may favour the emergence of fungicide-resistant isolate that may dominate the population.

Although a number of studies describe the biology of the infection process of grapevine by *P. viticola* (Allegre et al. 2007; Kortekamp 2005; Musetti et al. 2007; Unger et al. 2007; Werner et al. 2002) and the characterisation of defence reactions in susceptible and resistant grapevines (Dai et al. 1995; Gindro et al. 2003; Kortekamp 2006, Kortekamp and Zyprian 2003) or their response to inducers of defence reactions (Aziz et al. 2003; Hamiduzzaman et al. 2005; Trouvelot et al. 2008), a large-scale analysis of *P. viticola* and grapevine interaction has not been reported.

Upon recognition of an invader, signal transduction events ultimately leading to the induction of both chemical and physical defence responses are activated (Scheel 1998). The specific recognition of a pathogen often leads to a programmed cell death termed the hypersensitive response (HR). It is a rapid and efficient plant resistance mechanism leading to cell death at the site of infection, usually in gene-for-gene interactions with obligate biotrophic pathogens (Heath 2000). The cellular events taking place during the HR comprise among others the generation of reactive oxygen species (ROS) at the site of infection, cell wall reinforcements through callose-rich deposits, ion fluxes, and the synthesis of anti-microbial compounds and enzyme inhibitors (Glazebrook 2005). In grapevine, these include low-molecular antimicrobial compounds (phytoalexins), deposition of phenolics, lignin and callose and increased activity of pathogenesis-related (PR) proteins with hydrolytic activity (chitinases and glucanases; Derckel et al. 1999).

Since the difference between a resistant and susceptible plant is often just a question of timing and amplitude of the adequate defence response (Polesani et al. 2010), investigating a compatible interaction not only gives information on the availability of adequate defence mechanisms but will also assist in the development of new control strategies and lead to the identification of pathogen and host factors needed for disease progression. Investigating the molecular basis of plant-pathogen interactions often uncovers novel aspects of plant cell biology and signalling mechanisms. Expression profiles investigating compatible interactions in cultivated grapevine genotypes is limited and downy mildew has received very little attention compared to diseases caused by other biotrophic pathogens, such as powdery mildew and rusts (Polesani et al. 2008).

Functional genomic approaches provide powerful tools for identifying expressed genes. Among these techniques, expressed sequence tags (EST) (Adams et al. 1991), serial analysis of gene expression (SAGE) (Velculescu et al. 1995) and massively parallel signature sequencing (MPSS) (Brenner et al. 2000), have been successfully employed. Subtractive suppression hybridization (SSH) is an effective method that can be used to maximise the identification of genes that are involved in host responses to pathogen infection and disease development. This technique has been used to isolate plant genes that are expressed in response to infection (Birch et al. 1999; Degenhardt et al. 2005; Lu et al. 2004; Van den Berg et al. 2007). The SSH cDNA library approach reduces the cloning of abundantly expressed housekeeping genes or genes commonly expressed in both control and treated plants, therefore significantly enhancing the chances of cloning differentially expressed genes. This is particularly important because many PR genes are expressed at low levels. These genes are less likely to be represented in a library if standard EST cloning methods are used. The aim of this study was to construct a SSH cDNA-library to identify and verify genes that are differentially expressed during a susceptible response of *V. vinifera* cv. Chasselas to *P. viticola*. By this approach, *P. viticola*-responsive genes were identified and classified into functional groups and the expression profiles of some of the genes were examined by quantitative real time PCR following inoculation with *P. viticola*.

Material and methods

Biological materials

Grapevine plants were grown from *V. vinifera* cv Chasselas seeds (susceptible to *P. viticola*), in soil at Syngenta (Stein, Switzerland) in a glasshouse at day/night temperature of 23°C/ 18°C with a photoperiod of 16 h and relative humidity (RH) of 65±10%. Seedlings at the five-to six-leaf stage (6 weeks) were used for infection experiments.

Pathogen and inoculation of plants

P. viticola (house isolate, obtained from Syngenta, Stein, Switzerland) was maintained on susceptible Chasselas seedlings in a glasshouse. Weekly, the abaxial surfaces of young leaves were inoculated with an aqueous suspension containing ~4×10^4 sporangia ml^{-1}. The inoculated plants were kept at 100% RH overnight. After incubation for 6 days under ambient glasshouse conditions, the plants were again maintained overnight at 100% RH to induce sporulation. The sporangia were harvested and used for inoculation.

Six week old grapevine seedlings were inoculated at Syngenta by spraying a freshly prepared sporangia suspension (6×10^4 sporangia ml^{-1}) onto the abaxial leaf surfaces. For the mock-inoculation, seedlings were sprayed with water. Inoculated and mock-inoculated seedlings were placed overnight in a humid chamber. Samples were collected from inoculated and mock-inoculated leaves 24 h after inoculation (for library construction) or at 0, 8, 12, 24 and 48 h after inoculation (for quantitative RT-PCR) and immediately frozen in liquid nitrogen and stored at −80°C. For the next 6 days the plants were placed in a glasshouse under the conditions described above and finally placed in a humid chamber overnight to induce sporulation. On day 7 plants were assessed for visual disease symptoms (Fig. 4a–c). For the library construction, 15 inoculated and 15 mock-inoculated plants were used. Twenty 4 h after inoculation, the third, fourth and fifth leaves from inoculated and mock-inoculated plants were harvested and immediately frozen in liquid nitrogen and stored at −80C.

For gene expression analysis, at each time point a pool of 3 inoculated and 3 mock-inoculated plants was used. As for the library construction, the third, fourth and fifth leaves from the bottom of inoculated and mock-inoculated plants were harvested. Two independent experiments were carried out.

RNA isolation and SSH library construction

Total RNA was isolated from frozen leaf tissues using a modified CTAB extraction and lithium chloride precipitation method according to (Iandolino et al. 2004). The mRNA was isolated using Poly(A)Purist™ mRNA Purification Kit (Ambion) according to the manufacturer's protocol and purified mRNA quality was then determined on an Agilent 2100 Bioanalyser RNA 6000 Pico LabChip. Two micrograms of high quality mRNA were used for the construction of the libraries.

SSH was carried out using the PCR-Select cDNA Subtraction Kit (Clontech, Palo Alto, CA). cDNA prepared from the inoculated samples was used as the tester and that from the mock-inoculated samples as the driver for the forward subtraction to isolate fragments corresponding to genes whose expression level was increased following infection. The reverse library was carried out with the mock-inoculated samples as tester and inoculated samples as driver in order to isolate fragments corresponding to genes whose expression level decreased following infection. An Advantage PCR cloning kit (Clontech, USA) was used to selectively amplify the cDNA fragments preferentially present in the tester from the subtraction hybridization products. Secondary PCR products from the forward and reverse subtracted libraries were cloned into pGEM-T Easy vector (Promega) according to the manufacturer's protocol. Transformation was performed using Library Efficiency DH5α Competent Cells (Invitrogen). A total of 1344 clones form the forward SSH library and 1152 from the reverse SSH library were randomly picked and stored in 96-well plates.

SSH cDNA microarray screening

In preparation for screening with microarrays, following the method of Van den Berg et al. (2004) and modifications of Berger et al. (2007), PCR reaction products from all the clones in the forward and reverse SSH libraries were purified using the 96-well Multiscreen HTS, FB plates (Millipore). An aliquot of the purified PCR products were visualised on a 1% agarose Electro-Fast Stretch gel to check that a single band was amplified and to estimate the yield. The purified PCR products were dried in a vacuum centrifuge and resuspended in 20 µl 3X SSC, 1.5 M betaine. Libraries containing in total 2496 grapevine clones were arrayed in triplicate onto aldehydesilane-coated slides (NexterionTM Slide AL, Schott Nexterion, Jena, Germany) using an Omnigrid 300 contact-printing robotic micro-arrayer (Genomic Solutions, Ann Arbor, MI) equipped with SMP3 pins (TeleChem International Inc.). Spike controls (Lucidea Universal Scorecard, GE Healthcare) were included for each sub-grid of the microarray. Spot and printing quality were assessed visually after printing and the DNA was cross-linked to the slides by baking at 80°C for 1 h. The slides were post-processed with sodium borohydride using the protocol recommended by the manufacturer (Schott Nexterion, Jena, Germany).

Preparation of fluorescent targets and hybridization

Forward subtracted (FS), forward unsubtracted (FU) and reverse unsubtracted (RU) cDNA targets were used to screen the forward SSH library, whereas reverse subtracted (RS), reverse unsubtracted (RU) and forward unsubtracted (FU) cDNA targets were used to screen the reverse SSH library on glass slide microarrays (Berger et al. 2007). cDNA targets of FS, FU, RS and RU, with the adaptors removed were labelled by incorporation of Cy5 and Cy3-dUTP using Klenow enzyme according to Berger et al. (2007). The labelled targets were purified using the Wizard SV gel and PCR clean-up system (Promega) and quantified using a Nanodrop spectrophotometer.

Eight slides were arrayed with the cDNA probes from the forward and reverse library separately. Before hybridization, equimolar amounts (300 pmol) of each pair of appropriate labelled targets and 10 μg yeast tRNA were mixed and concentrated through a Microcon YM-30 filter (Millipore). The target combinations for screening of the forward library were as follows: Cy3-labelled FS and Cy5-labelled FU; dye swap: Cy3-labelled FU and Cy5-labelled FS (to calculate Enrichment ratio 2); Cy3-labelled FU and Cy5-labelled RU; dye swap: Cy3-labelled RU and Cy5-labelled FU (to calculate Enrichment ratio 3) (Berger et al. 2007). For the screening of the reverse library, the target combinations were: Cy3-labelled RS and Cy5-labelled RU; dye swap: Cy3-labelled RU and Cy5-labelled RS (to calculate Enrichment ratio 2); Cy3-labelled RU and Cy5-labelled FU; dye swap: Cy3-labelled FU and Cy5-labelled RU (to calculate Enrichment ratio 3). For hybridizations, the labelled probes were combined with 3X SSC and 0.4% SDS in a final volume of 30 μl. The solution was boiled for 1 min, centrifuged and applied to the microarray under a cover slip. The microarray slides were sealed in a hybridization chamber (TeleChem) and submerged in a water bath at 64°C overnight. After hybridization, slides were washed twice for 5 min in 2X SSC/ 0.1% SDS, twice for 1 min in 0.2X SSC and twice for 1 min in 0.1X SSC. Slides were dried by centrifugation at 900 g for 2 min and scanned with an Agilent DNA microarray scanner (Agilent Technologies). Genepix Pro 5.1 was used to localise and integrate every spot on the array. Analysis of microarray data was done according to Berger et al. (2007) by using the SSHscreen 1.0.4 package downloaded from http://microarray.up.ac.za/SSHscreen/. SSHscreen was used to implement the following statistical approaches: the normexp method was used to correct for the background, in order to normalise within arrays the printiploess method was used and finally to normalise between arrays the Aquantile method was used.

Sequence analysis

Sequencing of the inserted cDNA fragments was done using the T7 and SP6 primers at Microsynth (Switzerland). Sequences were compared to the UniProt database as well as NCBI database using BLAST. Functional categorisation of sequences was performed by comparison with sequences in the NCBI database as well as through the use of the MIPS functional catalogue database (FunCatDB). Grapevine cDNA sequences have been deposited in GenBank dbEST with the accession numbers GO652854-GO653046.

Reverse transcriptase quantitative real-time PCR (RT-qPCR) analysis

To remove any contaminating DNA in the RNA preparations, RNA was treated with TURBO DNA-free DNase I (Ambion) according to the manufacturer's protocol. After DNase treatment, the RNA concentration and quality was checked with the NanoDrop ND-1000 spectrophotometer (NanoDrop Technologies, Wilmingon, DE, USA). cDNAs were synthesised from 1 μg of total RNA using the Superscript III first strand synthesis system (Invitrogen) according to the manufacturer's instructions.

Real-time PCR reactions were performed in 96-well plates with a BioRad iQ iCycler Detection system (BioRad Laboratories, Ltd) using SYBR Green to detect dsDNA synthesis. The amount of transcripts of selected genes in *P. viticola*-inoculated versus mock-inoculated grapevine plants was compared by using the reference gene Elongation Factor 1-α (*EF1-α*) as endogenous reference for data normalization. Reactions were done in a total volume of 20 μl containing 200 nM of each primer, 2 μl of

1:5 diluted cDNA and 10 µl 2X SYBR Green Master Mix Reagent (Bio-Rad). Reactions were run using the following cycling parameters: 95°C for 3 min, 40 cycles of 95°C for 30 s, 57°C for 35 s and 72°C for 35 s followed by a melt cycle from 60°C to 95°C. No-template controls were included for each primer pair to assess its specificity and each PCR reaction was done in triplicate. Primer sequences (Table 1) were designed with Primer Design 4 based on the sequences of the selected clones. The software programme Gene-X was used to calculate the mean normalised expression of the genes (Vandesompele et al. 2002).

Microscopic observation

Leaf disks from *P. viticola*-inoculated and mock-inoculated grapevine plants were collected at two time points (4 and 7 days). Leaf disks were then used for different staining methods.

Lactophenol-trypan blue staining

To observe the structure of *P. viticola,* leaf disks were stained with lactophenol trypan blue according to (Keogh et al. 1980). The disks were incubated for 48 h in the staining solution and then destained with chloral hydrate (2.5 $g ml^{-1}$). Destained disks were kept in chloral hydrate solution and visualized with a light microscope. Hyphae, sporangiophores and sporangia stained blue.

Lignin staining

Leaf disks were boiled for 10 min in 95% ethanol and then treated with 10% phloroglucinol for 20 min and washed in 25% HCl. They were then mounted in glycerol (75%) and examined with a light microscope (Dai et al. 1995).

Flavonoid staining

Staining for the presence of flavonoids was performed according to (Pina and Errea 2008). Briefly, decolorized leaf disks were stained for 10 min in 1% (w/v) Naturstoff reagent A (2-Aminoethyl diphenylborinate, Fluka) in ethanol and then washed 4 times with ethanol. Visualization was done under an epifluorescence microscope with an UV filter (BP, 340–380 nm; LP, 450 nm)

Results

Construction of the SSH library

For the construction of cDNA libraries enriched for *P. viticola* responsive genes, pooled Chasselas plants inoculated with *P. viticola* or mock-treated plants were used for the extraction of mRNA and cDNA synthesis. Reciprocal subtractions of cDNAs derived from *P. viticola*-infected and non-infected plants yielded two cDNA libraries, one being enriched for *P. viticola*-responsive genes (forward library) and the other enriched for *P. viticola*-repressed genes (reverse library). In total 1344 clones form the forward SSH library and 1152 from the reverse SSH library were used.

Screening of the SSH library using glass slide microarrays

According to the PCR-select differential screening kit user manual (Clonetech), 1–4% of the clones identified by differential screening with inverse dot blots, turn out to be false positives. In order to select clones for sequencing and reduce the number of false positives, a more quantitative, efficient, rapid and high-throughput technique was used, namely screening the forward and reverse SSH library on glass slide microarrays (as described in Van den Berg et al. 2004). In addition an analysis R software package named SSHscreen was used, which employs functions of limma (linear models for microarray data) to analyse spot intensity data, thereby screening clones in the libraries to identify those that are significantly differentially expressed (Berger et al. 2007; Coetzer et al. 2010). Furthermore, this method is able to identify whether the clones were derived from rare or abundant transcripts in the treated sample. Quantitative screening was carried out by calculating different "SSH enrichment ratios (ER)" for each clone. Calculations were based on the intensity values of each cDNA clone as a result of hybridization by the Cy dye-labelled cDNA targets. After hybridization, scanning and normalization of the data to account for intensity biases within and between slides, the SSH enrichment ratios were calculated and a statistical test was applied with a false discovery rate (FDR) of 5% to account for multiple testing. SSH enrichment ratio 3 (ER3) values were calculated as $\log_2(FU/RU)$ (for

Table 1 Primers used for quantitative real-time RT-PCR

Gene name	Forward primer	Reverse primer
Chalcone isomerase (CHI)	5′-AACTTCTGGTAGGGACCCATCT-3′	5′-GAAGGATGAAACCTTCCCACCA-3′
Chalcone synthase (CHS)	5′-GAGAGCAACTCGACACGTTCTG-3′	5′-GTGTAGCAAGGCTGTGCAACAA-3′
Chitinase III (Chit III)	5′-ATCGTCTCGGCCATTAGGTGAT-3′	5′-ATACCTTCCTTCCACGCTTGCT-3′
Cinnamate-4-hydroxylase (C4H)	5′-AGTCCAAGTCACCGAGCCTGAT-3′	5′-TAGCAAGCCACCATGCGTTCAC-3′
EF1-α	5′-GAACGTTGCTGTGAAGGATCTC-3′	5′-CGCCTGTCAACCTTGGTCAGTA-3′
Flavonone-3-hydroxylase (F3H)	5′-GAGAGCAACTCGACACGTTCTG-3′	5′-CTGAAGGAGCAGCGTGATGGTT-3′
β-1,3-Glucanase (Gluc)	5′-CCAGTACAGCATCCGAAGAATGAG-3′	5′-AACCGCAACGTATCGGAATCTGAC-3′
Heavy-metal associated protein (HMAP)	5′-CATCCACTGCCATGGAGCTT-3′	5′-CTTCTGCCACCGACCTAATC-3′
Metallothionein (MT)	5′-TGCATGGATCGCAGTTGCAGTT-3′	5′-TCGCACCAGTGAAGATGCACT-3′
PR-1	5′-AACGACACTGTAGCTGCCTACG-3′	5′-GCATCGGTGCCTGTCAATGAAC-3′
PR-5	5′-TGTTGTCGTGGGATCTCACCTG-3′	5′-CTTCAAAGGGTCAGGCAAGTGT-3′
PR-10	5′-CCAACCAATCCTCCTTGTCTT-3′	5′-GAATCGAGGATAGAGGCCTAGC-3′
PR-17	5′-ATCGCCTGAGTAGCTTCCGATG-3′	5′-CGCTAGTATCTGCCACCGACTT-3′

the forward library) and $\log_2$(RU/FU) (for the reverse library) and gave an indication whether cDNAs were up- or down-regulated by *Plasmopara* infection (Fig. 1).

From the forward library screening, 441 grapevine clones were significantly up-regulated in response to *Plasmopara* infection with a positive ER3 value (and positive B statistics at FDR of 5%) and are represented in the top two quadrants of Fig. 1a. A number of clones in the forward library (161) had negative ER3 values and thus fell in the bottom two quadrants of Fig. 1a, indicating that they are "down-regulated" by *Plasmopara* infection, which means they are false positives and thus escaped the subtraction. SSHscreen produces a similar plot from the screening of the reverse library, but in this case positive ER3 values indicate grapevine clones that were down-regulated in response to *Plasmopara* infection (Fig. 1b). Screening of the reverse library resulted in a total of 64 clones that were down-regulated by *Plasmopara* infection (positive ER3 values, see top two quadrants of Fig. 1b) whereas 418 clones had negative ER values (i.e. escaped subtraction in the revere library Fig. 1b). These results indicate that the efficiency of the SSH process was far better in the forward library compared to that of the reverse library (73% vs. 13%). The SSHcreen analysis thus enabled the selection of clones for DNA sequencing out of the 441 forward library clones that were scored as up-regulated, and out of the 64 reverse library clones that were scored as down-regulated.

One of the main advantages of SSH is that it normalizes the cDNA abundance so that cDNAs encoded by genes that are expressed at low levels, but are nonetheless differentially regulated, can be identified readily. SSH enrichment ratio 2 (ER2) values were calculated as $\log_2$(FS/FU) (for the forward library) and $\log_2$(RS/RU) (for the reverse library) to give an indication whether cDNAs have been enriched or reduced relative to levels in the tester sample by the normalization process. Rare transcripts in *Plasmopara* infected plants will have negative inverse ER2 values and fall in the top left quadrant of Fig. 1a, while abundant transcripts will fall in the top right quadrant (Fig. 1a). In the forward library, half of the grapevine clones up-regulated by *Plasmopara* infection were rare transcripts, and half were abundant (48% vs. 52%). In the reverse library, the majority (75%) of the grapevine clones down-regulated by *Plasmopara* infection were rare transcripts (top left quadrant of Fig. 1b), while only 25% were abundant.

Sequencing analysis and functional classification of differentially expressed clones

No *P. viticola* genes were identified since the time point chosen to construct the SSH library was early,

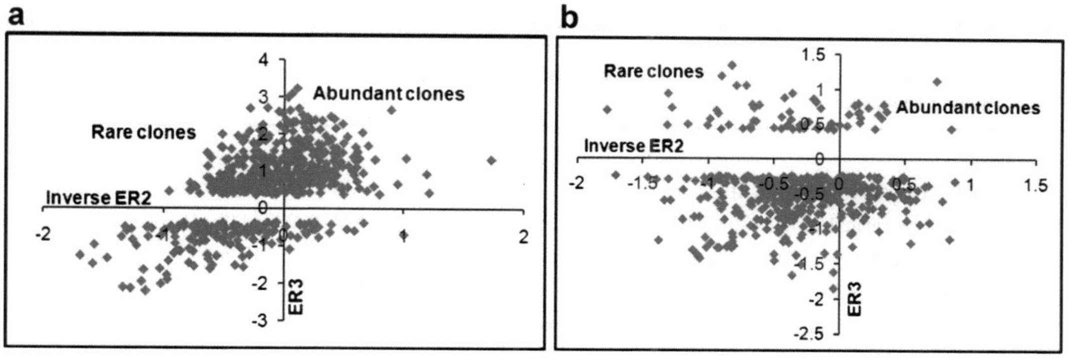

Fig. 1 Screening of forward (**a**) and reverse (**b**) SSH grapevine libraries on glass slide microarrays. An up/down regulation plot of enrichment ratio 3 (ER3) vs. inverse enrichment ratio 2 (ER2) is produced by the SSHscreen analysis. The clones that were calculated in SSHscreen to be statistically significantly up-regulated (ER3>0, positive B statistic, false discovery rate of 5%) after *P. viticola* inoculation in the forward library (**a**) are shown in the top two quadrants, while those that are down-regulated in the forward library (false positives) are shown in the bottom two quadrants. The inverse is true for the reverse library (**b**), where the top two quadrants contain clones that are significantly down-regulated after *P. viticola* inoculation, and the bottom two quadrants contain false positive clones that are up-regulated. The inverse ER2 values reflect the relative abundance of the transcripts for each clone in the unsubtracted cDNA samples. Thus transcripts that were rare in the cDNA from plants inoculated with *P. viticola*, but also were up-regulated relative to control plants fall in the top left quadrant (**a**), whereas those that were abundant in the cDNA from plants inoculated with *P. viticola*, but also were up-regulated relative to control plants fall in the top right quadrant (**a**). For the reverse library, the top left and top right quadrants show transcripts that were down-regulated after inoculation with *P. viticola* and that were rare and abundant, respectively, in the control plants prior to subtraction (**b**). No clones are observed in the region of the x-axis (ER3=0), since this region would contain genes that have similar transcript levels in inoculated and mock-inoculated samples, that would have been removed by the subtraction or not plotted on Fig. 1a and b since they are not calculated to be differentially expressed when applying a FDR of 5%

so that the amount of pathogen biomass (RNA) was very low compared to plant biomass.

A total of 223 clones were sequenced but 29 of those resulted in poor quality sequences or the sequenced insert was too short to give any BLAST results and were therefore discarded. Of the remaining 194 clones, 138 were up- and 56 down-regulated.

UniProt analysis of the cDNA sequences was carried out in order to identify their putative functions. Sequences without significant homology in the UniProt database were further analysed with BLASTx and BLASTn to predict their putative functions. Details of putative gene identities for the sequenced clones from the forward and reverse library are summarised in Table 2. Sequencing results for the 194 subtracted clones revealed only 3 clones (1.5%) that showed no similarities with known sequences within the UniProt and NCBI databases, and 6% corresponded to hypothetical proteins. Some of the clones were redundant and in these cases only one representative is shown. Redundancy resulted from the use of *RsaI*-restricted cDNA fragments in the SSH procedure. Because of the restriction step, two or more different cloned cDNA fragments can represent a single transcript.

The clones that had significant sequence homologies with known sequences in the UniProt and NCBI databases were divided into seven categories through the use of the MIPS functional catalogue database (FunCatDB, Ruepp et al. 2004) and extensive search of scientific literature. The distribution of the functional classification for genes that were both up- and down-regulated is represented in Fig. 2. The largest number of induced genes was classified under the category of defence response (24%), followed by genes involved in phenylpropanoid and lignin biosynthesis (11%), and in metabolism (10%), whereas the largest number of repressed genes was classified under photosynthesis and energy (12.5%) followed by those involved in metabolism (9%). Induced genes involved in photosynthesis and energy and transport constituted the smallest group, comprising between 1 and 3% of the genes, respectively, while those genes involved in response to stimulus (8%) and those genes encoding hypothetical proteins (7%), were more abundant.

Table 2 Identification of *P. viticola*-responsive genes isolated from the SSH library in grapevine after glass slide microarray screening

Clone name	Putative function and (accession number of original matching sequence)	Gene identifiers[b]	Up/ Down[a]	E-value	N° of clones
Defense Response					
Fw1-C9	Beta 1-3 glucanase (Q9M3U4)	XM_002277475.1	+	9.00E-67	6
Fw2-A7	Pathogenesis-related protein 10 (Q2I305)	XM_002274447	+	3.00E-57	21
Fw3-E10	Putative pathogenesis related protein 1 (Q7XAJ6)	XM_002274239	+	1.74E-66	10
Fw6-B8	Chitinase III (Q84S31)	XM_002264708	+	7.00E-96	3
Fw9-F7	F23N19.16 (Plant lipid transfer protein/Par allergen) (Q1SBU5)	XM_002264045	+	5.00E-20	5
Fw6-C6	Pleiotropic drug resistance protein 1 (NpPDR1) (Q949G3)	XM_002285142	+	4.12E-59	1
Fw11-B2	NtPRp27 (Q9XIY9) (PR-17)	XM_002269434	+	3.00E-70	2
Fw13-A12	Thaumatin-like protein (Q7XAU7)	XM_002283046	+	4.00E-80	1
Rv8-F12	Germin-like protein 6 (Q0PWM4)	XM_002279223	−	1.09E-50	1
Metabolism					
Fw1-D6	26 S proteasome regulatory particle triple-A ATPase subunit2b (Q8W422)	XM_002263298	+	6.00E-68	1
Fw11-H8	Maturase (Q6TNE4)	XM_002283754	+	0.71	1
Fw2-H3	Ubiquitin-conjugating enzyme (Q2I309)	XM_002281744	+	1.15E-57	2
Fw3-E5	40 S ribosomal protein S17 (Q6RJY3)	XM_002268961	+	3.00E-44	1
Fw1-E2	Monooxygenase (Monooxygenase 2) (O81816)	XM_002272572	+	4.27E-49	1
Fw5-G1	Highly similar to developmental protein DG1118 (Q9SSM4)	XM_002281246	+	6.85E-21	2
Fw6-C12	60 S ribosomal protein L13a-4 (R13A4)	XM_002268865	+	2.03E-15	2
Fw6-C8	Putative ribosomal protein S10 (Fragment) (Q0VJB2)	XM_002273250	+	2.72E-09	1
Fw5-G3	Aldehyde dehydrogenase (NAD+) (EC 1.2.1.3) (P93344)	XM_002263443	+	6.00E-57	4
Fw12-G5	Amine oxidase (Q9SW88)	XM_002263313	+	2.00E-109	5
Rv6-G2	Putative ribosomal protein S14 (Q6H7T1)	XM_002276598	−	4.97E-62	3
Rv7-F2	Putative tumor differentially expressed protein 1 (Q6K688)	XM_002270574	−	7.00E-48	2
Rv11-C12	Fiber protein Fb11 (Q8GT82)	XM_002265077	−	3.00E-24	2
Rv14-E1	IMP dehydrogenase/GMP reductase (Q1RSI8)	XM_002274242	−	7.00E-47	1
Rv7-E12	Galactinol synthase, isoform GolS-1 (Q9XGN4)	XM_002265911	−	3.80E-41	1
Rv3-E1	Cyclin-like F-box; Galactose-binding like (Q1S0J2)	XM_002279550	−	9.10E-35	2
Rv9-E3	High mobility group protein (Q4ZH67)	XM_002280048	−	1.13E-29	1
Rv14-E4	Aldo/keto reductase AKR (Q52QX9)	XM_002265991	−	5.00E-87	1
Rv9-G6	DnaJ-like protein (P93499)	XM_002265808	−	6.15E-30	1
Rv2-F6	60 S ribosomal protein L8-3 (Q42064)	XM_002269373	−	7.02E-58	1
Rv11-B2	Putative ankyrin-repeat protein (Q6TKQ6)	XM_002283462	−	1.06E-101	1
Rv5-D6	Hydrolase, alpha/beta fold family protein (Q10QA5)	XM_002285272	−	3.73E-20	1
Photosynthesis and Energy					
Fw7-F12	Oxygen-evolving enhancer protein 2, chloroplast precursor (OEE2) (Q9SLQ8)	XM_002283012	+	1.09E-100	1
Fw11-D2	Photoreceptor-interacting protein-like (Q9FJY3)	XM_002272516	+	4.00E-82	1
Fw11-D6	Chlorophyll A-B binding protein (CAB) (Q1EPK5)	XM_002274906	−	1.11E-77	10
Fw13-C7	Plastid ribosomal protein L19, putative (Q2A9H1)	XM_002265826	−	8E-18	4
Rv6-A5	Photosystem II 10 kDa polypeptide, chloroplast precursor (PII10) (Q40519)	XM_002271755	−	6.74E-45	2
Rv1-G1	Hypothetical chloroplast RF2 (Q0ZIV6)	DQ424856	−	1.44E-80	1
Rv10-F10	Type 26 kD CP29 polypeptide (Q00321)	XM_002264259	−	2.00E-47	2

Table 2 (continued)

Clone name	Putative function and (accession number of original matching sequence)	Gene identifiers[b]	Up/Down[a]	E-value	N° of clones
Rv4-H8	Rubisco activase (O82428)	XM_002282200	−	2.38E-90	3
Rv8-C7	Ribulose-1,5-bisphophate carboxylase/oxygenase small subunit (Q2I314)	XM_002276955	−	1.92E-47	2
Response to stimulus					
Fw11-A4	Methionine sulfoxide reductase A (Q6QPJ5)	XM_002278222	+	3.33E-52	1
Fw7-F4	Cysteine protease (O50002)	XM_002278588	+	3.30E-90	2
Fw1-E4	GTPase activating protein-like (Q9LVH4)	XM_002275199	+	1.46E-60	1
Fw8-B5	AtRab18 (Putative GTP-binding protein) (O23657)	XM_002267351	+	1.14E-41	1
Fw12-B9	Harpin-induced protein 1 (Q10NQ5)	XM_002265754	+	3.00E-13	1
Fw12-C9	Heavy-metal-associated domain-containing protein (AT5g14910) (Q93VK7) (HMAP)	XM_002275240	+	2.00E-13	3
Fw2-G10	Glutathione S-transferase GST (Q948X4)	XM_002262806	+	2.75E-06	2
Fw13-H6	Metallothionein-like protein 1 (MT-1)	XM_002283754	+	3.00E-04	5
Rv10-G3	AUX/IAA protein (Q1SP83)	XM_002284246	−	9.00E-70	1
Rv7-D1	Aminomethyltransferase, mitochondrial precursor (P49363)	XM_002277870	−	3.55E-71	1
Rv9-E4	Metallothionein-like protein type 3 (MT-3) (Q40256)	XM_002284942	−	1.83E-26	1
Rv8-A6	Putative metallothionein-like protein (Q9M4H3)	XM_002284942	−	2.89E-35	1
Rv2-E11	Metallothionin 2 (Q1W3C8)	XM_002265022	−	6.09E-17	1
Rv8-F5	Thioredoxin domain 2 (Q1T0K3)	XM_002266314	−	1.01E-32	1
Rv11-A12	1-aminocyclopropane-1-carboxylic acid oxidase 1 (Q84X67)	XM_002273394	−	5.41E-18	1
Rv10-B3	AOBP (Ascorbate oxidase promoter-binding protein) (Q39540)	XM_002281958	−	7.92E-39	1
Phenylpropanoid and Lignin Biosynthesis					
Fw1-F6	Chalcone synthase (O80407)	XM_002276910	+	1.00E-98	7
Fw7-B2	Pectin methylesterase inhibitor isoform (Q0Z7S4)	XM_002264168	+	1.15E-12	1
Fw2-D8	Expansin 45, endoglucanase-like (Q1RYX8)	XM_002284960	+	3.03E-13	1
Fw2-E10	Cinnamate-4-hydroxylase (Q3HM04)	XM_002266202	+	9.00E-106	3
Fw3-C10	Naringenin,2-oxoglutarate 3-dioxygenase (Flavonone-3-hydroxylase)	XM_002267604	+	1.00E-90	2
Fw9-E1	Putative pectin methylesterase (Q8H223)	XR_077457	+	3.00E-06	1
Fw11-G10	Chalcone isomerase	XM_002282072	+	2.00E-70	7
Signal Perception and Transduction					
Fw5-G7	Expressed protein (Leucine Rich Repeat family protein) (Q53PD8)	XM_002268472	+	2.36E-06	1
Fw5-F6	Sucrose responsive element binding protein (Q1XAN1)	XM_002283311	+	8.66E-63	1
Fw2-F11	Putative receptor-like protein kinase 1 (Q5JN27)	XM_002266170	+	9.19E-68	1
Fw2-A10	Putative peroxisomal membrane protein(22-kDa)(PMP22) (Q6Z1N7)	XM_002269300	+	3.00E-47	4
Fw11-A10	Membrane protein (Q2M5F0)	XM_002277120	+	2.90E-19	1
Fw14-B7	Zinc finger, C2H2 type family protein, expressed (Q2QX40)	AM482928	+	3.3	1
Fw8-E7	SOS2-like protein kinase (Q8LK24)	XM_002285460	+	5.00E-24	1
Fw1-C1	Cyclin-like F-box; Serine/threonine protein phosphatase, BSU1 (Q1RZ31)	XM_002265077	+	1.19E-49	1
Rv12-G10	Phosphate translocator precursor (Q40568)	XM_002278793	−	3.00E-56	1
Transport					
Fw5-C10	Sorbitol transporter (Q84KI7)	XM_002284527	+	4.29E-14	1
Fw2-B9	Putative peptide transporter protein (Q8VZN7)	XM_002266869	+	4.22E-20	2
Fw11-D5	Copper transporter 1 (Q547P9)	XM_002280052	+	4.00E-17	2

Table 2 (continued)

Clone name	Putative function and (accession number of original matching sequence)	Gene identifiers[b]	Up/ Down[a]	E-value	N° of clones
Unclassified					
Fw9-C9	Hypothetical protein T6H20.190 (Q9STF2)	XM_002282694	+	1.10E-38	1
Fw2-E9	T17H3.3 protein (Hypothetical protein) (Similar to CGI-126 protein) (Q9SXC8)	XM_002285545	+	3.00E-29	1
Fw6-D5	Hypothetical protein (At3g15810) (Q9LVZ8)	XM_002266546	+	1.00E-22	1
Fw6-E1	Hypothetical protein T16K5.140 (Hypothetical protein At3g49790) (Q9M2X9)	AM438165	+	2.2	1
Fw9-G4	Hypothetical protein rps12 (Q3BAI4)	AM483961	+	3.00E-27	1
Fw9-G9	Hypothetical protein At2g39140 (O80967)	XM_002270150	+	1.02E-24	1
Fw4-H6	Hypothetical protein (Q56ZH1)	XM_002275456	+	1.24E-06	1
Fw10-D4	Hypothetical protein (Q3LSN3)	XM_002279821	+	3.00E-60	1
Fw2-C7	Unnamed protein product (CAO38708.1)	XM_002265072	+	4.00E-19	1
Fw1-F10	Hypothetical protein ORF1 (Q9FSX3)	XM_002263298	+	7.00E-18	1
Fw1-F2	No homology		+		1
Fw2-C10	No homology		+		1
Fw1-E6	No homology		+		1
Rv7-H8	Expressed protein (Q10Q41)	XM_002271976	–	2.77E-28	1
Rv3-E12	Hypothetical protein (Q8LDM9)	XM_002283263	–	2.54E-07	1
Rv8-D10	Hypothetical protein (Q1SFB2)	XM_002270765	–	1.45E-10	1
Rv4-A11	Hypothetical protein (CAN65189)	XM_002266869	–	4.00E-94	1
Rv9-G5	No homology		–		1

[a] "+" means gene was induced by *P. viticola* from screening of the SSH libraries, and comes from the forward library, "–" means gene was suppressed by *P. viticola* from screening of the SSH libraries and comes from the reverse library

[b] Grapevine gene identifiers were annotated according to the NCBI database

Expression analysis of differentially responsive genes to *P. viticola* infection by RT-qPCR

To analyse the expression patterns of a number of genes during the interaction of *V. vinifera* with *P. viticola,* reverse transcriptase quantitative real-time PCR (RT-qPCR) was performed on inoculated and mock-inoculated samples 0, 8, 12, 24 and 48 h after *P. viticola* inoculation. The selected genes corresponded to a range of functional categories: defence response (6 genes), phenylpropanoid and lignin biosynthesis (4 genes) and response to stimulus (2 genes) (Fig. 3). RT-qPCR was carried out with two biological replicates. Total RNA extracted from grapevine seedlings was reverse-transcribed into first-strand cDNA, and used as template for quantitative RT-PCR analysis. Results of the time-course experiment showed that PR-genes were highly induced compared to those involved in phenylpropanoid and lignin biosynthesis and in response to stimulus, which was in good agreement with the library verification results (data not shown). Among the PR-genes, a β-1,3-glucanase (*Gluc*) and a pathogenesis-related protein 10 (*PR-10*) showed the highest fold increase in expression level. The expression level of *Gluc* increased transiently over the 48 h time course, whereas the expression level of *PR-10* reached a maximum level at 12 hpi, decreasing at 24 hpi and remaining constant up to 48 h. The expression pattern of a chitinase III (*Chit III*) was similar to that of *PR-10* with the exception that after a slight decrease in expression level at 24 hpi, it increased gradually until 48 hpi. Another PR gene, pathogenesis-related protein 1 (*PR-1*) was induced 9.5-fold at 12 hpi where it reached maximum level and then decreased reaching levels close to the non-infected samples. Like *Gluc* the expression level of a pathogenesis-related protein 17 (*PR-17*) was transiently induced over the 48 h time course. The only defence-related gene which did not show any significant change in expression level over

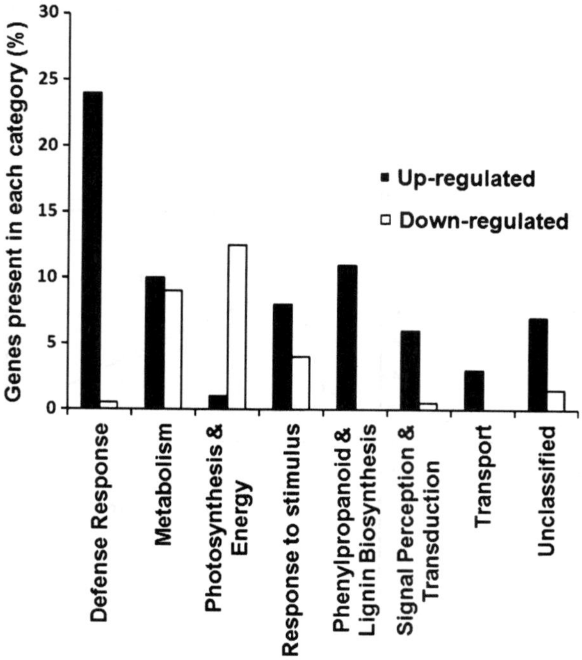

Fig. 2 Functional classification of genes up- and down-regulated in grapevine by *P. viticola* inoculation. Differentially expressed genes were classified into seven groups and the percentage of the genes within each category is shown

the time course was a thaumatin-like protein (*PR-5*). Similarly, one of the four genes involved in phenylpropanoid and lignin biosynthesis, namely a chalcone synthase (*CHS*) was down-regulated. Flavonone-3-hydroxylase (*F3H*) and cinnamate-4-hydroxylase (*C4H*) were not highly induced over the 48 h time course, both showed a 2-fold increase in expression level at 24 hpi, whereas chalcone isomerase (*CHI*) was induced 8-fold at 24 hpi. Of the two genes involved in response to stimulus, only a metallothionein-like protein (*MT*) showed a 2-fold increase in expression at 12 hpi after which the expression level decreased reaching levels of the non-infected samples at 48 hpi, whereas a heavy-metal-associated protein (*HMAP*) showed no significant change in expression level between infected and non-infected plants.

Microscopic observation of the response of the susceptible grapevine cv. Chasselas to *P. viticola* infection

To monitor response mechanisms and pathogen growth in leaves of a susceptible grapevine, leaf disks were sampled at different time points post inoculation. The development of *P. viticola* in the susceptible grapevine caused visible symptoms such as curling of a heavy infected leaf, oil spots which started to undergo necrosis and sporulation on the abaxial leaf surface within 7 days after inoculation (Fig. 4a, b & c). After 4 dpi hyphal growth was observed (Fig. 4d) and after 7 dpi extensive colonization of leaf tissues was observed by the presence of abundant branched hyphae and the formation of sporangiophores bearing a large number of sporangia (Fig. 4e & g). Aggregation of hyphae in the sub-stomatal cavity was observed (Fig. 4f). As expected mock-inoculated samples at 7 dpi showed no signs of infection (Fig. 4o).

Lignification of the leaf tissues was observed by phloroglucinol-HCI staining leading to a red colourization of lignin. In the inoculated samples 2 and 4 dpi very few red-stained cells were observed (Fig. 4h & i). However at 7 dpi prominent red-stained cells were observed (Fig. 4j) evenly spread throughout the host cells and in close proximity to a stoma (Fig. 4k). No lignification was observed in the mock-inoculated samples at 7 dpi (Fig. 4p)

Plant tissues stained with Naturstoff display a light yellow fluorescence that shows the presence of flavonoids. As expected, no accumulation of flavonoids was found in the mock-inoculated samples 7dpi (Fig. 4q) nor in the inoculated samples 2 dpi (Fig. 4l) but at 4 dpi the presence of the light yellow fluorescence could be seen in the inoculated samples (Fig. 4m) and by 7 dpi a predominant increase in the light yellow fluorescence could be observed (Fig. 4n).

Discussion

Compatibility describes the complementary relationship between a plant species and an adapted pathogen which ultimately results in disease. The progression of *P. viticola* culminating in the appearance of oil spots (Fig. 4b) and the formation of fully developed sporangiophores with sporangia (Fig. 4e) in *V. vinifera* cv. Chasselas represents a compatible interaction. Despite considerable recent progress, our understanding of many aspects of plant-biotroph compatibility is still in its early stages.

Obligate biotrophs, such as powdery and downy mildews and rust fungi, cannot be extensively cultured *in vitro,* since they have to form specialized infection structures, haustoria, within infected host cells. In contrast to necrotrophic and hemibiotrophic fungal

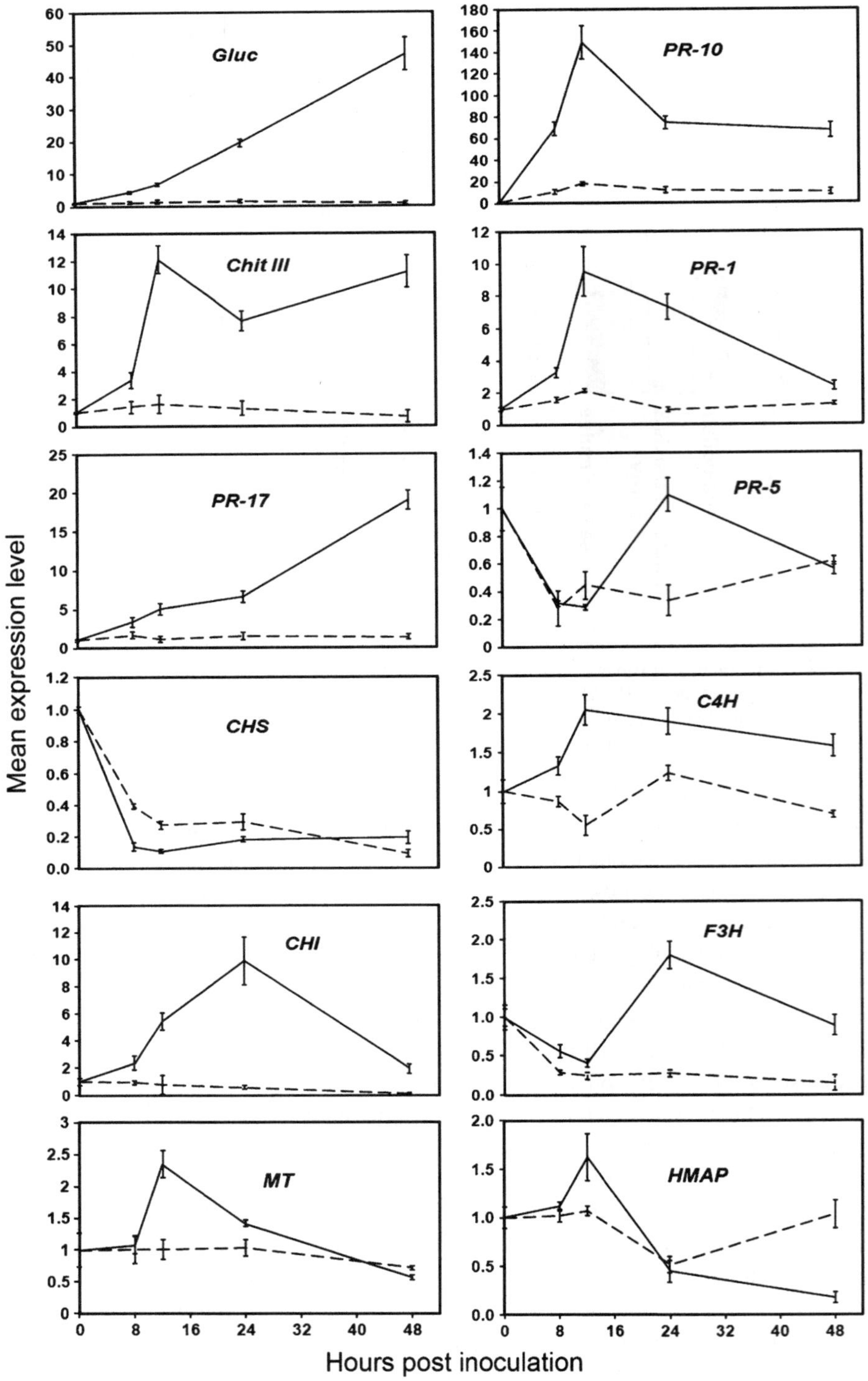

Fig. 3 Reverse transcriptase quantitative real-time PCR (RT-qPCR) analysis of transcript accumulation in grapevine in response to *P. viticola* challenge. Transcript accumulation of *Gluc*, *PR-10*, *Chit III*, *PR-1*, *PR-17*, *PR-5*, *CHS*, *C4H*, *CHI*, *F3H*, *MT* and *HMAP* genes was monitored in mock-inoculated (dashed lines) and *P. viticola*-inoculated (solid line) plants. Transcript levels of the twelve genes are expressed as relative values normalized to the transcript level of *EF1-α* gene, used as an internal reference. Results are means of triplicate data from one representative experiment among two independent repetitions

pathogens, obligate biotrophs are entirely dependent on living plant tissue for their growth and propagation. This lack of host cell death in plant-biotroph interactions might be accomplished by the pathogen's ability to avoid or durably suppress pre-formed and induced host defences (Mendgen and Hahn 2002). It is therefore probable that specific host genes and/or proteins are targeted by biotrophs to achieve these goals, and so be considered as compatibility factors that are essential for successful pathogenesis (Panstruga 2003).

Construction and screening of a SSH library at 24 h after *P. viticola* inoculation

Gene expression profiling has become a valuable tool in functional genomics. SSH is a powerful method to isolate genes that are specifically expressed under diverse sets of conditions. To better understand the molecular basis of compatibility and disease development in a grapevine-*Plasmopara* interaction, we used SSH to identify cDNAs that were differentially expressed in infected plants 24 h after inoculation. Recently, it was shown that rice defence genes induced in resistant plants were also induced in susceptible interactions (Vergne et al. 2007). Moreover, Jantasuriyarat et al. (2005) show that the transcriptional response of a susceptible plant at 24 hpi is most similar to the response of a resistant plant at 6 hpi. They also found that the most highly induced rice genes at 24 hpi (50-fold induction) in a susceptible interaction are also induced in resistant plants (6 and 24 hpi).

Downy mildew oomycetes enter the intercellular spaces of the leaf mesophyll via stomata where they form an infection vesicle with primary hypha and the first haustorium. This mode of entry is expected to have consequences on the timing of recognition of the pathogen by the plant cells and on the nature of induced defence responses. The outcome of plant-pathogen interaction seems to be determined at the time of the subsequent penetration of intact host cells by the haustorium. Recently, two microscopy studies of the infection process of *P. viticola* showed that elongated hyphae with haustoria invaded the intercellular space of the mesophyll by 24 hpi in the resistant *Vitis riparia* cv. Gloire de Montpellier and in the susceptible *V. vinifera* cv. Muscat Ottonel (Diez-Navajas et al. 2008; Unger et al. 2007). Since it is known that the first infection steps, occurring within 24 h after inoculation are essentially the same in a resistant and in a susceptible cultivar we decided to construct the SSH library at 24 hpi. Microscopically, the main difference in infection between the two mentioned cultivars occurs within 48–72 hpi.

Recently, Polesani et al. (2008) used a cDNA-AFLP method to investigate the compatible interaction between *P. viticola* and grapevine at the oil spot stage. They found that nearly 70% of the identified differentially expressed grapevine genes were down-regulated during infection due to the exploitation of cellular resources and the suppression of defence responses. We only found 30% and 13% of the differentially expressed genes to be down-regulated in the forward and reverse library, respectively, the main difference between these two studies being that Polesani and associates carried out their analyses at the oil spot stage, where infection has already been established, whereas our analyses was done at 24 h after inoculation, which mirrors the early stages of infection. At this time point we still expect the susceptible cultivar to try to mount a defence response.

Since the majority of the methods used to screen a SSH library involves visual inspection of inverse dot blots, this constitutes a potential source of false positives. We chose a quantitative screening method that was developed to evaluate the quality of the SSH libraries (Van den Berg et al. 2004; Berger et al. 2007). This approach facilitates the choice of clones from the library for further analysis, such as DNA sequencing, RT-qPCR, Northern blotting or detailed expression profiling using a custom cDNA microarray (Berger et al. 2007). Another advantage of using this screening method with the software SSHscreen is the inclusion of statistical tests of the microarray data to provide further confidence in the choice of genes.

Defence-related genes

The differentially expressed cDNAs isolated from the SSH were classified into several categories and were involved in a number of physiological and molecular events including: metabolism, photosynthesis and energy, response to stimulus, phenylpropanoid and lignin biosynthesis, signal perception and transduction, transport and defence response. A significant number of genes identified in this study that were involved in defence included PR proteins (PR-1, PR-5, PR-10 and PR-17). PR proteins were identified several years ago as being associated with resistance of plants to various pathogens and abiotic stresses (van Loon et al. 2006). Members of the PR-1, and the thaumatin-like PR-5 families have been associated with activity against oomycetes. PR-5-type proteins are thought to create trans-membrane pores and exhibit antifungal activities *in vitro* by blocking spore germination and germ tube growth of *Uncinula necator*, *Phomopsis viticola* and *Botrytis cinerea* (Monteiro et al. 2003). Interestingly, we found *PR-1* to be induced as early as 12 hpi whereas *PR-5* showed no induction over the 48 h time course after inoculation with *P. viticola*. The latter is in agreement with Kortekamp (2006), who found no difference in PR-5 expression in a susceptible and resistant cultivar after inoculation with *P. viticola* indicating that the PR-5's role in defence against *P. viticola* remains unclear. The role of *PR-1* in defence against *P. viticola* also remains uncertain. Interestingly, *P. viticola* did not induce the expression of *PR-1* in the susceptible cultivar Chardonnay according to Chong et al. (2008) but *PR-1* was detected 24 h following inoculation with *P. viticola* in *V. vinifera* cv. Riesling (Wielgoss and Kortekamp 2006). These differences in *PR-1* expression may be related to the use of different cultivars, different isolates of *P. viticola*, semi-quantitative versus quantitative RT-PCR, or even different time points after inoculation. Chong and associates looked at *PR-1* expression between 18 and 96 hpi and we found that at 48 hpi the expression level of *PR-1* was already returning to basal levels close to those of the non-inoculated plants.

A large number of PR-10-related sequences have been identified in many plants. Sequence comparison indicated that PR-10 proteins are structurally related to ribonucleases. The inducible expression of PR-10 genes in response to attacks by pathogens has been widely investigated in a number of plant species. Pathogens triggering a PR-10 response include viruses, bacteria and fungi (Liu and Ekramoddoullah 2006). Some PR-10 proteins display constitutive expression patterns unrelated to the pathogen response. It has been shown that PR-10 proteins are also developmentally regulated in different plant tissues and organs, therefore it is not reasonable to assign a unique function to PR-10 proteins throughout the plant kingdom. Nevertheless, *PR-10* expression was induced in grapevine after challenge with *P. viticola*. A strong induction of *PR-10* was observed at 12 hpi which is in agreement with a number of other studies that found transcript accumulation of *PR-10* in the susceptible grapevine cv. Riesling 12 hpi with *P. viticola* (Kortekamp 2006), in rice within 12 hpi with *Magnaporthe grisea* (McGee et al. 2001) and in sorghum, inducible *PR-10* transcripts reached a maximum level at 12 hpi (Lo et al. 1999). When taking these results into account, the induction of PR-10 may play an important role in the defence of grapevine against *P. viticola*.

Fig. 4 Development of *P. viticola* and accumulation of lignin ▶ and flavonoids in the susceptible cultivar Chasselas. Plants were inoculated with sporangia of *P. viticola* (4×10^4 sporangia ml^{-1}) (**a–n**), sampled 2 dpi (**h** and **l**), 4 dpi (**d**, **i** and **m**) and 7 dpi (**a–c**, **e**, **f**, **g**, **i**, **j**, **k**, and **n**). Mock-inoculated plants were sampled at 7 d (**o–q**). Infected leaves were stained with lactophenol-trypan blue and analyzed by light microscopy (**d–g**), stained with phloroglucinol-HCL and analyzed by light microscopy (**h–k** and **p**), or stained with Naturstoff reagent A and analyzed by epifluorescence microscopy (**l–n** and **q**). **a** White sporulation of *P. viticola* on the abaxial leaf surface; **b** heavily infected leaf showing typical curling symptoms; **c** oil spots starting to undergo necrosis on the adaxial leaf surface. **d** Initial hyphal growth, **e** extensive colonization of leaf tissues and abundant branched hyphae, **f** aggregation of hyphae (hy) in the sub-stomatal cavity (st), **g** sporangiophores bearing a large number of sporangia. **h** Very little amount of lignin, **i** presence of small amount of lignin as shown by red-coloured tissue, **j** increased amount of red-stained lignin in tissues, **k** prominent lignin deposition close to a stoma. **l** No presence of flavonoids at 2 dpi, **m** presence of flavonoids as indicated by yellow fluorescence (4 dpi), **n** accumulation of yellow fluorescence. **o** No presence of infection in mock-inoculated plants at 7d, **p** and **q** no presence of lignin and flavonoids in mock-inoculated plants respectively

Many pathogenesis-related proteins were first characterised in tobacco, NtPRp27, is constitutively expressed in tobacco roots but can be induced by tobacco mosaic virus, wounding, drought and by the application of ethylene, methyl jasmonate, salicylic acid and abscisic acid (Okushima et al. 2000). Recently NtPRp27 became the prototype for the new PR family PR-17. Accumulation of mRNA encoding PR-17 proteins has been reported after benzothiadiazole treatment, inoculation with *Erysiphe graminis* f. sp. *tritici* in wheat (Gorlach et al. 1996) and inoculation with *Blumeria graminis* f. sp. *hordei* (*Bgh*) in barley (Christensen et al. 2002). The transient accumulation of *PR-17* after *P. viticola* inoculation was in close agreement with both transcript and protein accumulation in the compatible and incompatible interactions in barley in response to *Bgh*, diversions in these interactions occured after 72 hpi.

Two other PR proteins namely a *chitinase III* (*PR-8*) and a *β-1,3-glucanase* (*PR-2*) were induced after *P. viticola* inoculation. Both PR proteins are thought to provide defence against fungal pathogens by hydrolysing components of the cell walls of invading fungal hyphae (Mauch et al. 1988). It has been shown that cell walls of *P. viticola* consist of chitin and β-1,3-glucan as seen on the basis of histochemical staining with specific fluorochromes, (Kortekamp 2005) therefore expression of both *chitinase III* and

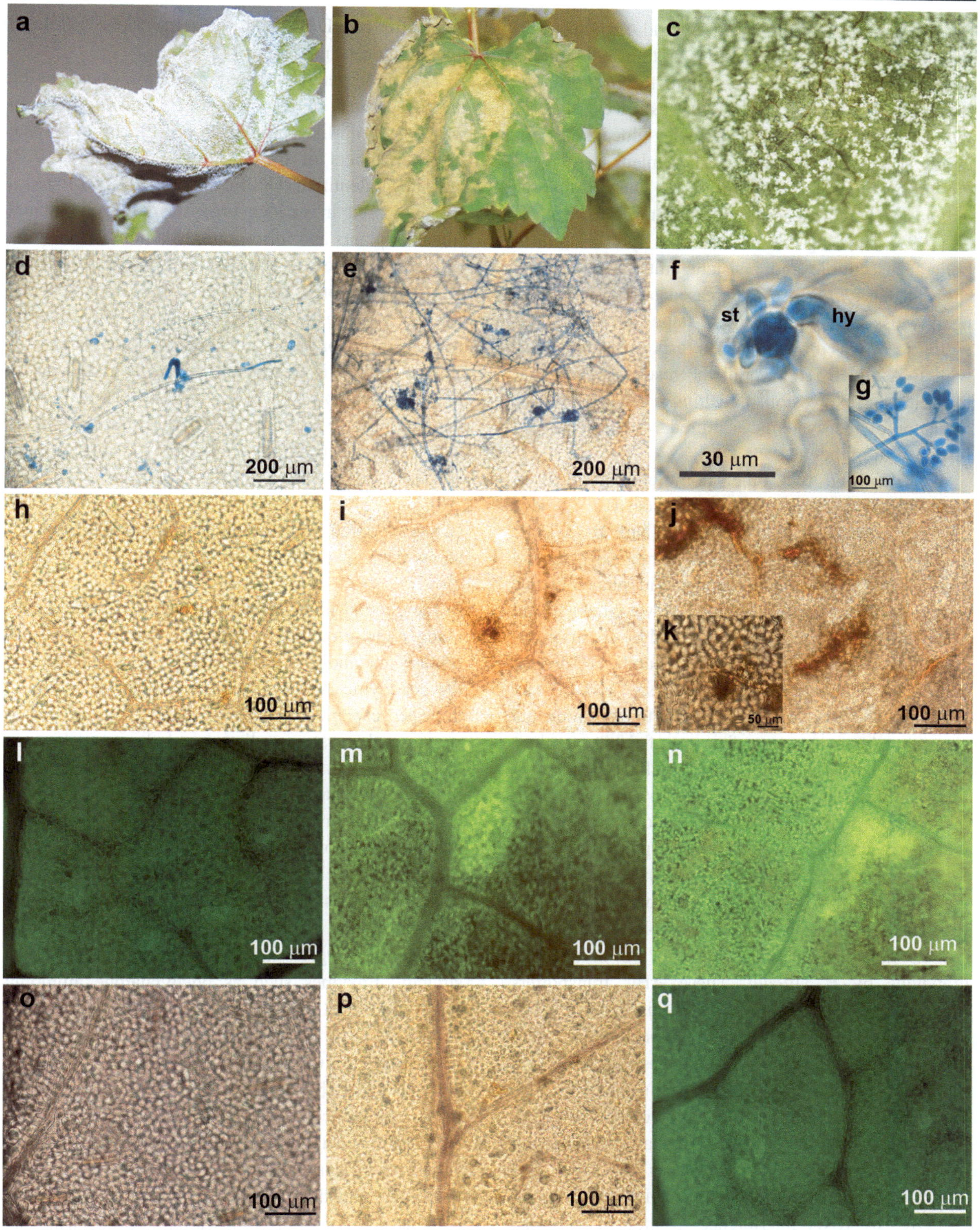
a
b
c
d
200 μm
e
200 μm
f
st
hy
30 μm
g
100 μm
h
100 μm
i
100 μm
j
k
50 μm
100 μm
l
100 μm
m
100 μm
n
100 μm
o
100 μm
p
100 μm
q
100 μm

β-1,3-glucanase may inhibit fungal growth. The induction of *β-1,3-glucanase* was transient and to much higher levels than that found for *chitinase III*. Interestingly, Kortekamp (2005) showed that *P. viticola* septa are composed of β-1,3-glucans but not chitin, which could trigger the host plant to increase the production of β-1,3-glucanase. Our results for *chitinase III* are in accordance with those of Busam et al. (1997), showing that its induction by *P. viticola* in both a susceptible and resistant cultivar. They also showed that *chitinase III* increased transiently in the healthy tissue of younger, next-stage leaves and that *chitinase III* was up-regulated in response to treatment with INA and BTH. Overall the results suggest that the selective expression of *chitinase III* is a reliable indicator of the systemic acquired resistance (SAR) response in grapevine.

We identified two other homologs involved in defence, namely a lipid transfer protein (LTP) and a pleiotropic drug resistance protein 1 (NpPDR1). LTPs are members of the family of pathogenesis-related proteins (PR-14) that are involved in plant defence responses. Plant LTPs also have developmental and tissue specificities, they are modulated by environmental and stress conditions and have a number of diverse functions, including cutin biosynthesis, surface wax formation and adaptation of the plant to environmental changes. An interesting feature of LTPs is their structural similarity with elicitin, which is a small lipid-binding protein that is secreted by the phytopathogenic oomycetes *Phytophthora* and *Pythium*, and which triggers the hypersensitive response (HR) and SAR in tobacco (Ponchet et al. 1999). A homologous gene encoding a protein of the pleiotropic drug resistance (PDR) subfamily of ATP-binding cassette (ABC) transporters was found to be induced in grapevine by *P. viticola*. These ABC transporters mediate the translocation of a diverse array of compounds across biological membranes, including toxins, drugs, glutathione conjugates, peptides and secondary metabolites (Yazaki 2006). There is also increasing evidence for the involvement of ABC transporters in defence. Stukkens et al. (2005) found that a PDR protein in tobacco leaves (NpPDR1) was induced following pathogen attack and silencing of NpPDR1 by RNA interference resulted in increased sensitivity to the plant pathogen *Botrytis cinerea*. More recently Kobae et al. (2006) demonstrated that a loss of an *Arabidopsis* plasma membrane-localized PDR ABC transporter (AtPDR8) lead to hypersensitive cell death upon pathogen infection.

The only defence-related gene we found to be down-regulated in response to *P. viticola* infection was a germin-like protein 6 (GLP). In grapevine seven members of the GLP gene family (*VvGLP*) have been identified (Godfrey et al. 2007). *GLP* genes are expressed in various organs and developmental stages in plants and in response to a number of biotic and abiotic stresses (Bernier and Berna 2001). In grapevine, expression analysis revealed that the *VvGLP* genes exhibited diverse and highly specific patterns of expression in response to a variety of abiotic and biotic treatments, including challenge by *Erysiphe necator*, *P. viticola* and *B. cinerea*. Down regulation of *VvGLP6* in leaf tissue infected with *P. viticola* has been reported previously (Godfrey et al. 2007).

Genes involved in the phenylpropanoid and lignin biosynthesis pathway

In the general phenylpropanoid pathway, three enzymatic transformations redirect the carbon flow from primary metabolism, transforming phenylalanine into the coenzyme A (CoA)-activated hydroxycinnamoyl thioester capable of entering the two major downstream pathways: monolignol and flavonoid biosynthesis. The pathway that provides the lignin-building monolignol units is strongly activated after infection by pathogens or treatments with elicitors. In infected plants, deposition of phenylpropanoid compounds is a part of the cell wall reinforcement that restricts pathogen invasion. Accumulation of soluble and cell-wall bound phenolics in plant tissues challenged by fungal pathogens has been suggested as a defence strategy. Enzymes such as phenylalanine ammonia lyase (PAL; EC 4.3.1.5), cinnamate-4-hydroxylase (C4H; EC 1.14.13.11) and 4-coumarate:coenzyme A ligase (4CL, EC 6.2.1.12) are crucial to phenylpropanoid metabolism. We found the expression of *C4H* to be induced upon *P. viticola* challenge peaking at 12 hpi. Sewalt et al. (1997) showed that reduced C4H activity correlated with reduced levels of lignin indicating that there could be a correlation between *C4H* expression and lignification. Along with the induced expression of *C4H* we found an increase in lignin deposition as shown through histochemical staining (Fig. 4h–k).

All flavonoids are derived from the chalcone scaffold, which is biosynthesized by the enzyme chalcone synthase (CHS). Following CHS, an assemblage of isomerases (CHI), reductases, hydroxylases (F3H), glycosyltransferases and acyltransferases make up the basic flavonoid skeleton leading to a number of flavonoid chemical subclasses such as flavones, isoflavanoids, flavonols, flavandiols, proanthocyanidins and anthocyanins. Even though we found the expression of *CHS* to be down-regulated, the other two enzymes which are involved down-stream from *CHS*, namely *CHI* and *F3H*, were induced by *P. viticola* challenge. The expression level of both *CHI* and *F3H* peaked at 24 hpi. The expression of these two genes coincided with the production of flavonoids as seen by histochemical staining (Fig. 4l–n).

In grapevine, among the best characterised defence reactions upon fungal infection, is the accumulation of stilbene phytoalexins. Resistant grape cultivars artificially inoculated with *P. viticola* showed high amounts of stilbenic phytoalexins at the site of infection, and these phytoalexins have significant inhibitory effects on the mobility of *P. viticola* zoospores (Pezet et al. 2004). Chalcone synthase and stilbene synthase (*STS*) are further key branch-point enzymes in the phenylpropanoid pathway and their enzymatic activities and gene expression are related to phytoalexin biosynthesis in grapevine. We expect to find *STS* present in the SSH library since in separate studies, expression of STS in grapevine was induced after *P. viticola* infection (Chong et al. 2008; Kortekamp 2006). We have consistently seen that the expression of *STS* in *P. viticola*-inoculated grapevine seedlings decreases drastically at 24 hpi and remains low up to 48 hpi (Slaughter et al., unpublished results). Interestingly, Chong et al. (2008) found the expression of *STS* to be induced 48 h after *P. viticola* inoculation and not before and for that reason we were not able to identify *STS* in our library. We found the expression of *CHS* to be down-regulated and we did not find the presence of *STS* and since both are key enzymes in the production of stilbene phytoalexins, they could play an important role in the resistance of grapevine to *P. viticola* but since the SSH library was constructed from a susceptible cultivar to *P. viticola,* the down-regulation of *CHS* and the decrease in STS expression could contribute to the susceptibility of this cultivar to *P. viticola*.

Response to stimulus-related genes

Plant metallothioneins (MTs) are small, highly conserved cysteine-rich heavy-metal-binding proteins, typically classified into four categories (types 1 to 4). Plant MT gene expression is regulated by various factors, including metal ions, developmental stages, symbiotic interaction, and various stress responses. Stresses such as wounding and virus infection were shown to induce the expression of MT in tobacco, while differential expression of MT was detected in *Arabidopsis* following infection of compatible and incompatible strains of *Peronospora* spp. In this respect plant MTs have been proposed to function as reactive oxygen species (ROS) scavengers that protect plant DNA from oxidative damage caused by free radicals released during the oxidative burst (Wong et al. 2004). Although the majority of expressed MT genes belong to type 2, type 1, 3 and 4 MT have also been reported to be involved in various plant processes. Transcript accumulation of type 1 MT was shown in various organs and also in response to pathogen attack (Butt et al. 1998). In this study, induction of a type 1 MT was observed 12 hpi followed by a decline at 24 hpi. A plausible explanation is that the plant accumulated large amounts of MT transcripts in response to *P. viticola* within early stages of infection, but mRNA levels rapidly returned to levels comparable with those expressed in control plants. A similar trend of MT differential gene expression in tobacco plants following *Tobacco mosaic virus* (TMV) infection over a period of 72 h was observed (Choi et al. 1996). Relative amounts of MT mRNA were highly induced at 12 h, sharply dropped after 24 h following TMV infection and remained the same thereafter. We found the expression of a type 2 MT to be down-regulated in response to *P. viticola* challenge, which is in accordance with Wong et al. (2004) who found in rice the expression of *OsMT2b* to be synergistically down-regulated by *OsRac1* and rice blast-derived elicitors. They also demonstrated increased susceptibility to bacterial blight and blast fungus in transgenic plants overexpressing *OsMT2b*. We also found a type 3 MT to be down-regulated in response to *P. viticola* challenge. So far there has only been one case in plants where a type 3 MT was inducible in response to pathogen attack, in velvetleaf leaves in response to *Colletotrichum coccodes*. (Dauch and Jabaji-Hare,

2006). Up-regulation of plant type 3 MT so far has been mostly observed as part of senescence-related responses in plants. In the case of a heavy-metal-associated domain-containing protein we did not find any induction between inoculated and control plants.

Suppression of photosynthesis-related genes

In this study, the largest group of genes suppressed in grapevine during *P. viticola* challenge were photosynthesis-related, namely chlorophyll a-b binding protein, photosystem II polypeptide and the CP29 subunit of photosystem II. Transcriptional down-regulation of photosynthesis-related genes has previously been reported for incompatible and compatible interactions (Moy et al. 2004; Restrepo et al. 2005; Vergne et al. 2007). A down-regulation of genes related to photosynthesis was shown by Polesani et al. (2008) in grapevine at the oil spot stage after challenge with *P. viticola*. The reason for the reduction in photosynthesis during the compatible interaction between grapevine and *P. viticola* is unknown, but Fung et al. (2008) have suggested that in a compatible interaction of grapevine with powdery mildew, there is an up-regulation of invertases, which are involved in degradation of carbon reserves into hexoses, resulting in the reduction of photosynthetic rates. This metabolic shift seems to be coordinated with up-regulation of genes involved in the synthesis of secondary metabolites (flavonoids and lignin) via the oxidative pentose phosphate and shikimate pathway. Increased invertase activity leads to carbohydrate accumulation, which in turn may inhibit the Calvin cycle, which also limits photosynthesis. Two genes encoding enzymes involved in the Calvin cycle, namely rubisco activase and ribulose-1,5-bisphosphate carboxylase/oxygenase small subunit were down-regulated.

Metabolism-associated genes

Proteolysis of important regulatory proteins is a key aspect of cellular regulation in eukaryotes and there is evidence that the ubiquitin-proteasome pathways are important in implementation of the plant defence response. In response to pathogen attack, the Ub/26 S proteasome pathway initiates programmed cell death to localise pathogen spread (Smalle and Vierstra 2004). We found homologs of a 26 S proteasome and an ubiquitin-conjugating enzyme to be up-regulated in grapevine in response to *P. viticola*.

In summary, the SSH method has allowed us to generate a differential cDNA library enriched for *Plasmopara*-responsive genes from grapevine seedlings. This is the first large-scale investigation into the early response (24 hpi) of compatibility between *V. vinifera* and the biotrophic pathogen *P. vitocola*. Polesani et al. (2008) performed cDNA-AFLP analysis on a compatible interaction of grapevine with *P. viticola* on the oil spot stage of infection which follows 2–3 weeks after inoculation; this could be defined as a late response to pathogen challenge, whereas we were interested in analyzing the early response to *P. viticola* challenge. Even though we are dealing with a susceptible cultivar to *P. viticola* challenge, the host plant was still able to mount a defence response as seen by the number of genes identified which are involved in defence mechanisms. Further characterization and functional analysis of the genes generated from the library will facilitate our understanding of the defence mechanisms in grapevine plants to *P. viticola* infection.

Acknowledgements This project was funded by the National Centre of Competence in Research (NCCR) Plant Survival, a research programme of the Swiss National Science Foundation. We thank Mrs. M. Waldner (Syngenta, Stein, Switzerland) for the grapevine seedlings and Dr J. Weber (Lausanne DNA Array Facility, University of Lausanne) for the assistance in the making of the SSH slides for the microarray screening.

References

Adams, M. D., Kelley, J. M., Gocayne, J. D., Dubnick, M., Polymeropoulos, M. H., Xiao, H., et al. (1991). Complementary DNA sequencing: expressed sequence tags and human genome project. *Science, 252*, 1651–1656.

Allegre, M., Daire, X., Heloir, M. C., Trouvelot, S., Mercier, L., Adrian, M., et al. (2007). Stomatal deregulation in Plasmopara viticola-infected grapevine leaves. *The New Phytologist, 173*, 832–840.

Aziz, A., Poinssot, B., Daire, X., Adrian, M., Bezier, A., Lambert, B., et al. (2003). Laminarin elicits defense responses in grapevine and induces protection against Botrytis cinerea and Plasmopara viticola. *Molecular Plant-Microbe Interactions, 16*, 1118–1128.

Berger, D. K., Crampton, B. G., Hein, I., & Vos, W. (2007). Screening of cDNA libraries on glass slide microarrays. *Methods in Molecular Biology, 382*, 177–203.

Bernier, F., & Berna, A. (2001). Germins and germin-like proteins: Plant do-all proteins. But what do they do exactly? *Plant Physiology and Biochemistry, 39*, 545–554.

Birch, P. R. J., Avrova, A. O., Duncan, J. M., Lyon, G. D., & Toth, R. L. (1999). Isolation of potato genes that are induced during an early stage of the hypersensitive response to Phytophthora infestans. *Molecular Plant-Microbe Interactions, 12*, 356–361.

Brenner, S., Johnson, M., Bridgham, J., Golda, G., Lloyd, D. H., Johnson, D., et al. (2000). Gene expression analysis by massively parallel signature sequencing (MPSS) on microbead arrays. *Nature Biotechnology, 18*, 630–634.

Busam, G., Kassemeyer, H. H., & Matern, U. (1997). Differential expression of chitinases in Vitis vinifera L. responding to systemic acquired resistance activators or fungal challenge. *Plant Physiology, 115*, 1029–1038.

Butt, A., Mousley, C., Morris, K., Beynon, J., Can, C., Holub, E., et al. (1998). Differential expression of a senescence-enhanced metallothionein gene in Arabidopsis in response to isolates of Peronospora parasitica and Pseudomonas syringae. *The Plant Journal, 16*, 209–221.

Choi, D., Kim, H. M., Yun, H. K., Park, J. A., Kim, W. T., & Bok, S. H. (1996). Molecular cloning of a metallothionein-like gene from Nicotiana glutinosa L. and its induction by wounding and tobacco mosaic virus infection. *Plant Physiology, 112*, 353–359.

Chong, J., Le Henanff, G., Bertsch, C., & Walter, B. (2008). Identification, expression analysis and characterization of defense and signaling genes in Vitis vinifera. *Plant Physiology and Biochemistry, 46*, 469–481.

Christensen, A. B., Cho, B. H., Naesby, M., Gregersen, P. L., Brandt, J., Madriz-Ordenana, K., et al. (2002). The molecular characterization of two barley proteins establishes the novel PR-17 family of pathogenesis-related proteins. *Molecular Plant Pathology, 3*, 135–144.

Coetzer, N., Gazendam, I., Oelofse, D., & Berger, D. K. (2010). SSHscreen and SSHdb, generic software for microarray based gene discovery: application to the stress response in cowpea. *Plant Methods, 6*, 10.

Dai, G. H., Andary, C., Mondolot-Cosson, L., & Boubals, D. (1995). Histochemical studies on the interaction between three species of grapevine *Vitis vinifera, V. rupestris and V. rotundifolia* and the downy fungus, *Plasmopara viticola. Physiological and Molecular Plant Pathology, 46*, 177–188.

Dauch, A. L., & Jabaji-Hare, S. H. (2006). Metallothionein and bZIP Transcription Factor Genes from Velvetleaf and Their Differential Expression Following Colletotrichum coccodes Infection. *Phytopathology, 96*, 1116–1123.

Degenhardt, J., Al-Masri, A. N., Kurkcuoglu, S., Szankowski, I., & Gau, A. E. (2005). Characterization by suppression subtractive hybridization of transcripts that are differentially expressed in leaves of apple scab-resistant and susceptible cultivars of Malus domestica. *Molecular Genetics and Genomics, 273*, 326–335.

Derckel, J. P., Baillieul, F., Manteau, S., Audran, J. C., Haye, B., Lambert, B., et al. (1999). Differential Induction of Grapevine Defenses by Two Strains of Botrytis cinerea. *Phytopathology, 89*, 197–203.

Diez-Navajas, A. M., Wiedemann-Merdinoglu, S., Greif, C., & Merdinoglu, D. (2008). Nonhost versus host resistance to the grapevine downy mildew, Plasmopara viticola, studied at the tissue level. *Phytopathology, 98*, 776–780.

Fung, R. W., Gonzalo, M., Fekete, C., Kovacs, L. G., He, Y., Marsh, E., et al. (2008). Powdery mildew induces defense-oriented reprogramming of the transcriptome in a susceptible but not in a resistant grapevine. *Plant Physiology, 146*, 236–249.

Gindro, K., Pezet, R., & Viret, O. (2003). Histological study of the responses of two Vitis vinifera cultivars (resistant and susceptible) to Plasmopara viticola infections. *Plant Physiology and Biochemistry, 41*, 846–853.

Glazebrook, J. (2005). Contrasting mechanisms of defense against biotrophic and necrotrophic pathogens. *Annual Review of Phytopathology, 43*, 205–227.

Godfrey, D., Able, A. J., & Dry, I. B. (2007). Induction of a grapevine germin-like protein (VvGLP3) gene is closely linked to the site of Erysiphe necator infection: a possible role in defense? *Molecular Plant-Microbe Interactions, 20*, 1112–1125.

Gorlach, J., Volrath, S., Knauf-Beiter, G., Hengy, G., Beckhove, U., Kogel, K. H., et al. (1996). Benzothiadiazole, a novel class of inducers of systemic acquired resistance, activates gene expression and disease resistance in wheat. *The Plant Cell, 8*, 629–643.

Hamiduzzaman, M. M., Jakab, G., Barnavon, L., Neuhaus, J. M., & Mauch-Mani, B. (2005). beta-Aminobutyric acid-induced resistance against downy mildew in grapevine acts through the potentiation of callose formation and jasmonic acid signaling. *Molecular Plant-Microbe Interactions, 18*, 819–829.

Heath, M. C. (2000). Nonhost resistance and nonspecific plant defenses. *Current Opinion in Plant Biology, 3*, 315–319.

Iandolino, A. B., Goes da Silva, F., Lim, H., Choi, H., Williams, L. E., & Cook, D. R. (2004). High-quality RNA, cDNA, and dervided EST libraries from grapevine (*Vitis vinifera* L.). *Plant Molecular Biology Reporter, 22*, 269–278.

Jantasuriyarat, C., Gowda, M., Haller, K., Hatfield, J., Lu, G., Stahlberg, E., et al. (2005). Large-scale identification of expressed sequence tags involved in rice and rice blast fungus interaction. *Plant Physiology, 138*, 105–115.

Keogh, R. C., Deverall, B. J., & McLeod, S. (1980). Comparison of histological and physiological responses to *Phakoopsora pachyrhizi* in resistant and susceptible soybean. *Transactions of the British Mycological Society, 74*, 328–333.

Kobae, Y., Sekino, T., Yoshioka, H., Nakagawa, T., Martinoia, E., & Maeshima, M. (2006). Loss of AtPDR8, a plasma membrane ABC transporter of Arabidopsis thaliana, causes hypersensitive cell death upon pathogen infection. *Plant & Cell Physiology, 47*, 309–318.

Kortekamp, A. (2005). Growth, occurrence and development of septa in Plasmopara viticola and other members of the Peronosporaceae using light- and epifluorescence-microscopy. *Mycological Research, 109*, 640–648.

Kortekamp, A. (2006). Expression analysis of defence-related genes in grapevine leaves after inoculation with a host and a non-host pathogen. *Plant Physiology and Biochemistry, 44*, 58–67.

Kortekamp, A., & Zyprian, E. (2003). Characterization of Plasmopara-resistance in grapevine using in vitro plants. *Journal of Plant Physiology, 160*, 1393–1400.

Liu, J. J., & Ekramoddoullah, A. K. M. (2006). The family 10 of plant pathogenesis-related proteins: Their structure, regulation, and function in response to biotic and abiotic

stresses. *Physiological and Molecular Plant Pathology, 68*, 3–13.
Lo, S. C., Hipskind, J. D., & Nicholson, R. L. (1999). cDNA cloning of a sorghum pathogenesis-related protein (PR-10) and differential expression of defense-related genes following inoculation with Cochliobolus heterostrophus or Colletotrichum sublineolum. *Molecular Plant-Microbe Interactions, 12*, 479–489.
Lu, G., Jantasuriyarat, C., Zhou, B., & Wang, G. L. (2004). Isolation and characterization of novel defense response genes involved in compatible and incompatible interactions between rice and Magnaporthe grisea. *Theoretical and Applied Genetics, 108*, 525–534.
Mauch, F., Mauch-Mani, B., & Boller, T. (1988). Antifungal hydrolases in pea tissue: II. Inhibition of fungal growth by combinations of chitinase and beta-1, 3-glucanase. *Plant Physiology, 87*, 936–942.
McGee, J. D., Hamer, J. E., & Hodges, T. K. (2001). Characterization of a PR-10 pathogenesis-related gene family induced in rice during infection with Magnaporthe grisea. *Molecular Plant-Microbe Interactions, 14*, 877–886.
Mendgen, K., & Hahn, M. (2002). Plant infection and the establishment of fungal biotrophy. *Trends in Plant Science, 7*, 352–356.
Monteiro, S., Barakat, M., Picarra-Pereira, M. A., Teixeira, A. R., & Ferreira, R. B. (2003). Osmotin and thaumatin from grape: a putative general defense mechanism against pathogenic fungi. *Phytopathology, 93*, 1505–1512.
Moy, P., Qutob, D., Chapman, B. P., Atkinson, I., & Gijzen, M. (2004). Patterns of gene expression upon infection of soybean plants by Phytophthora sojae. *Molecular Plant-Microbe Interactions, 17*, 1051–1062.
Musetti, R., Marabottini, R., Badiani, M., Martini, M., di Toppi, L. S., Borselli, S., et al. (2007). On the role of H2O2 in the recovery of grapevine (Vitis vinifera cv. Prosecco) from Flavescence doree disease. *Functional Plant Biology, 34*, 750–758.
Okushima, Y., Koizumi, N., Kusano, T., & Sano, H. (2000). Secreted proteins of tobacco cultured BY2 cells: identification of a new member of pathogenesis-related proteins. *Plant Molecular Biology, 42*, 479–488.
Panstruga, R. (2003). Establishing compatibility between plants and obligate biotrophic pathogens. *Current Opinion in Plant Biology, 6*, 320–326.
Pezet, R., Gindro, K., Viret, O., & Richter, H. (2004). Effects of resveratrol, viniferins and pterostilbene on *Plasmopara viticola* zoospore mobility and disease development. *Vitis, 43*, 145–148.
Pina, A., & Errea, P. (2008). Differential induction of phenylalanine ammonia-lyase gene expression in response to in vitro callus unions of Prunus spp. *Journal of Plant Physiology, 165*, 705–714.
Polesani, M., Bortesi, L., Ferrarini, A., Zamboni, A., Fasoli, M., Zadra, C., et al. (2010). General and species-specific transcriptional responses to downy mildew infection in a susceptible (*Vitis vinifera*) and a resistant (*V. riparia*) grapevine species. *BMC Genomics, 11*, 117.
Polesani, M., Desario, F., Ferrarini, A., Zamboni, A., Pezzotti, M., Kortekamp, A., et al. (2008). cDNA-AFLP analysis of plant and pathogen genes expressed in grapevine infected with *Plasmopara viticola*. *BMC Genomics, 9*, 142.
Ponchet, M., Panabieres, F., Milat, M. L., Mikes, V., Montillet, J. L., Suty, L., et al. (1999). Are elicitins cryptograms in plant-Oomycete communications? *Cellular and Molecular Life Sciences, 56*, 1020–1047.
Restrepo, S., Myers, K. L., del Pozo, O., Martin, G. B., Hart, A. L., Buell, C. R., et al. (2005). Gene profiling of a compatible interaction between Phytophthora infestans and Solanum tuberosum suggests a role for carbonic anhydrase. *Molecular Plant-Microbe Interactions, 18*, 913–922.
Ruepp, A., Zollner, A., Maier, D., Albermann, K., Hani, J., Mokrejs, M., et al. (2004). The FunCat, a functional annotation scheme for systematic classification of proteins from whole genomes. *Nucleic Acids Research, 32*, 5539–5545.
Scheel, D. (1998). Resistance response physiology and signal transduction. *Current Opinion in Plant Biology, 1*, 305–310.
Sewalt, V. J. H., Ni, W. T., Blount, J. W., Jung, H. G., Masoud, S. A., Howles, P. A., et al. (1997). Reduced lignin content and altered lignin composition in transgenic tobacco down-regulated in expression of L-phenylalanine ammonia-lyase or cinnamate 4-hydroxylase. *Plant Physiology, 115*, 41–50.
Smalle, J., & Vierstra, R. D. (2004). The ubiquitin 26 S proteasome proteolytic pathway. *Annual Review of Plant Biology, 55*, 555–590.
Stukkens, Y., Bultreys, A., Grec, S., Trombik, T., Vanham, D., & Boutry, M. (2005). NpPDR1, a pleiotropic drug resistance-type ATP-binding cassette transporter from Nicotiana plumbaginifolia, plays a major role in plant pathogen defense. *Plant Physiology, 139*, 341–352.
Trouvelot, S., Varnier, A. L., Allegre, M., Mercier, L., Baillieul, F., Arnould, C., et al. (2008). A beta-1, 3 glucan sulfate induces resistance in grapevine against Plasmopara viticola through priming of defense responses, including HR-like cell death. *Molecular Plant-Microbe Interactions, 21*, 232–243.
Unger, S., Buche, C., Boso, S., & Kassemeyer, H. H. (2007). The Course of Colonization of Two Different Vitis Genotypes by Plasmopara viticola Indicates Compatible and Incompatible Host-Pathogen Interactions. *Phytopathology, 97*, 780–786.
Van den Berg, N., Berger, D. K., Hein, I., Birch, P., Wingfield, M. J., & Viljoen, A. (2007). Tolerance in banana to Fusarium wilt is associated with early up-regulation of cell wall-strengthening genes in the roots. *Molecular Plant Pathology, 8*, 333–341.
Van den Berg, N., Crampton, B. G., Hein, I., Birch, P., & Berger, D. K. (2004). High-throughput screening of suppression subtractive hybridization cDNA libraries using DNA microarray analysis. *Biotechniques, 37*, 818–824.
van Loon, L. C., Rep, M., & Pieterse, C. M. (2006). Significance of inducible defense-related proteins in infected plants. *Annual Review of Phytopathology, 44*, 135–162.
Vandesompele, J., De Preter, K., Pattyn, F., Poppe, B., Van Roy, N., De Paepe, A., et al. (2002). Accurate normalization of real-time quantitative RT-PCR data by geometric averaging of multiple internal control genes. *Genome Biology, 3*, 7.

Velculescu, V. E., Zhang, L., Vogelstein, B., & Kinzler, K. W. (1995). Serial analysis of gene expression. *Science, 270*, 484–487.

Vergne, E., Ballini, E., Marques, S., Sidi Mammar, B., Droc, G., Gaillard, S., et al. (2007). Early and specific gene expression triggered by rice resistance gene Pi33 in response to infection by ACE1 avirulent blast fungus. *The New Phytologist, 174*, 159–171.

Werner, S., Steiner, U., Becher, R., Kortekamp, A., Zyprian, E., & Deising, H. B. (2002). Chitin synthesis during in planta growth and asexual propagation of the cellulosic oomycete and obligate biotrophic grapevine pathogen Plasmopara viticola. *FEMS Microbiology Letters, 208*, 169–173.

Wielgoss, A., & Kortekamp, A. (2006). Comparison of PR1 expression in grapevine cultures after inoculation with a host- and a non-host pathogen. *Vitis, 45*, 9–13.

Wong, H. L., Sakamoto, T., Kawasaki, T., Umemura, K., & Shimamoto, K. (2004). Down-regulation of metallothionein, a reactive oxygen scavenger, by the small GTPase OsRac1 in rice. *Plant Physiology, 135*, 1447–1456.

Yazaki, K. (2006). ABC transporters involved in the transport of plant secondary metabolites. *FEBS Letters, 580*, 1183–1191.

Eur J Plant Pathol (2011) 129:171–182
DOI 10.1007/s10658-010-9697-7

Quantitative studies on downy mildew (*Peronospora destructor* Berk. Casp.) affecting onion seed production in southern Uruguay

Pablo H. González · Paula Colnago · Sebastián Peluffo · Héctor González Idiarte · Javier Zipitría · Guillermo A. Galván

Accepted: 7 October 2010 / Published online: 22 October 2010

Abstract Onion downy mildew (Dm) symptoms and damage on seed production fields in southern Uruguay were quantified during two seasons as the progress of incidence, severity, and as the effect of the level of seed-stalks infections on seed yield and quality. In addition, the effects of two plantation dates and two plant densities on Dm were studied in a factorial experiment. Maximum incidence along the season ranged from 15 to 65% in four commercial fields in 2005 and two fields in 2006. Maximum severity ranged from 4.5 to 9.3% of leaf area affected in 2005, and 0.35 to 1.17% in 2006. Whereas Dm incidence varied among studied fields, disease progress as Dm severity varied mainly between studied years. Crop rotation, crop vigour and plant density were identified as major factors affecting Dm variation in the field. Intensive fungicide schedules did not control Dm when other conditions favoured the disease. A high level of Dm severity defined as large necklace spots on seed-stalks significantly reduced seed yield in comparison with healthy seed-stalks in 2005, as well as seed yield and germination in 2006. Weight of 1,000 seeds was not significantly reduced by Dm infections on seed-stalks. Six genera of fungi were recovered from seeds harvested on highly infected seed-stalks (*Stemphylium, Fusarium, Alternaria, Penicillium, Aspergillus* and *Botrytis*), but not *Peronospora*. Late planting date and low plant density had significantly lower Dm incidence and severity. This finding questioned the early planting dates and high densities previously recommended in order to achieve high yielding crops. The combination of several practices to reduce initial inoculum, susceptibility status of the host, and environmental conditions promoting the disease is discussed as forming a basis for effective disease control.

Keywords *Allium cepa* · Epidemiology · Integrated disease management · Seed-borne pathogens

P. H. González
Plant Protection Department, Facultad de Agronomía, Universidad de la República, Progreso, Canelones, Uruguay

P. Colnago · S. Peluffo · H. González Idiarte · J. Zipitría · G. A. Galván (✉)
Plant Production Department, Centro Regional Sur (CRS), Facultad de Agronomía, Universidad de la República, Camino Folle km 36, Progreso, Canelones, Uruguay
e-mail: horticrs@fagro.edu.uy

Introduction

Onion (*Allium cepa* L.) is one of the main vegetable crops in farming systems in southern Uruguay, with an annual acreage in the order of 2,000 ha. Production of onion seeds in small households has been a traditional practice that led to the development and maintenance of germplasm adapted to local conditions (Galván et al. 1997, 2005). Although seed production is a widespread activity, on-farm produced seed lots have dissimilar physiological and sanitary qualities (Cardani et al. 1990). Recently, onion breeding activities in Uruguay released cultivars (Vilaró et al. 2005) which currently hold more than 50% of the cultivated onion area (MGAP 2006). The uptake of these cultivars depended on their availability in the required quantity and quality. Consequently, there was the development of a seed certification programme involving seed growers (González et al. 2007).

Downy mildew (Dm) caused by *Peronospora destructor* Berk. (Casp.) is a major leaf disease for the production of onion bulbs, as well as for the production of onion seeds in the second year of this biennial species (Schwartz and Mohan 2008). Yield reductions of onion bulbs due to Dm outbreaks may range from 30% up to 70% when the environment is conducive for the disease (Van Doorn 1959; Maude 1990). Yields are strongly reduced by *P. destructor* during seed production, where 30–97% losses have been reported in Brazil (Verona et al. 1996) and 60–70% in India (Sharma et al. 2002). The spraying with (not always effective) chemicals adds to the economic losses. Forecast systems like 'Downcast' (Sutton 1986) reduce the number of chemical interventions during the season (Lorbeer et al. 2002), although not significantly in Uruguay because environmental conditions for Dm sporulation and infection frequently occur (Maeso 2005). A high level of host resistance does not occur within *A. cepa* (Kofoet and Zinkernagel 1989). Introgression of Dm resistance from *A. roylei* was recently achieved, and the development of onion resistant cultivars is continuing (Scholten et al. 2007). However, use of this resistance in onion cultivars adapted to diverse regions is a long term process. Knowledge on cultural practices influencing Dm leading to an integrated disease management is still important.

Typical symptoms of *P. destructor* are pale green to yellowish areas on leaves and seed-stalks, ovate to cylindrical in shape, which become covered by masses of gray fruiting bodies under appropriate environmental conditions (Schwartz and Mohan 2008). *P. destructor* may have high sporulation rates capable of destroying crop foliage quickly. As infected leaf areas turn necrotic and leaf tips collapse, plant growth and crop yields are negatively affected (Maude 1990). Loss of foliage results in thick onion necks and reduction in the keeping quality of onion bulbs (Schwartz and Mohan 2008). Stored bulbs can remain infected, and act as a primary source of inoculum for seed production (Virányi 1974). Dm is especially harmful for seed production because the supply of water and nutrients to the umbel can be interrupted when seed-stalks are infected, which may weaken and break down (Schwartz and Mohan 2008). Moreover, seeds may be infected by *P. destructor*, acting as primary inoculum for the next season (Rondomanski 1971).

Peronospora destructor occurs all over the world but is important in temperate regions (CMI 1990). Nevertheless, it requires very specific environmental conditions and therefore the timing and relevance of Dm outbreaks in a location may vary as a consequence of weather differences between years and crop management practices influencing micro-environment. Sporulation of Dm requires relative humidity above 95% or dew, temperatures from 4–7°C up to 22–25°C and darkness, though a previous period of intense light promotes sporulation (Hildebrand and Sutton 1982). As a consequence, sporulation occurs during the night when optimal conditions prevail for at least 6 h. Rainfall prevents sporulation. Release of sporangia takes place during the day when relative humidity decreases (Leach et al. 1982). New infections occur with a film of water present on the leaves for at least two to 6 h and temperature between 3 and 14°C. Latency period up to the development of new sporulation lasts from 9 to 16 days (Maude 1990; Schwartz and Mohan 2008).

The effects of onion downy mildew on yields and quality of onion seed in Uruguay have not been evaluated previously, although the presence of seed-stalk diseases and the lack of epidemiological information are major constraints in seed production. Plant diseases add complexity to crop management and uncertainty about the level of production. This research aims to quantify symptoms and damage caused by *P. destructor* in onion seed production in southern Uruguay. By surveying commercial fields across two seasons, Dm incidence and severity were evaluated, as well as the effect of seed-stalk infection levels on yield and quality of onion seeds. In addition, an experiment was set up to study the effects of two planting dates and two plant densities on the development of the disease. Factors influencing Dm differences between commercial farm fields were analyzed.

Materials and methods

Disease incidence and severity

The prevalence of downy mildew was assessed by monitoring commercial onion seed producing fields in the seasons 2005 and 2006 in Canelones, Uruguay (Table 1). Cultivar 'Pantanoso del Sauce CRS' was grown in all cases. Planting date was between August 1st and 12th, onset of anthesis took place in the second half of November, and harvesting in the first half of January. In the studied fields, weeds were controlled by herbicides and manual clearing, and did not influence the development of diseases, nor seed yields. Production as yield per plant, yield per hectare and seed germination are presented in Table 1. Farmers used similar combinations of fungicides and 7 to 10 days spraying schedule during October to December, except for DP in 2005 and JL in 2006, who sprayed fewer times after monitoring a low presence of Dm (Table 1).

Disease incidence was quantified as the percentage of plants with Dm symptoms, and disease severity was estimated as the percentage of leaf area infected per plant. Only Dm typical pale green to yellowish spots, or leaf areas covered by typical Dm sporulation were assessed for incidence and severity. Necrotic areas caused by old infections on leaf areas were disregarded, with only Dm active infections considered. Atypical symptoms of Dm were assessed on leaves and seed stalks by removing samples from the field and placing in a humid chamber before observation using a binocular microscope.

In 2005, Dm incidence was evaluated in four commercial fields that differed in crop rotation, plant density, use of fertilizers and fungicides (Table 1). The observed level of Dm disease in the previous season as perceived by farmers also varied and was recorded, although the piece of land within the farm exploited for onion seed production differed year to year. Each crop was divided into four quadrant sections, and five plants were randomly chosen within each section. Plants were assessed every 2 weeks from October 18th till December 27th (six times). Only symptoms developed on leaves were assessed for incidence and severity.

In 2006, Dm incidence and severity were evaluated in two fields selected for their contrasting prevalence of Dm disease in the farm previously(Table 1). One percent of the plants were randomly chosen within each quadrant (five and eight plants per quadrant per farm). The plants were assessed weekly from the 4th October to 22nd November 2006 (nine times). Dm severity on leaves and severity on seed-stalks were estimated separately.

Effects of seed-stalk infections on seed yield and quality

The effects of the level of Dm infections on seed-stalks were investigated by sampling inflorescences in commercial fields. A field with intermediate to high infestation was selected in the season 2005, and another in 2006 (Table 2), which permitted the differentiation of three infection levels on seed-stalks, defined as follows: (0) healthy stems; (1) 25–50% of infected tissue; and (2) 25–50% of infected tissue, with necklace spots extending over the circumference of the stem. Plants corresponding to each infection level were randomly selected and marked at the end of anthesis (early December). Six replications by infection level were set, each consisting of three stems (18 stems in total). In the season 2005, the sanitary status of seed-stalks was evaluated again at harvest (January 12th 2006), as well as the diameter of the umbels. Seed yield per plant and yield per hectare were calculated (with 4.5 stems per plant in 2005 and 4.0 in 2006). Seed quality was evaluated following ISTA protocols (1976) as percentage of seed germination at 12 days from the beginning of the test, seed vigour as percentage of germinated seedling at 6 days, and weight of 1,000 seeds. Genera of fungi that developed on the seed surface during germination tests were identified and quantified.

Effect of planting date and plant density

An experiment aiming to evaluate the effects of two planting dates and two plant densities on Dm incidence and severity was set at the Centro Regional Sur (Progreso, Canelones). Onion bulbs cv. 'Pantanoso del Sauce CRS' were planted on July 27th and August 29th 2005 at two plant densities (4.45 and 8.90 plants·m^{-1}), following a factorial split-plot design with four replications. Planting dates were the large experimental plots (6.0×3.2 m), and plant densities were assessed as small plots (3.0×3.2 m) within planting dates. Four

Table 1 Description and productive results of onion seed producing fields in southern Uruguay surveyed for downy mildew in 2005 and 2006

Characteristics	Season 2005				Season 2006	
	JL	CM	JM	DP	JL	NZ
Description						
Location	Cuchilla de Sierra	Cañada Grande	Cañada Grande	San Jacinto	Cuchilla de Sierra	Paso Garúa
Field size (m^2)	500	600	1032	3394	620	1700
Plant density (m^{-2})	4.00	2.80	1.55	2.80	4.00	3.20
Plant distribution	Simple rows 0.25 m between plants	Double rows 0.45 m between plants	Simple rows 0.40 m between plants	Double rows 0.45 m between plants	Simple rows 0.25 m between plants	Simple rows 0.25 m between plants
Manures and Fertilizers	Chicken manure and Superphosphate	Ammonium phosphate	Ammonium phosphate	No fertilizers	Chicken manure and Superphosphate	NPK 15-15-15 and Urea
Fungicides sprayed	Copper oxichloride, mancozeb, Propamocarb metil, metalaxil	Copper oxichloride, mancozeb, Propamocarb metil, metalaxil	Copper oxichloride, mancozeb, Propamocarb metil, metalaxil	Mancozeb +Metalaxil	Copper oxichloride, mancozeb, Propamocarb metil, metalaxil	Copper oxichloride, mancozeb, Metalaxil
Nr. of spraying times and period	6 Oct–Dec	6 Oct–Dec	6 Oct–Dec	1 Nov	2 Oct–Dec	6 Oct–Dec
Previous crop	Onion	Sweet potato	Maize	Tomato	Sweet pepper	Potato
Downy mildew in the previous season	Low	High	No detected	Low	High	Low
Productive results						
Yield per plant (g)	26	18	18	19	18.1	18.6
Yield (kg ha^{-1})	1040	504	279	532	744	579
Germination (%)[a]	96	93	86	89	93	92

[a] Test performed after seed cleaning

Table 2 Description of onion seed producing-fields in which the effects of the level of downy mildew infection on seed-stalks were investigated

Characteristics	Season 2005–06 AR	Season 2006–07 MR
Location	Canelón Chico, Canelones	La Paloma, Canelones
Field size (m^2)	495	1650
Plantation date	July 25–30	August 4
Plant distribution	Double rows 0.3 m between plants	Double rows 0.6 m between plants
Plant density (m^{-2})	3.85	2.18
Fertilization	Chicken manure 10 Mg ha^{-1}	Ammonium phosphate 200 kg ha^{-1}
Fungicides sprayed	Copper oxichloride, mancozeb +metalaxil	Copper oxichloride, mancozeb +metalaxil
Nr. of spraying times and period	3 Oct–Nov	4 Oct–Nov
Previous crops	Potato	Fallow
Downy mildew in the previous season	No detected	High
Productive results		
Yield per plant (g)	8.33	5.83
Yield (kg ha^{-1})	320	127
Germination (%)[a]	87	66

[a] Test performed after seed cleaning

plants per plot were assessed weekly for Dm incidence and severity (16 plants per treatment-combination). The effects of these treatment-combinations on seed yield, seed germination, seed vigour, weight of 1,000 seeds, and the presence of fungal infections on seeds were also assessed.

Data analysis

For the survey of commercial fields, standard deviations for Dm incidence means *(p)* were calculated using the four quadrants as replicates (n) and normal approximation for a binomial distribution, as: $SD = \sqrt{(p) \times (1-p)/n}$ (Box et al. 1978). Pearson correlation and Spearman rank correlation between Dm variables and onion seed yield in commercial fields were performed. For the experiment designed to assess two plantation dates and two plant densities, Dm incidence and severity were analyzed for each evaluation time as variables with respectively binomial and multinomial distributions, using residual maximum likelihood analysis in Genstat Discovery Edition 7.2.0.220 (VSN International Ltd., Lawes Agricultural Trust, UK). Seed yield and seed quality traits were studied by analysis of variance, and significant differences were established with Fischer's-protected test (LSD). Before analysis, seed germination, seed vigour and the presence of rotten seeds expressed as percentages were transformed by natural logarithm.

Results

Disease incidence and severity

In 2005, Dm incidence in the four studied fields increased until November and decreased in December (Fig. 1). Maximum incidences of 50 and 65% in two farm fields (JL, JM) contrasted with 15 and 20% reached in the other two fields (CM, DP). Dm severity increased from October 18th until November 29th. Evaluations of severity during December were disregarded because necrosis caused by natural leaf senescence and by *Stemphylium vesicarium* and *Alternaria porri* infections become predominant. Maximum Dm severity ranged from 4.45 up to 9.25% of leaf area affected in the four fields. The farm field JM with maximum Dm incidence also had maximum severity. The farm field CM had disease distribution in patches, and consequently standard deviation values were larger than other fields and as large as severity mean values (data not shown).

In 2006, maximum Dm incidence for two studied fields was 30 and 45%, and severity was respectively 0.35 and 1.17%. Whereas disease progress in the former field was erratic, in the latter field disease kept increasing until November 29th, and in both cases decreased in December (Fig. 1). Dm symptoms on seed-stalks were observed less frequently than symptoms on leaves (data not shown).

The range of variation between fields for Dm incidence was similar between years, but Dm severity was higher in 2005 than that observed in 2006. Maximum Dm incidence, maximum Dm severity, and cumulative Dm severity in the surveyed fields were not (negatively) associated with seed yields per plant, yield per hectare, nor with seed germination (correlation tests not shown).

Fig. 1 Downy mildew evaluated in commercial onion seed producing fields along the season. Only typical symptoms caused by new infections and leaf areas covered by sporulation were accounted for. (**a**) Incidence and (**c**) severity in four fields surveyed in 2005. For simplicity, standard deviations were drawn only for the lowest and highest incidence means within each evaluation time, and JM and DP severity means. (**b**) Incidence and (**d**) severity in two fields surveyed in 2006

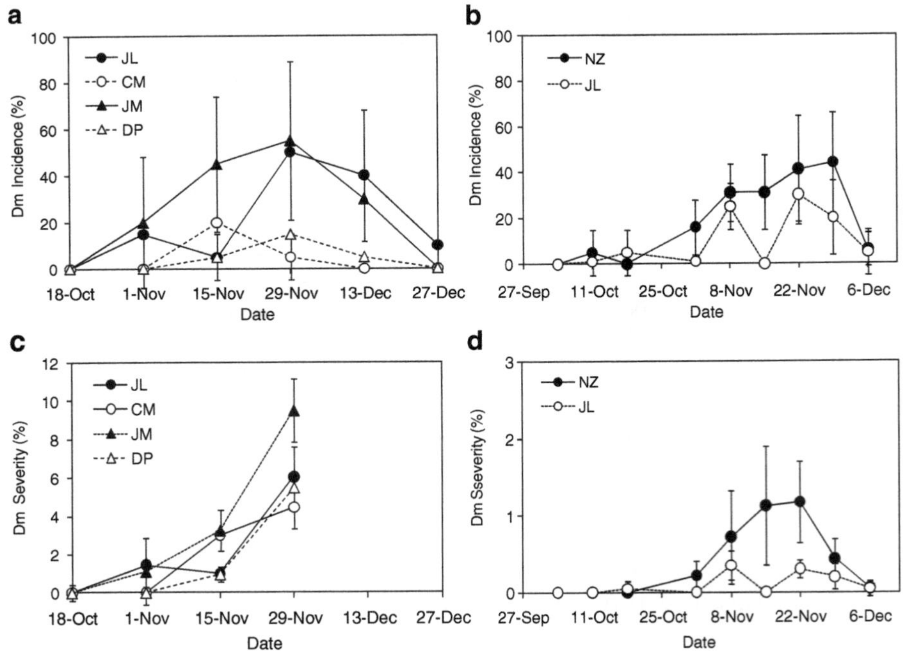

Effects of downy mildew on seed yields and quality

In 2005, onion seed yield obtained with severity levels 2 (25–50% of the seed-stalk infected with necklace spots) and 1 (25–50% of the seed-stalk infected without necklace spots) were lower than yields obtained from healthy stems ($p<0.05$) (Fig. 2). At harvest, seed-stalk area affected by Dm was estimated as 51, 44 and 0% for respectively infection levels 2, 1 and 0. Among stalks marked as infection level 1 at the end of anthesis, 56% of the stalks presented necklace spots at harvest due to secondary infections caused by *Stemphylium vesicarium* or *Alternaria porri* on lesions initially caused by *P. destructor*. Average diameters of umbels were 6.9, 7.2 and 7.3 cm for infection levels 2, 1 and 0 respectively, which were not significantly different.

In 2006, seed yields in the studied field were approximately one quarter of the yields obtained in 2005. Under these conditions, seed yield of seed-stalks with severity level 2 was significantly lower than yields obtained with infection levels 1 and 0 (Fig. 2).

Seed germination, vigour and weight of 1,000 seeds were not significantly reduced by the level of Dm infection on the seed-stalk in the season 2005 (Table 3). The effects of Dm infection levels on seed quality were more severe in 2006. For instance, germination means ranged from 76 to 71% in 2005 and from 55 to 42% in 2006 for the different infection levels. Germination tests produced sub-normal growth of seedling radicles and/or epicotyls, as well as the presence of other pathogens (rotten seeds). In the season 2006, seed germination for seed-stalks with infection level 2 was lower than levels 1 and 0, whereas seed vigour was lower for infection level 1 in comparison with levels 0 and 2 ($p<0.05$). Weight of 1,000 seeds was not significantly affected by Dm infection levels on seed-stalks (Table 3).

Six genera of fungi were identified from infected seeds on blotters prepared to evaluate seed germination (Table 4). Fungal infections increased from 8% for seeds produced on healthy stems to 23.67% for seeds collected on seed-stalks with the higher Dm infection level. *Stemphylium* and particularly its perfect state *Pleospora* (perithecia observed on seed surface and radicle) was the most frequently observed genus. *Fusarium*, *Alternaria*, *Penicillium*, *Aspergillus* and *Botrytis* were found at lower frequencies. No *Peronospora destructor* infection was detected following this methodology.

Effects of planting date and plant density

Overall, early planting date (July 27th) originated higher Dm incidence and severity than late planting

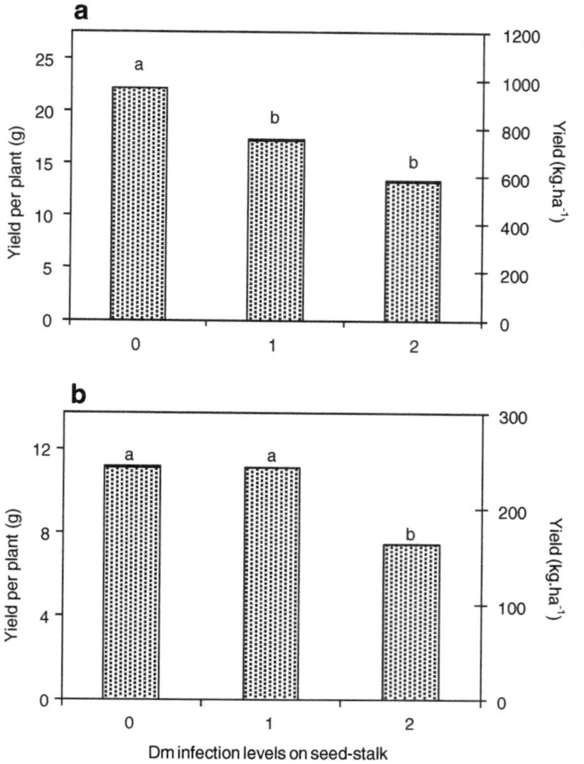

Fig. 2 Onion seeds yields per plant and per hectare for three levels of downy mildew infections on onion seed-stalks, defined as (0) stems without symptoms, (1) 25–50% of infected tissue, and (2) 25–50% of infected tissue with necklace spots extended over the circumference of the stem, in the seasons (**a**) 2005 and (**b**) 2006. Bars with the same letter on top do not differ (Fischer LSD test, $p<0.05$)

date (August 29th), but the effects of plant density were less relevant (Fig. 3). Dm incidence (active symptoms) for early planting date treatments were detected from September 20th, and increased until the end of October, before decreasing during the second half of November. A counterintuitive result was that Dm incidence increased earlier for the less dense cultivation and reached maximum incidence 1 week earlier. Nevertheless, severity was higher for the dense cultivation from October 25th until November 8th, with significant interaction between planting date and plant density within this three-week period (Fig. 3). At the end of the season (December), estimations of severity were disregarded because of high infestation of *Alternaria porri* and *Stempylium vesicarium,* and natural senescence of the leaves.

The combination of later planting date and lower plant density led to no Dm symptoms. In addition, development of *Alternaria porri* and *Stempylium vesicarium* infections at the end of the season were lesser in this treatment combination in comparison with the other three treatment combinations.

For early planting date, a seed yield of 20.4 g plant^{-1} for the less dense cultivation was higher than the yield obtained with the dense cultivation (14.8 g plant^{-1}) (Table 5). These yields were significantly higher than yields obtained for later planting dates (9.72 and 9.10 g plant^{-1} respectively for low and high density, without significant difference between them). Early planting and denser cultivation produced higher seed yield per hectare, without interaction between

Table 3 Effects of downy mildew infection levels on seed-stalks on germination, vigour and weight of 1000 onion seeds in the seasons 2005 and 2006

Infection levels[a]	Season 2005			Season 2006		
	Germination[b] (%)	Vigour[c] (%)	Weight of 1000 seeds (g)	Germination (%)	Vigour (%)	Weight of 1000 seeds (g)
0	76.4ns	26.6ns	4.33ns	55.2a	25.6a	3.25ns
1	72.0	23.0	4.21	51.8a	19.4b	3.13
2	70.7	21.2	4.02	41.6b	25.3a	2.73
C.V.%	4.4[d]	11.2[d]	7.4	4.8[d]	13.3[d]	14.4

[a] (0) stems without symptoms; (1) 25–50% of infected tissue; and (2) 25–50% of infected tissue, with necklace spots extended over the circumference of the stem

[b] Normal germinated seedlings 12 days after the beginning of the test, following ISTA rules (1976). Test performed before seed cleaning

[c] Normal germinated seedlings 6 days after the beginning of the test

[d] Data were transformed by natural logarithm before analysis of variance

Table 4 Fungi infecting onion seeds for the different downy mildew infection levels on seed-stalks in the seasons 2005 and 2006

Identified genera	Season 2005 Infection levels on seed-stalk[a]		Season 2006 Infection levels on seed-stalk[a]		
	1	2	0	1	2
Stemphylium sp.[b, c]	21.9a	22.2a	4.33a	15.33b	17.00b
Fusarium sp.	2.9	5.4	1.67	1.67	4.00
Alternaria sp.	2.4	1.7	0.67	1.00	1.33
Penicillium sp.	3.4	5.4	0.67	0.33	1.00
Aspergillus sp.	–	–	0.67	0	0
Botrytis sp.	–	–	0	0	0.33
Others	5.5	4.4	–	–	–
Total[b]	22.8a	24.7a	8.00a	18.33b	23.67b

[a] (0) Stems without symptoms, (1) 25–50% of infected tissue, (2) 25–50% of infected tissue with necklace spots extended over the circumference of the stem. Infection level 0 was only evaluated in the season 2006

[b] Within each season and row, means followed by the same letter do not differ ($p<0.05$)

[c] *Stemphylium* and its perfect state *Pleospora* sp

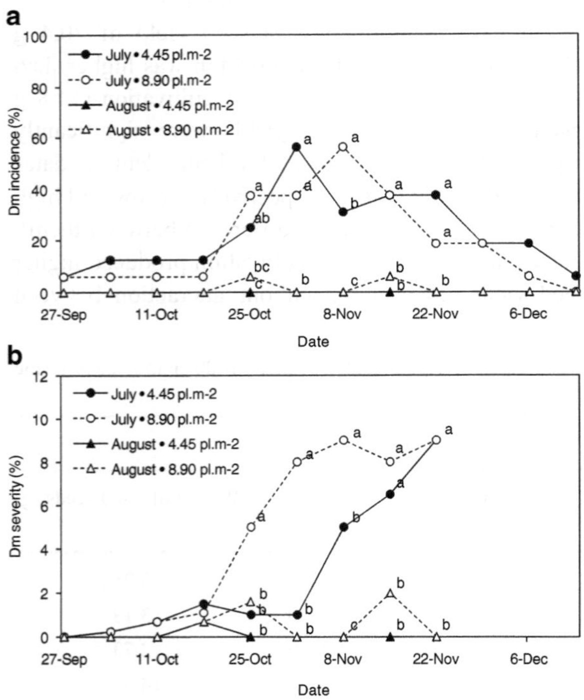

Fig. 3 (**a**) Mean incidence and (**b**) severity of downy mildew along the season in four treatment-combinations of planting date and plant density in onion seed production. Only typical symptoms caused by new infections and areas covered by sporulation were accounted for. Within each evaluation time, means labelled with the same letter do not differ (REML analysis, $p<0.05$)

these two factors. Conversely, weight of 1,000 seeds, seed germination and seed vigour were significantly larger for the late planting date in comparison with values obtained for the early planting date (Table 5). The presence of rotten seeds on blotter tests increased with the combined effect of early planting date and low plant density.

Discussion

Uruguay has 1,100 to 1,300 l m^{-2} as an average annual rainfall, that contributes to create a favourable environment for onion Dm caused by *Peronospora destructor*. Several seed companies from Europe and United States of America have established onion seed production abroad in search of dry climates. When this strategy is not possible, Dm may be particularly harmful in the production of onion seeds. This research carried out a case study on commercial fields in southern Uruguay, as knowledge on the pathogen, farming history and the effects of diverse factors within the agro-ecosystem are critical elements in designing effective disease management (Lorbeer et al. 2002).

Downy mildew was found as a significant disease for onion seed production, and symptoms developed mainly in the period October–November when plants had maximum vegetative development and seed-stalks evolved from onset until full anthesis. Nevertheless, Dm

Table 5 Seed yield per plant and seed quality traits as a function of planting date and plant density

Planting date	Plant density (m^{-2})	Seed yield (g $plant^{-1}$)[a]	Germination (%)[b]	Vigour (%)[c]	Rotten seeds (%)	Weight of 1000 seeds (g)
July 27	4.45	20.4a	77.25b	17.5b	6.35c	4.433b
	8.90	14.8b	80.25b	17.8b	4.83b	4.585b
August 29	4.45	9.7c	82.00a	24.0a	4.53b	4.882a
	8.90	9.1c	83.75a	25.8a	0.50a	4.800a
C.V.%		13.7	1.6	17.8[d]	14.2[d]	3.0

[a] Within each column, means followed by the same letter do not differ significantly

[b] Normal germinated seedlings 12 days after the beginning of the test, following ISTA rules (1976). Test performed before seed cleaning

[c] Normal germinated seedlings 6 days after the beginning of the test

[d] Data were transformed by natural logarithm before analysis of variance

incidence and/or severity differed between studied years and farm fields (Fig. 1). These differences could be caused by differences in environmental and micro-environmental circumstances. Either for infection, for the progress of the disease and sporulation, *P. destructor* requires the occurrence of specific conditions of relative humidity, temperature and darkness for a minimum period of time (Hilderbrand and Sutton 1982; Gilles et al. 2004). Differences between years in the fulfilment of these particular combinations of environmental factors probably gave rise to the observed variation, particularly for Dm severity. The number of days per year in which these conditions are achieved would cause variation in timing and severity of Dm outbreaks. A distinct result was observed for Dm incidence, as the range among fields was similar for the studied years. General conditions required for infection of *P. destructor* seemed to be more regularly achieved, but affected by factors at field level.

Downy mildew incidence and severity differed among farm fields within the same year (Fig. 1). Rudolph (1990) reported 15% Dm incidence on seed stalks as average among surveyed fields in Germany, with a range between 0 and 82%. This range is wider than the one found in the present research for Dm incidence scored on leaves. The fields JL and DP contrasted for their respectively high and low levels of disease in 2005, either in incidence or severity. This result was in line with the potential difference in initial inoculum (Table 1): JL had a previous onion crop, however, DP was a piece of land without an onion crop history. Plant debris in the soil may act as reservoir of *P. destructor* oospores, as well as onion bulbs used as planting material for seed production which could be infected in the previous season. Therefore, effective disease control of the crop in which bulbs are produced (Palti 1989), and long rotation with non-host crops would contribute to reduce Dm outbreaks during seed production. Elimination of volunteer onion plants is another proposed action to reduce Dm (Palti 1989).

Initial inoculum may explain the difference between JL and DP, but does not add to an understanding of the differences between the other two farm fields surveyed in 2005, nor between the fields surveyed in 2006 (Fig. 1). The observed relative levels of Dm incidence and severity certainly were not related with the background of disease in the farm, as subjectively reported by the farmers. Plant density, an agronomic practice that changes air circulation and relative humidity within the canopy, tended to be positively associated with maximum Dm incidence (r^2=0.59, n=6, n.s.). However, plant density does not explain a high Dm incidence for JM with 1.55 plants m^{-2}, and low incidence for CM with 2.80 plants m^{-2} (Table 1). Nor did the number of chemical interventions explain the differences between fields, as a more frequent spraying schedule was applied by those farmers having significant Dm outbreak. Fungicides did not eradicate Dm when other factors were conducive to the disease, but probably delayed disease development during the season in comparison with untreated fields. Many factors may contribute to explaining variation in Dm outbreaks among fields, such as micro-environmental conditions differing from farm to farm (topography, neighbouring forests, etc.), differences in initial inoculum, chemical interventions, and differences in agronomic practices (Rudolph 1990). For instance, Sharma et al. (2002) observed that higher density of weeds increased the presence of Dm. A single factor could not explain differences between fields in incidence or severity, but the combination of several factors determines Dm disease development.

For some of the studied cases (e.g., farm fields in 2006), an initial small maximum peak value for incidence was observed early in the season (October) before the main maximum disease outbreak that usually took place in November (Fig. 1). The occurrence of an early small incidence peak was also observed in the screening of onion germplasm to Dm (Colnago 2009), and may represent the polycyclic behaviour of the disease. In such a way, initial Dm infections probably occur during winter or early spring, and two main cycles of sporulation may be completed during spring in southern Uruguay. Using local onion germplasm and a local *P. destructor* isolate for inoculations in a growing chamber, latency period lasted 20 to 22 days (Colnago 2009). This time-span corresponds with epidemiological observations carried out in commercial fields in this research, and is longer than the period reported by Hildebrand and Sutton (1982). Monitoring crops for early detection of Dm symptoms and/or modelling early sporulation and infection phases would be additional relevant elements in disease management, in order to reduce maximum incidence and severity afterwards during the season (Lorbeer et al. 2002).

Seed yield is affected by agronomic practices and environmental conditions that determine plant growth, leaf development, differentiation of primary and secondary inflorescences, and the number of flowers per inflorescence finally developed. Among these conditions affecting seed yield, Dm reduces active leaf area and plant growth, as well as weakening seed-stalks causing yield losses. In contrast with this theoretical concept, yields obtained in the surveyed farm fields were not (negatively) related with Dm incidence or severity during the season. This lack of relationship would be the consequence of the complexity of factors involved in yield formation, and the measures applied by farmers to stop and overcome the potential negative effects of Dm disease.

Nevertheless, when a disease outbreak was out of control and numerous Dm infections developed on seed-stalks, seed yield and quality were negatively affected (Fig. 2). That was the case for the fields where the effects of the level of seed-stalk infections were studied (Table 2). Seed-stalks with necklace infections yielded 40% and 33% less seeds than healthy seed-stalks respectively in 2005 and 2006. These yield losses for severely infected seed-stalks are lower than those reported in Brazil (Verona et al. 1996) and India (Sharma et al. 2002) for seed producing fields. Furthermore, observed yield reductions cannot be strictly assigned to Dm infections, as apparently these Dm infections assist the occurrence of *Stemphylium* and *Alternaria* diseases late in the season. As a result, even in those farm fields where disease control was not effective at all, Dm in Uruguay seemed not to reach the aggressiveness and effects reported in other regions, probably because of differences in environmental conditions between locations.

Seed quality was also negatively affected by seed-stalk infections (Table 3). Seed germination tended to decrease, though only significantly in 2006, and the presence of diverse potentially pathogenic fungi on seed surface increased, particularly *Stemphylium* and its perfect state *Pleospora* (Table 4). Previous studies of commercial onion seed lots also detected *Alternaria, Aspergillus, Fusarium* and *Stemphylium* on seed surfaces (Cardani et al. 1990; Verona et al. 1996; Sharma et al. 2002), *Penicillium* (Cardani et al. 1990), and *Botrytis squamosa* (Ellerbrock and Lorbeer 1977), although *Aspergillus niger*, *Botrytis aclada* and *Fusarium oxysporum* f.sp. *cepae* were outlined as prevalent species (Özer and Köycü 2004). Seed cleaning aiming to remove the share of seeds with lower weight and seed treatments with fungicides can improve seed germination values when these are below acceptable standards (Cardani et al. 1990; Sharma et al. 2002; Özer and Köycü 2004). In addition, regarding the influence of seed-stalk infection on seed quality (Table 3), rejection of inflorescences from seed-stalks severely infected during hand-harvesting would be another action to improve seed quality.

Even when *P. destructor* was reported as a seed-borne pathogen (Rondomanski 1971), the importance of this phase in initiating the disease is not well documented. As Dm infections occur only on adult plants, when leaves have more than 20 cm in length (Maeso 2005), disease development occurs only late in the season. It is not straightforward to recognize Dm infections at seed or seedling stages. During the production of onion bulbs, the presence of dwarf plants with curved leaves and fully covered by Dm sporulation, also observed in Uruguay, allows speculation about systemic Dm infections from the beginning of the season (Virányi 1981; Maeso 2005). Recent and ongoing development of immunoassay-based detection of *P. destructor* (Kennedy and Wakeham 2008) and molecular tests as developed for other *Peronosporales* (Scott et al. 2004; Tsay et al. 2006) will contribute to elucidate the role of seed transmission in the dynamics of this disease.

Infections of *P. destructor* on seed-stalks helped ulterior infections by the *Stemphylium–Alternaria* complex. Indeed, healthy stems (infection level 0) remained as such at the end of the season, whereas half of the stems with infection level 1 moved to infection level 2 (necklace spots) caused by *Stemphylium* and/or *Alternaria* infections that developed on the lesions caused by *P. destructor*. Such interaction was also observed in the experiment in which planting dates and plant densities were studied. Early planting date treatments that were more affected by *P. destructor* developed later on *Stemphylium* and *Alternaria* lesions that increased general disease severity up to 100% of leaf surfaces. In contrast, the treatment combining late planting date and low plant density did not have Dm nor *Stemphylium–Alternaria* infections. The latter ones are weak opportunistic pathogens that, although they can also infect intact tissues, usually infect plant tissues through wounds caused by *Thrips tabaci* and necrotic spots caused by other diseases or abiotic factors (Maude 1990; Schwartz and Mohan 2008). Further research is needed to understand interactions between pathogens and their effects, a poorly studied topic of disease dynamics.

Planting in June–July was previously recommended in order to obtain maximum onion seed yields in Uruguay (Arboleya et al. 1987). Early planting date permits larger leaf development and adequate vernalization before emission of seed-stalks. However, a trade-off between seed yield and the development of Dm occurs, in such a way that early planting date is associated with larger development of Dm and other leaf diseases, as well as an increase in the presence of seed-borne pathogens (Table 5). This trade-off is the consequence of larger plants increasing relative humidity within the canopy, and holding larger proportions of older leaves that are more susceptible to Dm infections. Planting date and leaf age associated susceptibility is also observed in the production of onion bulbs (Maeso 2005).

Early August was already chosen in commercial seed production in Uruguay as a planting date that combines profitable yield and moderate conditions for development of Dm. Although results from a single experiment, planting at the end of August as studied in this research seemed to produce markedly lower yields regardless of the elevated sanitary status of the crops and the obtained seeds, and can still be an attractive alternative for the production of organic seeds. Plant density was found as another factor influencing seed yield and Dm development because ventilation within leaf canopy is modified, as well as the extent of the period in which conditions for infection and/or sporulation are fulfilled. Nevertheless, plant density effects were less evident than those observed for planting date.

Adaptation of Dm forecast systems applied in bulb production for the seed production phase would be a valuable element in disease management and spraying decisions. However, this cannot be the single element. An effective control of Dm in onion seed producing fields would be the result of the combination of several practices into an integrated disease management. These practices should be directed to either reduce initial inoculum, susceptibility status of the host, and environmental conditions promoting disease development.

Acknowledgements This research was part of a Project coordinated by H. González Idiarte and founded by CSIC-SP, Universidad de la República, Uruguay. We thank onion seed-growers belonging to CALSESUR who kindly contributed to the surveys and experiments for this research. Thanks are due also to O. Costa, N. Curbelo, V. Ferreira, and N. Jara for their technical assistance during installation and evaluation of field experiments and seed quality tests.

References

Arboleya, J. E., Villamil, J. M., & Itoh, M. (1987). Efecto de la fecha de plantación de bulbos de cebolla sobre el rendimiento y la calidad de la semilla. *Horticultura Argentina, 6*, 49–65.

Box, G. E. P., Hunter, W. G., & Hunter, J. S. (1978). *Statistics for experimenters: An introduction to design, data analysisi and model building*. New York: Wiley Interscience. 653p.

C. M. I. (1990). Distribution maps of plant diseases, Map Nr. 76, Onion downy mildew. Kew, Surrey, United Kingdom.

Cardani, A., Giménez, G., & Verdier, E. (1990). *Calidad de semilla de cebolla (Allium cepa L.)*. Montevideo: Tesis Ing. Agr. Facultad de Agronomía, Universidad de la República.

Colnago, P. (2009). Evaluación de la respuesta a Peronospora destructor (Berk.) Casp. en el germoplasma local de cebolla. Tesis M. Sc. Facultad de Agronomía, Universidad de la República. Montevideo, Uruguay.

Ellerbrock, L. A., & Lorbeer, J. W. (1977). Sources of primary inoculum of *Botrytis squamosa*. *Phytopathology, 67*, 363–372.

Galván, G., González, H., & Sollier, S. (1997). Productive adaptation of onion (*Allium cepa* L) landraces used for post-harvest storage. *Acta Horticulturae, 433*, 165–170.

Galván, G., González, H., & Vilaró, F. (2005). Estado actual de la investigación en poblaciones locales de hortalizas en Uruguay y su utilización en el mejoramiento. *Agrociencia, 9*, 115–122.

Gilles, T., Phelps, K., Clarkson, J. P., & Kennedy, R. (2004). Development of MILIONCAST, an improved model for predicting downy mildew sporulation on onions. *Plant Disease, 88*, 695–702.

González, H., Peluffo, S., & Galván, G. (2007). Caracterización del Programa de Producción de semilla certificada del cultivar de cebolla "Pantanoso del Sauce CRS" (abstract). Congreso Nacional de Horti-Fruticultura, 11th. Montevideo, Uruguay.

Hilderbrand, P. D., & Sutton, J. C. (1982). Weather variables in relation to an epidemic of onion downy mildew. *Phytopathology, 72*, 219–224.

ISTA. (1976). International rules for seed testing. International Seed Testing Association. *Seed Science and Technology, 4*, 3–177.

Kennedy, R., & Wakeham, A. J. (2008). Development of detection systems for the sporangia of *Peronospora destructor*. *European Journal of Plant Pathology, 122*, 147–155.

Kofoet, A., & Zinkernagel, V. (1989). Resistance to downy mildew (*Peronospora destructor* (Berk.) Casp.) in *Allium* species. *Journal of Plant Diseases and Protection, 97*, 13–23.

Leach, C. M., Hilderbrand, P. D., & Sutton, J. C. (1982). Sporangium discharge by *Peronospora destructor*: influence of humidity, red-infrared radiation and vibration. *Phytopathology, 72*, 1052–1056.

Lorbeer, J. W., Kuhar T. P., & Hoffmann M. P. (2002). Monitoring and forecasting for disease and insect attack in onions and *Allium* crops within IPM strategies. InH. D. Rabinowitch & L. Currah (Ed.), Allium crop science: recent advances (pp. 293–310). CABI Publishing.

Maeso, D. (2005). Enfermedades del cultivo de cebolla. In J. Arboleya (Ed.), Tecnología para la producción de cebolla. INIA Boletín de divulgación 88, Uruguay, pp. 151–188.

Maude, R. B. (1990). Leaf diseases of onions. In H. D. Rabinowitch & J. L. Brewster (Eds.), *Onion and allied crops* (pp. 173–212). Boca Raton: CRC.

MGAP, DIEA-DIGEGRA. (2006). Encuesta Hortícola Zonas Sur y Litoral Norte año 2006. Montevideo, Uruguay.

Özer, N., & Köycü, N. D. (2004). Seed-borne fungal diseases of onion and their control. In K. G. Mukerji (Ed.), *Disease management of fruits and vegetables* (pp. 281–306). Dordrecht-Boston-London: Kluver Academic.

Palti, J. (1989). Epidemiology, prediction and control of onion downy mildew caused by *Peronospora destructor. Phytoparasitica, 17*, 31–48.

Rondomanski, W. (1971). The sources of primary infection of onion downy mildew (*Peronospora destructor* (Berk.) Fries). *Tagunsbericht Deutsche Akademie der Landwirtschaftswissenschaften zu Berlin, 115*, 157–171.

Rudolph, M. (1990). On disease infection in onion seed plants (*Allium cepa* L.). *Nachrichtenblatt Pflanzenschutz, 44*, 180–182.

Scholten, O. E., Van Heusden, A. W., Khrustaleva, L. I., Burger-Meijer, K., Mank, R. A., Antonise, R. G. C., et al. (2007). The long and winding road leading to the successful introgression of downy mildew resistance into onion. *Euphytica, 156*, 345–353.

Schwartz, H. F., & Mohan, S. K. (2008). *Compendium of onion and garlic diseases*. St. Paul: APS. 127p.

Scott, J. B., Hay, F. S., & Wilson, C. R. (2004). Phylogenetic analysis of the downy mildew pathogen of oilseed poppy in Tasmania, and its detection by PCR. *Mycological Research, 108*, 198–205.

Sharma, R. C., Gill, S. S., & Kohli, N. (2002). Pathological problems in production and storage of onion seed in Punjab and their remedial measures. *Seed Research, 30*, 134–141.

Sutton, J. C. (1986). Forecasting onion diseases: a key to efficient fungicide use. *Highlights, 9*, 13–16.

Tsay, J. G., Chu, C., Chuang, Y. H., & Chen, R. S. (2006). Specific detection of *Peronospora tabacina* by PCR-amplified rDNA sequences. *Plant Pathology Journal, 5*, 378–382.

Van Doorn, A. M. (1959). Onderzoekingen over het optreden en de bestrijding van valse meeldauw (*Peronospora destructor*) bij uien. *Tijdschriven Plantenziekten, 65*, 193–255.

Verona, L. A. F., Pacheco, A. C., Zanini Neto, J. A., Gandin, C. L., & Thomazelli, L. F. (1996). Qualidade e produtividade de sementes de cebola na regiao Oeste Catarinense - safra 1994/95. *Agropecuaria Catarinense, 9*, 29–32.

Vilaró, F., Vicente, E., Pereyra, G., & Rodríguez, G. (2005). Cultivares y mejoramiento genético en cebolla. In J. Arboleya (Ed.), Tecnología para la producción de cebolla. INIA, Uruguay. Boletín de Divulgación 88. pp. 31–42.

Virányi, F. (1974). Studies on the biology and ecology of onion downy mildew (*Peronospora destructor* (Berk.) Fries) in Hungary. I. Onverwintering of the pathogen in onion bulbs. *Acta Phylopathologica Academiae Scientiarum Hungaricae, 9*, 3–4.

Virányi, F. (1981). Downy mildew of onion. In D. M. Spencer (Ed.), *The downy mildews* (pp. 463–472). Littlehampton: Academic.

Eur J Plant Pathol (2011) 129:183–197
DOI 10.1007/s10658-010-9711-0

Diversity and fitness of *Plasmopara viticola* isolates resistant to QoI fungicides

Marie-France Corio-Costet · Marie-Cécile Dufour · Jérémy Cigna · Pierre Abadie · Wei-Jen Chen

Accepted: 20 October 2010 / Published online: 17 November 2010

Abstract The effectiveness of Quinone outside Inhibitor (QoI) fungicides against grape downy mildew in European vineyards has significantly decreased in the last decade. One nucleotide polymorphism, G143A in the cytochrome *b* gene of *Plasmopara viticola,* is involved in resistance to QoIs. Previous genetic examination on the mitochondrial genomes showed four major haplotypes (IR, IS, IIR, IIS) coexisting in European vineyards. A resistant allele (G143A) was present in IR and IIR haplotypes. The purpose of the present study was to estimate the diversity of the different mitochondrial haplotypes and their distribution in QoI-resistant populations before evaluating the potential cost of the resistant mutation G143A in *P. viticola* population. From 2000 to 2004, the frequencies of resistant isolates ranged from 0% to 23.25% with an average of 4.64 % among the populations examined. To evaluate the fitness of sensitive and resistant isolates, a comparison of different biological parameters including latent period, spore production and infection frequency was performed, enabling a fitness index (F_I) to be determined. Resistant isolates exhibited greater infection frequency than sensitive isolates, whereas no significant difference was found in sporulation ability and latent period between sensitive and resistant isolates. To further investigate competitiveness among isolates, an assay including two resistant isolates in different proportion with a sensitive isolate was conducted on eight asexual growing cycles in the absence of a QoI fungicide. The competitiveness of resistant isolates varied according to their fitness parameters, suggesting that there is no noticeable cost of QoI resistance in controlled conditions in *Plasmopara viticola*.

Keywords Cytochrome b · Fitness · Fungicide resistance · Grapevine downy mildew · Mitochondrial variability · qPCR

Abbreviations

CAPS	Cleaved Amplified Polymorphic Sequence
F_I	fitness index
QoI	Quinone outside Inhibitor
ARMS	Amplification refractory mutation system
SNPs	Single Nucleotide Polymorphisms

M.-F. Corio-Costet (✉) · M.-C. Dufour · J. Cigna
UMR Santé Végétale, 1065, ISVV, IFR 103, INRA,
Rue Edouard Bourleaud, BP 81, 33883 Villenave d'Ornon, France
e-mail: coriocos@bordeaux.inra.fr

Present Address:
P. Abadie
UMR 1202 BIOGeCo,
Laboratoire de génétique des arbres forestiers, INRA,
69 route d'Arcachon,
33612 Cestas, France

Present Address:
W.-J. Chen
Institute of Oceanography, National Taiwan University,
No.1 Sec. 4 Roosevelt Rd.,
Taipei 10617, Taiwan

Introduction

Downy mildew caused by the oomycete *Plasmopara viticola* is a major grape disease worldwide. The pathogen is a native species from North America that was reported in Europe in 1878 in Coutras in (Bordeaux region) by Planchon (Galet 1977). Grapevine downy mildew (*P. viticola*) attacks all important cultivars for vine and table grape production. Depending on epidemic pressure and cultural practices, production losses of up 100% can occur. For effective control of the disease, various fungicides, including Quinone outside Inhibitor fungicides (QoIs), have been widely used. However, strong selection pressure following repeated QoI applications can result in the development of QoI-resistant populations which has limited the success of these fungicides in grapes and other crops (Gisi et al. 2002; Grasso et al. 2006). Thus, only a few years after the introduction of QoI fungicides in 1996, *P. viticola*-resistant populations were detected not only in French vineyards, (Magnien et al. 2003) but also in the vineyards of other European countries and the United States (Baudoin et al. 2008; Corio-Costet et al. 2008; Collina et al. 2005; Heaney et al. 2000; Sierotzki et al. 2005).

The QoI fungicides inhibit the cytochrome bc1 enzyme complex of the respiration chain in mitochondria (Jordan et al. 1999; Bartlett et al. 2002). Due to their site-specificity, QoI fungicides have been reported as being at high risk for encouraging the development of resistance in several target pathogens (Grasso et al. 2006). In most pathogens, two major mutations in the cytochrome *b* gene have been described (G143A, F129L) as being involved in resistance mechanisms. The mutation resulting in the replacement of glycine by alanine at amino acid codon position 143 (G143A) has been widely recognized (Gisi et al. 2002; Grasso et al. 2006).

To understand the mechanism leading to the evolution of fungicide resistance in a plant-pathogen system, Chen et al. (2007) conducted a phylogenetic analysis of a large mitochondrial DNA fragment including the cytochrome b gene (2.281 bp) across a wide range of *P. viticola* isolates. Four major haplotypes were detected belonging to two genetically distinct groups (I and II), in which the resistance allele carrying the G143A mutation can be present. This indicated at least two origins of fungicide resistance in the grapevine downy mildew population. In France, haplotype group I (IR and IS) and group II (IIR and IIS) reached a frequency of 75% and 25% respectively among populations tested, but their distribution may vary (Chen et al. 2007; Corio-Costet et al. 2008). For instance, in the Champagne (France) an average of 10% of haplotype group II was detected and the significance of the two mitochondrial haplotypes is unclear (Corio-Costet et al. 2006). As regards the resistant allele proportion IR and IIR within a population, the resistant haplotype IR ranged from 67.7% to 98.3% of resistant alleles depending on the location (Chen et al. 2007; Corio-Costet et al. 2006). The first objective of this work was to estimate the distribution of every haplotype in European *P. viticola* populations.

Although a biological test (Wong and Wilcox 2000; Genet et al. 1997) or molecular tools have been used to quantify the resistance alleles using real-time q-PCR, ARMS PCR or PCR, (Collina et al. 2005; Grasso et al. 2006; Toffolatti et al. 2007; Sirven and Beffa 2003) or to detect resistant haplotypes using the CAPS method, (Corio-Costet et al. 2006; Baudoin et al. 2008; Furuya et al. 2009) little is known about the fitness of QoI-resistant isolates. One study suggested that an artificial *P. viticola* population containing 5% of resistant isolates tended to increase QoI sensitivity (Genet et al. 2006). The second objective of our study was to try to understand how fungicide resistance develops by measuring the fitness of *P. viticola* iolates, without fungicide pressure. The fitness, or the selective value, of pathogens (Antonovics and Alexander 1989; Pringle and Taylor 2002) can be defined as the survival and reproductive success of an allele, individual or group, or as an organism's ability to contribute to the subsequent gene pool (Vanderplank 1982), which is measured as a function of its ability to grow, reproduce and survive through repeated life cycles. It is estimated either by measuring components of fitness such as reproductive ability and pathogenicity, or by assessing the actual population after competing with each other on a susceptible host. Usually, the general ability of pathogens to accommodate resistance mutations without a harmful loss of fitness is largely unknown, but in early studies the appearance of fungicide resistance was not generally accompanied by an immediate fitness cost in the absence of fungicides (Anderson 2005). The rapidity of the appearance of QoI resistance and the behaviour of resistant isolates in populations could suggest a good fitness value of resistant isolates that is assessed here.

The final objective of the study was to determine whether QoI resistance in *P. viticola* was associated with a resistance cost. For this, we performed fitness and competitiveness studies to investigate the potential of dispersal and the preservation of *P. viticola* isolates resistant to QoI fungicides. In particular, we assessed the following: i) the diversity of QoI haplotype in European countries: ii) the fitness of sensitive and resistant isolates of *P. viticola*, by comparing latent period, sporulation, infection frequency and competitiveness: and iii) the cost of QoI resistance in grapevine downy mildew.

Materials and methods

Plant material

All tests were performed with grapevine leaves of Cabernet Sauvignon cultivar. Grapevine plants (*Vitis vinifera* cv. Cabernet Sauvignon) were propagated from wood cuttings in a greenhouse. All tests were performed with grapevine leaves of 2-month-old plants (ten to 12 leaves) for experimentation on foliar discs.

P. viticola sampling and collection

A total of 1,366 downy mildew lesions on *Vitis vinifera* were collected between 2003 and 2004 or were donated by a Swiss colleague (D. Gobbin) for 2000 and 2001. Each sample from 2003 and 2004 came from a sparse plot on leaves collected between June and July during the growing season (Table 1).

For isolate collection, a total of 11 QoI-sensitive isolates and 12 QoI-resistant isolates were collected in 2003 at the beginning of the growing season on grapevine leaves in the Bordeaux vineyard in France (11 + 11) and one in the Freiburg vineyard in Germany (1) (Table 2). All monosporangial isolates were preserved on grapevine leaves at–20°C.

Fungicide sensitivity tests

To characterize the sensitivity of *P. viticola* isolates to QoI fungicides, single lesions were multiplied onto leaves of *Vitis vinifera* cv. Cabernet Sauvignon. Sporangia were collected and suspended in sterile water at 4°C. Inoculation was performed by depositing fifteen 10 μl droplets onto the abaxial face of each leaf. One day later the water droplets were removed by aspiration with a water pump. The leaves were incubated in Petri dishes under controlled conditions (22°C) with a 16 h light day (35 μm m^{-2} s^{-1}) for 7 days. Freshly produced sporangia were harvested to inoculate the test units. Assays were described in Chen et al. (2007). Briefly, Cabernet Sauvignon leaf discs with abaxial surface exposed were prepared as described above. Excised discs were randomized and placed lower surface exposed on moist filter paper into Petri dishes, with eight repetitions for each concentration. Experiments were also duplicated independently. The QoI tested was famoxadone (IUPAC name: 3-anilino-5-methyl-5-(4-phenoxyphenyl)-1, 3-oxazolidine-2,4-dione) (WG 50%, Dupont de Nemours). Famoxadone was applied by spraying in water solution at desired concentrations. The solutions were applied using a hand sprayer in order to spray 10 μl per cm^2. The fungicide test was repeated at least twice with a range of eight or ten concentrations adapted to the response of each isolate in order to obtain a dose-response curve. After using a discriminatory dose of 10 mg l^{-1}, according to the sensitivity of each isolate, ranges of concentrations for resistant isolates (0, 50, 100, 200, 500, 800, 1,000, mg l^{-1}) or for sensitive isolates (0, 0.02, 0.05, 0.08, 0.1, 0.15, 0.2, 0.5, 1, 2 mg l^{1}) were prepared to determine the EC_{50} of isolates. After drying surfaces, the leaf discs were inoculated with three droplets of 10 μl per disc of inoculum at 20.000 sporangia per ml. Petri dishes were incubated for 7 days under the conditions described above. Seven days after inoculation, development of mildew on leaf discs was determined by assessing the proportion of sporulation per unit of leaf area (Genet et al. 1997). The average score for each fungicide concentration was converted to a percentage of inhibition by comparison with the untreated disks. The fungicide concentration inhibiting 50% of growth (EC_{50}) discriminated two classes of isolates according to their sensitivity to famoxadone: famoxadone-sensitive isolates (EC_{50} and MIC <10 mg l^{-1}) and highly resistant isolates (EC_{50} and MIC $>1{,}000$ mg l^{-1}). Dose response curves for individual isolates were generated by plotting the relative growth values against $\log_{10}$ of the fungicide concentration used. The $\log_{10}$ effective dose to reduce growth by 50% (EC_{50}) was calculated from the

Table 1 Average of mitochondrial haplotype distribution in European populations of *Plasmopara viticola* collected between 2000 and 2004

Country	Location	N	Year	% of European mitochondrial haplotypes			
				IS	IR	IIS	IIR
France	Bordeaux	506	2003	50.20	23.90	22.70	3.20
	Champagne	122	2003	73.78	22.13	4.10	0.00
	Dordogne	84	2003	47.62	21.43	31.00	0.00
	Loire	121	2003	69.42	8.26	22.30	0.00
	Vallée du Rhône	86	2003	80.23	5.81	14.00	0.00
	Bourgogne	33	2004	36.37	30.30	12.10	21.20
	Jurançon-Gers	63	2004	38.10	34.90	27.00	0.00
Mean		1015		56.53	20.96	19.03	3.49
Germany	Rheingau	16	2000	68.75	0.00	31.25	0.00
	Rheingau	27	2003	70.37	0.00	29.67	0.00
	Baden	50	2003	76.00	16.00	8.00	0.00
Mean		93		71.71	5.33	22.97	0.00
Greece	Thessalie	21	2001	90.47	0.00	9.53	0.00
Italy	Piemont	8	2000	75.00	0.00	25.00	0.00
	Chianti	18	2000	77.78	0.00	22.22	0.00
	Haut Adige	15	2000	100.00	0.00	0.00	0.00
Mean		41		84.26	0.00	15.74	0.00
Portugal	Refois do lima	18	2004	89.00	0.00	11.00	0.00
	Corricha	25	2004	88.00	0.00	12.00	0.00
	Gaifar	25	2004	64.00	8.00	28.00	0.00
Mean		68		80.33	2.67	17.00	0.00
Romania	Timisoara	67	2004	97.02	0.00	2.98	0.00
Switzerland	Tessin	21	2000	90.48	0.00	9.52	0.00
	Vaud	22	2000	63.64	0.00	36.36	0.00
	East County	18	2000	88.89	0.00	11.11	0.00
Mean		61		81.00	0.00	19.00	0.00
Europe (Mean±Sem)		1366	2000–2004	80.19±4.98	4.14±2.91	15.18±2.56	0.5±0.5

regression equation generated through the linear proportion of the sigmoid curve.

Measurements of aggressiveness components

The following fitness components were measured comparatively in sensitive and resistant isolates: latent period, the number of spores obtained per deposited spore after 7 days of growth, and the infection frequency, using methods similar to those described by Tooley et al. 1986 and Kadish and Cohen 1988. All the measurements were performed on three plates, each containing five discs of grapevine leaves inoculated as described above with three droplets of 15 µl of sporangial suspension at 2,500 spores per ml.

The latent period (LP) was estimated by observing sporangia appearance daily, as the mean period (hours) to obtain 50% of the total sporulation. Three plates containing five discs of leaf abaxial face on the top were inoculated with three droplets of 15 µl at 2,500 sporangia per ml, leading to 45 points per isolate per experiment. The experiments were repeated at least twice independently. The onset of sporulation was measured daily on a notation scale as described in Chen et al. 2007.

Table 2 Origin and characteristics of monosporangial isolates used in fitness studies

Isolate	Locality of vineyard	Year	Mitochondrial haplotype (I or II)	QoI sensitivity	QoI CI_{50} (mg/ml±SEM)
AVI 02	Champagne	2003	II	S	0.90±0.01
COU 15	Bordeaux	2003	I	S	0.07±0.01
FEM 03	Bourgogne	2003	II	S	0.10±0.03
MOR 22	Bordeaux	2003	II	S	0.05±0.01
PIC 21	Bordeaux	2003	II	S	0.10±0.02
BOM 06	Bordeaux	2003	II	S	0.07±0.02
EPE 09	Champagne	2003	I	S	0.22±0.01
CAZ 22	Bordeaux	2003	II	S	0.13±0.01
MIC 103	Bourgogne	2003	II	S	0.04±0.01
MIC 123	Rhône Valley	2003	I	S	0.09±0.01
ALS 07	Alsace	2003	I	S	0.28±0.01
CGU 20	Bordeaux	2003	I	R	>1,000
COU 23	Bordeaux	2003	II	R	>1,000
EAU 13	Midi-pyrénées	2003	I	R	>1,000
FRE 01[1]	Freiburg	2003	I	R	>1,000
MIC 125	Midi-pyrénées	2003	I	R	>1,000
MIC 126	Midi-pyrénées	2003	I	R	>1,000
MIC 128	Midi-pyrénées	2003	I	R	>1,000
MOR 08	Bordeaux	2003	I	R	>1,000
PAR 29	Bordeaux	2003	I	R	>1,000
PIC 59	Bordeaux	2003	I	R	>1,000
PAU 32	Bordeaux	2003	II	R	>1,000
SAL 19	Bordeaux	2003	II	R	>1,000

Spore production (N_t / N_{t0}) was assessed by washing five inoculated discs of each plate (3) in a vial containing 16 ml of Isoton II (saline buffer) supplemented with a drop of dispersant type IIIA, and the number of sporangia produced per plate containing five discs inoculated by three drops was determined (N_t) with a Multisizer 3 Coulter counter (Beckman Coulter, France, Villepinte). Experiments were repeated at least twice independently, giving six measures each corresponding to the sporulation obtained for 562 deposited sporangia (N_{t0}).

Infection frequency (IF) was defined as the proportion of inoculated leaf disks on which lesions developed. Observations were quantified as the number of inoculation points with a suspension of 2,500 spores per mL giving sporulating lesions 7 days after inoculation. Three plates containing five leaf discs were inoculated by three drops of 15 µl per disc. Experiments were repeated twice independently.

From these different parameters, a composite fitness index (F_I) was calculated for each isolate using the formula $AI = Ln(N_t/N_{t0} \times IF \times 1/LP)$ derived from Montarry et al. 2008.

Competitiveness test

The relative competitiveness of two pairs of resistant/sensitive isolates MIC128-COU15 (R1-S1) and PAU32-COU15 (R2-S1) was compared at three initial ratios of spore concentrations, 20:80, 50:50, and 80:20. Spore suspensions of each isolate were prepared as described previously. Mixed inocula were generated by thoroughly mixing appropriate volumes of sporangia to obtain the desired R:S ratios on the basis of 40.000 spores per ml total in the different mixtures. Inoculation was carried out with a mixed spore suspension in the same way as described for infection tests. Four plates containing five discs were inoculated for each ratio and every 7 days, four new plates with five discs were inoculated again with the spores obtained at 40,000 spores per ml. Then, a biological test with famoxadone at 100 mg l^{-1} was performed as described above in fungicide sensitivity test section and a reference curve was established to evaluate the ratio of resistant isolates in the spore mixtures obtained after each asexual cycle. To confirm those ratios, a quantitative PCR assay was also performed to quantify the presence of resistant alleles from spore suspensions used to inoculate the new leaf discs at each cycle. Two independent experiments were performed for 8 weeks.

Statistical analysis

Data from two experiments were pooled together according to homogeneity of variance using the ANOVA procedure in Systat software (Systat 11. software, 2004, USA) to determine homogeneity of variance among repeated experiments. The values or groups were compared using the Tukey *t* test with $P<0.05$ for q-PCR assay or were compared using Fisher's LSD test with $P<0.05$ for the bioassay. All data for each fitness component were subjected to an analysis of variance (ANOVA). To analyze frequency haplotype distribution, a chi2 test was performed.

DNA extraction

P. viticola lesions were frozen. After freeze-drying overnight, DNA was extracted from each oil spot or spore suspension (containing infected grapevine leaf tissue and sporangia) according to the standard CTAB-phenol/chloroform method described by (Zolan et Pukkila 1986) with a slight modification (phenol-chloroform was replaced with isoamylic alcohol/chloroform). After extraction, DNA was precipitated with isopropanol, the pellet was washed with 70% ethanol and resuspended in 50 µl of sterile water.

Allele-specific restriction

The molecular method used to differentiate sensitive or resistant isolates was the amplification of the allele specific to the Cytochrome *b* gene of *P. viticola* (accession numbers DQ459464, DQ459466, DQ459467, DQ459469) exhibiting single nucleotide polymorphisms between sensitive and resistant isolates or between haplotype I or II (Chen et al. 2007). A CAPS analysis of cyctochrome *b* gene fragments was performed using the primers COB 279F and COB 865R (Table 3) under amplification conditions according to the method described by Baudoin et al. (2008) and Corio-Costet et al. (2006). Polymerase chain reaction (PCR) was performed in a 15 µl volume containing 1 µl of DNA diluted three-fold, 1.5 µl of 10x PCR buffer (Eurogentec), 1.5 mM $MgCl_2$, 0.13 µM of each dNTP, 0.13 µM of each primer (COB 279 F and CB 865R), 0.2 U (3.33 nkat) *Taq* silverstar DNA (Eurogentec). The PCR program was a first denaturation step of 4 min at 96°C, followed by 36 cycles of 45 s at 96°C, 50 s at 50°C, 1 min at 72°C. This DNA fragment was cut with the restriction enzymes *Sat1* and *Hinf1*, with the recognition sequence 5′-GCNGC-3′ and 5′–GANTC-3′. Cleaving reactions were performed in 10 µl volume with 1.2 µl of PCR product, 1 µl of 10X- buffer Tango (MBI fermentas), 1 U (16.67 nkat) of *Sat1* and 1U (16.67 nkat) of *Hinf1* enzymes (MBI Fermentas) at a temperature of 37°C and an incubation period of 2 h. Restriction fragments were visualized on 2% agarose after staining with ethidium bromide. *Sat1* enzyme was used to detect the resistant mutation and to cut the amplified fragment of 605 bp, generating two fragments of 147 and 458

Table 3 Sequence of primers and hybridization probes used in the real-time polymerase chain reaction assay on *Plasmopara viticola* isolates

Primer	Sequence (5′to 3′)	Nucleotide position	Sense
COB 279F	TAT ACA TAT TTT TAG GGG TTT G	1100	Forward
COB 865R	ACA TAG CGA TAA CAC CAC CAA	1704	Reverse
qPV-L	TTA GGG GTT TGT ATT ACG GAT CTT	1111	Forward
qPV-R	TCC CCA AGG CAA AAC ATC CC	1229	Reverse
qPV143R-L	GGA CAA ATG AGT TTT TGG GC	1227	Forward
qPV143RS-R	ATT ATC AAC GGC GAA TCC AC	1337	Reverse

pb each. *HinfI* enzyme was used to determine haplotypes and to cut the fragment, generating two fragments of 222 and 383 pb.

QoI real-time q-PCR

Genotyping of QoI-resistant mutations was performed with a real-time thermal iCycler iQ (BioRad, Hercules, CA, USA) using specific sets of primers and hybridization probes (Table 3). Primer sets qPV-L/qPV-R and qPV143R-L/qPV143RS-R (Table 3) were used to amplify preferentially one allele by matching the desired allele and mismatching the other allele at the 3′ end of the primers according to Sirven and Beffa (2003). Reverse primers were designed to have mismatches at the 3′ as described by Sirven and Beffa (2003). For each reaction, 1 µl of each primer at 10 nM and 7 µl of 2x-Mix blue SYBR green fluorescein Mix including Hot start DNA polymerase, DNTP and $MgCl_2$ (Abgene, France) and 5 µl of DNA diluted three times were used in compliance with the manufacturer's instructions. Assays (14 µl) were performed in ABgene PCR thermo-fast® 96, semi-skirted plates (Thermoscientific, France). Amplification and detection were performed in a real-time thermal iCycler iQ under the following conditions: one cycle at 94°C for 15 min, and 40 cycles at 94°C for 10 s, 60°C (primer set qPV-L/qPV-R) and 60°C (qPV143R-L/qPV143RS-R) for 10 s, and 72°C for 20 s. Increase in fluorescent emission signal from SYBR Green I (Δ*Rn*) was recorded at 72°C without normalization of the signal. The iCycler™ Q software version 3.0_a (BioRad) was used to analyze the data. For each sample, the amount of target DNA was quantified using appropriate calibration curves in which *Ct* was plotted against the amount of target DNA.

Results

Mitochondrial haplotype distribution in European vineyards

The survey of the geographical distribution of mitochondrial haplotypes from a panel of 1,366 *P. viticola* isolates collected between 2000 and 2004 in European vineyards was performed with an SNP typing assay. The frequency of the four main European haplotypes was estimated in four French, three German, one Greek, three Italian, one Romanian, three Portuguese and three Swiss populations (Table 1). Haplotype group I (IS and IR) predominated in all countries in a range from 77.44% to 97.02%, with an average of 84.33%. Haplotype group II (IIS and IIR) represented 2.98% to 22.56% depending on the country, with an average of 15.68%. With regard to QoI resistance, haplotype IR predominated, with levels depending on the country and locality in a range from 0% to 20.99% and an average of 4.14%. The resistant haplotype IIR was the most uncommon and was found only in two French vineyards in Bordeaux and Bourgogne. More resistant haplotypes were detected in 2003–2004 than in 2000–2001, with an average of 16.26% and 0% respectively.

QoI sensitivity and haplotype characteristics

Throughout the study, 11 sensitive isolates and 12 resistant isolates collected in 2003 were tested independently at least twice to evaluate their famoxadone sensitivity. EC_{50} values for resistant isolates were higher than 1,000 mg l^{-1} of famoxadone (Table 2). Sensitive isolates exhibited EC_{50} values ranging from 0.1 to 0.9 mg l^{-1}, with a mean of 0.19 mg l^{-1}. Seven sensitive isolates belonged to haplotype II and four belonged to haplotype I (COU15, COL07, MIC123 and EPE 09). By contrast, the resistant haplotype II was uncommon, while only three isolates (SAL19, PAU 32, COU 23) from the laboratory collection belonged to this haplotype. The level of QoI resistance between sensitive and resistant isolates expressed as the relative resistance factor varied in a range from 1111 to 25,000, with an average of 5263.

Latent period, sporangium production and infection frequency

Data on the fitness of 23 isolates inoculated separately onto grapevine leaf discs are given in Table 4. Three parameters (latent period, sporulation capacity and infection frequency) were recorded. A wide variation in fitness parameters was found in sensitive and resistant isolates. However, no difference was found for latent period between sensitive and resistant isolates (*P*=0.523) (Table 4, Fig. 1). Sensitive isolates had a mean of 102.7±1.57 h and resistant isolates

Table 4 Fitness components of sensitive and QoI-resistant isolates on grapevine leaf disks at 22°C (means±SEM)

Isolate	QoI sensitivity	Latent period (h±SEM)	Sporulation (Nt/Nt_0) (± SEM)	Infection frequency (± SEM)	AI (Ln (Nt/Nt_0 IF /LP)
ALS 07	S	140±2.93	537±70	0.94±0.01	1.28[a]
AVI 02	S	86.5±1.47	1022±310	0.94±0.03	2.41
BOM 06	S	96±3.56	332±73	0.86±0.03	1.09
CAZ 22	S	97±2.35	783±119	0.91±0.02	1.99
COU15	S	88±0.59	534±76	0.98±0.02	1.78
EPE 09	S	107.5±2.64	351±79	0.72±0.06	0.86
FEM 03	S	97±0	821±166	0.97±0.03	2.11
MIC 103	S	107±1.7	7 83±119	0.79±0.01	1.77
MIC 123	S	97.5±1.47	1267±238	0.97±0.01	2.53
MOR22	S	103±0.59	896±227	0.76±0.02	1.99
PIC 21	S	110±0.59	421±49	0.42±0.13	0.48
Mean		102.7±1.57	704±139	0.84±0.035	1.75±0.65
CGU 20	R	89.5±0.29	565±120	1.00±0.00	1.84
COU 23	R	108.5±0.29	555±139	0.96±0.04	1.59
EAU 13	R	91.5±2.05	1015±264	1.00±0.00	2.41
FRE 01	R	104.5±0.29	550±125	1.00±0.00	1.66
MIC 125	R	93.25±1.61	996±156	0.99±0.01	2.36
MIC 126	R	103±0.59	802±208	0.93±0.07	1.98
MIC 128	R	92.5±1.47	844±90	0.94±0.06	2.15
MOR 08	R	93.25±0.44	901±139	0.94±0.03	2.21
PAU32	R	139±5.28	693±133	0.96±0.04	1.57
PAR 29	R	95±0.59	1048±115	0.91±0.09	2.31
PIC 59	R	96±1.76	1508±407	1.00±0.00	2.75
SAL 19	R	105±1.17	484±160	0.68±0.14	1.14
Mean		100.9±1.32	830±171	0.94±0.02	2.05±0.45

[a] The variation of the fitness index varies from 5% to 18% of the value of Fi according to the isolates considered. For three isolates used for the experiment of competitiveness, we have F_{iPAU32}=1.57±0.28 (*a*), F_{iCou15}=1.78±0.22 (*ab*), $F_{iMIC128}$=2.15±0.15 (*b*), with a total variance analysis of *P*=0.015

needed 100.9±1.32 h to reach their latent period. The range of latent period varied from 86.5 h to 140 h in sensitive isolates and from 89.5 to 139 h in resistant isolates.

Concerning spore production, no significant difference was observed between sensitive and resistant isolates (*P*=0.137). Sensitive isolates produced an average of 704±139 sporangia and resistant isolates produced 830±171 sporangia per deposited sporangium (Table 4, Fig. 1). A wide variation was observed within the sensitive isolate group and the QoI-resistant group. Indeed, within a group some isolates produced 3.11- to 3.8-fold more sporangia than others in the resistant (PIC59/SAL19) and sensitive (BOM06/AVI01) groups, respectively. Sporangium size in sensitive and resistant isolates varied widely in a range from 8 to 33 μm in the sensitive and resistant groups, with an average of 18.47 μm± 0.03 μm. We also observed a significant increase in resistant sporangium size from 3% to 4.6% in the absence or in the presence of famoxadone (data not shown).

As regards infection frequency (IF) in our experimental conditions, a significant difference was obtained (*P*=0.02) between sensitive and resistant groups with 0.84±0.03 of sensitive isolates and 0.94± 0.02 of resistant isolates succeeding in infecting leaves (Table 4, Fig. 1). The range of infection frequency varied from 0.42 to 0.97 in the sensitive group and from 0.68 to 1 in the resistant group.

From these different parameters, a composite fitness index (F_I) was calculated for each isolate derived from F_I of Tooley et al. (1986) and from the aggressiveness index of Montarry et al. (2008). The mean F_I in the sensitive group was 1.75±0.65 and 2.05±0.45 in the resistant group, but no significant difference was observed (*P*=0.147) (Table 4). The fitness index varied considerably depending on

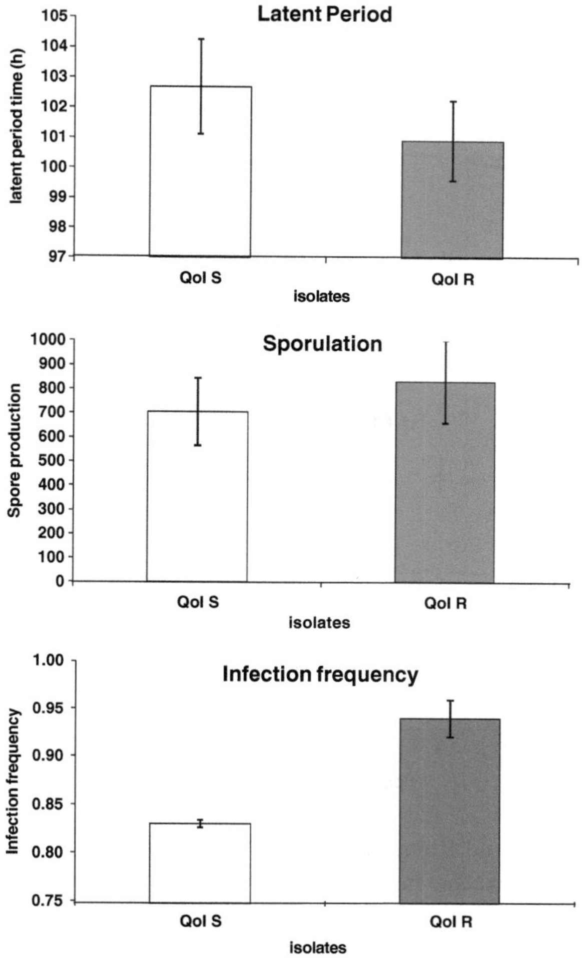

Fig. 1 Fitness component values (Latent period, Sporulation, Infection frequency means±SEM) as a function of sensitive (*white*) or QoI-resistant isolates (*grey*)

whether the isolates were sensitive or resistant (Fig. 2). A good correlation was obtained between spore production and the F_I with a regression coefficient R^2 equal to 0.86 ($P<0.001$). The F_I was then used to choose sensitive and resistant isolates for competitiveness tests having different fitness.

Competitive ability of resistant and sensitive isolates

The objective was to assess the competitive ability of resistant and sensitive isolates in mixed-isolate inoculations on leaf discs and to link these findings to the fitness parameters of isolates. The fitness index of the three isolates was different (Table 4, Fig. 2) with a sensitive isolate COU15 statistically intermediate (F_I= 1.78, *ab*) between both resistant isolates PAU 32 (F_I= 1.57, *a*) and MIC128 (F_I=2.15, *b*) (P=0.015).

The changes in the frequency of resistant isolates were related to their fitness parameters and their fitness index (Figs. 3 and 4). In one of the pairs used in this study R1-S1, after eight asexual cycles all ratios between sensitive and resistant isolates remained stable at 20%, 50% or 80% of R1 after quantification by q-PCR (Fig. 3). A slightly significant increase ($P<0.05$) in 20% and 50% of initial ratios was observed at the eighth cycle with the biological quantification method.

In the second pair (S1-R2), a decline in the frequency of R2 was observed after five cycles (Fig. 4) in all three initial ratios, indicating that the resistant isolate R2 was a weak competitor in comparison with its S1 partner. After five cycles, the proportion of resistant isolates decreased 20–25%

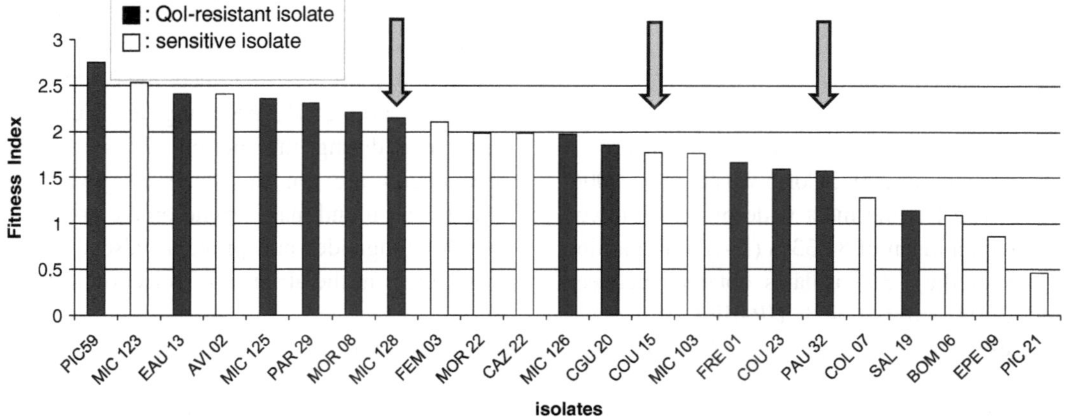

Fig. 2 Fitness index (F_I) of 11 sensitive (*white*) and 12 QoI-resistant isolates (*black*). The *arrows* designate the isolates used in competitiveness experiments

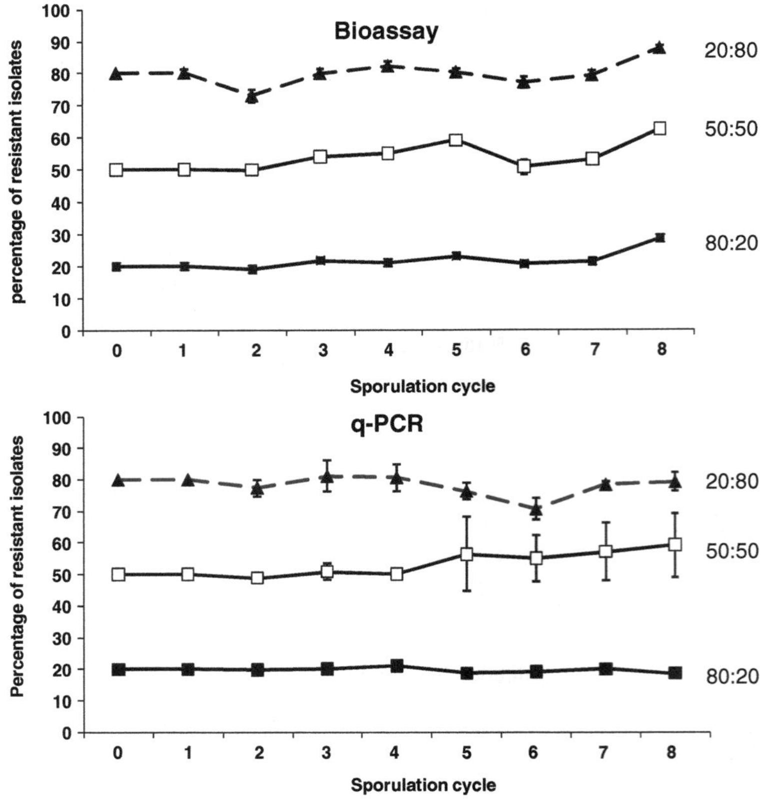

Fig. 3 Dynamic changes (means±SEM) in the frequency of QoI-resistant isolate sporangia in sporangial populations harvested from grapevine leaf discs inoculated with three mixtures of QoI-sensitive COU15 and QoI-resistant sporangia MIC128 in various proportions (20:80 (-■-), 50:50 (-□-) and 80:20 (–▲–) during eight asexual generations, assessed by bioassay or q-PCR method. The second sporulation cycle was initiated with sporangia resulting from the initial infection and subsequent sporulation cycles were conducted accordingly. Each point is mean of six replicates

compared to the initial ratio. With the q-PCR test, a significant and stronger decrease in the different ratios with R2 was observed from the third cycle, leading to a dramatic reduction during the eighth cycle, with a decrease of 5% (P=0), 49.56% (P=0.006) and 97.5% (P=0.004) at the ratios 80:20, 50:50 and 20:80, respectively (Fig. 4).

Discussion

Diversity

According to previous results obtained in French vineyards (Chen et al. 2007; Corio-Costet et al. 2006, 2008), haplotype I is the most widespread haplotype in Europe, with a mean of 84.33% (IS-IR). Regarding the analyses of the 1,366 isolates obtained between 2000 and 2004, only 4.54% (IR-IIR) of isolates were QoI-resistant, with haplotype IR representing 91.19% of resistant haplotypes. By contrast, French vineyards were significantly more resistant during this period than other vineyards exhibiting few resistant isolates (0–2.94%) such as Greek, Italian, Swiss and Romanian vineyards. Haplotype IIR was rare, except in Bourgogne where it represented 21.2%. The role of the presence of haplotypes I and II in *P. viticola* population remains unclear.

Our results on 1,015 mildewed lesions showed that in French vineyards in 2003 and 2004, at the beginning of the growing season and before any selection by a fungicide, 76.75% of *P. viticola* isolates were still sensitive. However, according to Sierotzki et al. (2005), who showed a sharp decrease of QoI sensitivity in 2002 and 2003 in European vineyards, we found more resistant haplotypes in 2003–2004 than in 2000–2001, revealing an increase of the resistance during this period. Thus, the resistance frequencies are not the same everywhere and the differences found may depend on the pressure exerted by QoI fungicides and probably also on the population fitness in the absence of selection pressure.

Fitness parameters

The development and the spread of fungicide resistance depends on the fitness of resistant isolates, and knowledge of processes such as diversity, sex

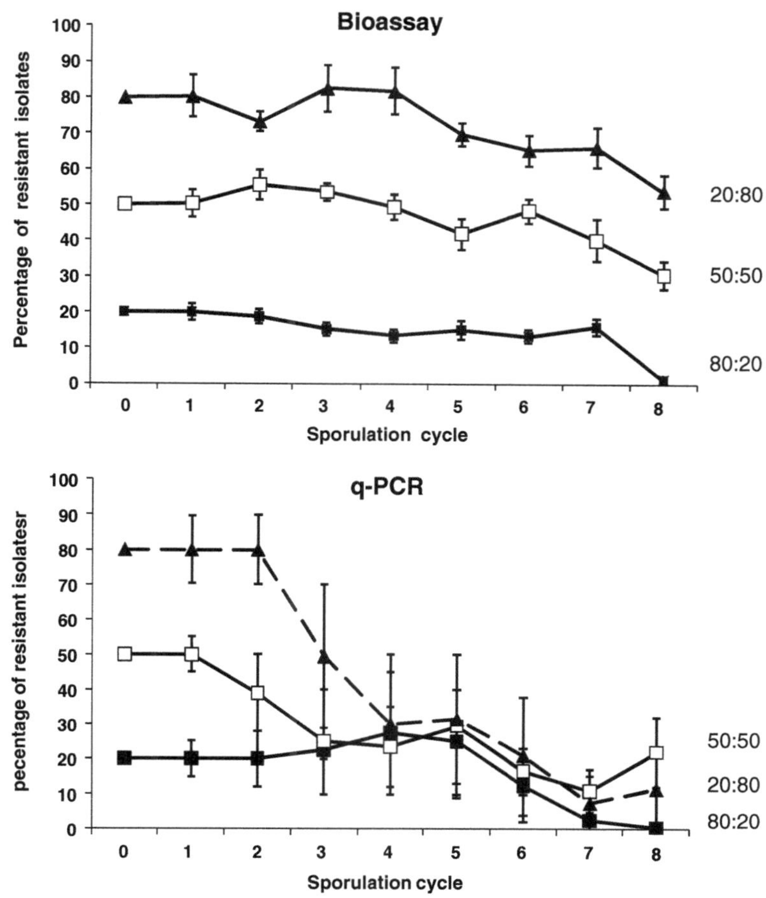

Fig. 4 Dynamic changes (means±SEM) in the frequency of QoI-resistant isolates sporangia in sporangial populations harvested from grapevine leaf disks inoculated with three mixtures of QoI-sensitive COU15 and QoI-resistant sporangia PAU 32 in various proportions (20:80 (-■-), 50:50 (-□-) and 80:20 (–▲–) during eight asexual generations, assessed by bioassay or q-PCR method. The second sporulation cycle was initiated with sporangia resulting from the initial infection and subsequent sporulation cycles were conducted accordingly. Each point is mean of six replicates

evolution and fungicide pressure is essential to try to prevent and/or to control the diseases better. For this, it is important to understand how fitness shapes the selection of individuals in nature under different selection pressure. From various sensitive or QoI resistant isolates belonging to different mitochondrial haplotypes, we measured fitness parameters without fungicide selection pressure. A major point should be made: resistant isolates are as fit as sensitive isolates, with regard to sporangia production, latent period and infection frequency in asexual cycles. Only infection frequency of resistant isolates was significantly higher than in sensitive isolates (P= 0.02). As the resistant isolates belonged essentially to haplotype I, it was possible that the difference was due in part to the belonging to the haplotype I group. A comparison between the fitness of haplotypes I and II inside sensitive group, or with the totality of isolates, including the resistant isolates, showed no significant difference in the spore production (P=0.75 and 0.24), the latent period (P=0.36 and 0.35), or the fitness index (P=0.86 and 0.10). Only infection frequency was significantly more important for haplotype I (P=0.03) and when QoI-resistant isolates were considered. As above, the significance of the infection frequency could well be related to the QoI resistance.

From this first study on the fitness parameters, we were thus able to conclude that the presence of G143A SNP in cytochrome b gene has no deleterious effect on growth parameters observed under controlled conditions and without selection pressure.

However, the overall fitness of one population relative to another would be difficult to predict solely on the basis of knowledge of fitness components. For this reason, the combination of different parameters would reasonably approximate overall fitness in calculating a composite fitness index. The fitness index described here, based on infection frequency, spore production and latent period enabled us to show that there was no significant difference between groups of haplotypes I and II and between sensitive and QoI resistant isolates. This index also revealed a good correlation with spore production suggesting that under our experimental conditions spore production could be a sufficient measure to assess the fitness of isolates. This F_I based on growth parameters seems adequate to standard conditions, but one of its weaknesses is certainly that it does not take in account survival capacity under less favourable environmental conditions, or fitness potential of the sexual cycle (*i.e.* oospore production, overwintering).

As with other studies on the fitness of oomycetes, such as *Phytophthora nicotianae* or *P. infestans* resistant to metalaxyl or mefenoxam (Hu et al. 2008; Kadish and Cohen 1988), QoI resistance did not show reduced fitness. However, it is possible that resistance cost might only become apparent under a specific set of environmental conditions which were not encountered in the experimental set-up, or that the associated antagonistic pleiotropy might not be detectable (Rosenberg 2001); or, yet again, that a fitness deficit linked to a resistance cost can be compensate for after numerous generations by different compensatory mechanisms (Anderson 2005; Cowen et al. 2001). One example of resistance cost due to environmental conditions was metalaxyl-resistant isolates of *Plasmopara viticola* that exhibited reduced fitness to grape leaves, but only at high temperature (Piganeau and Clerjeau 1985). A preliminary study performed under controlled conditions in our laboratory did not show differences between sensitive and resistant QoI isolates at 15°C or 28°C.

In connection with definite F_I, we estimated competitiveness of sensitive and resistant pairs of isolates. The follow-up during 8 cycles of asexual reproduction showed that, in the absence of selection pressure, resistant isolates can compete with sensitive isolates in mixed inoculation depending on their fitness parameters. Thus, a resistant isolates with a lower F_I (PAU32) than the sensitive isolates (COU 15) tends to disappear after eight cycles, while a resistant isolate (MIC 128) with a high F_I preserves even tends to increase after eight cycles. These results were reinforced by other isolate pairs tested previously, where a resistant isolate with a higher F_I (PIC59, F_I=2.75) than the sensitive isolate FEM103 (F_I=2.11) increased rapidly to reach 100% of resistant isolate in the whole mix after the fifth cycle. In the same way, another mix with PAU 32 (F_I=1.57) and a sensitive strain MIC 103 (F_I=1.77) also showed a decrease in the presence of PAU 32 from the fourth cycle (Abadie 2007).

Globally the competitiveness tests performed were coherent with the estimated parameters of the fitness and our F_I, although imperfect, enabled us to obtain a good idea of the capacities of isolate survival under controlled conditions. In accordance with Kadish and Cohen (1988), we showed that the pattern of changes in frequency of resistant isolates could be largely dependent on fitness parameters in laboratory conditions. During the competitiveness assays, we used the bioassay or the q-PCR method, the curves of competitiveness (Figs. 3 and 4) showed some differences, although tendencies were similar. The bioassay seemed to facilitate the presence of resistant isolates (R1 and R2). One possible explanation could be that the method of q-PCR is strictly quantitative and applies directly to spores harvested after each asexual cycle, while the bioassay required an additional cycle in the presence of famoxadone, facilitating the growth of resistant isolates and potentially integrating others fitness parameters. Thus, the quantification of the percentage of resistant isolates based on reference curves in the presence of famoxadone in the bioassay could lead to a slight overestimation of the presence of resistant isolates compared to q-PCR carried out without fungicide pressure. However, the faster, easier q-PCR method is an adequate method to estimate the proportion of resistant isolates.

As with the analysis of fitness parameters, the competitiveness assays strengthened the idea that the presence of G143A mutation would have no deleterious effect on *P. viticola* under controlled conditions without fungicide selection pressure. As all resistant isolates used in this study have already been sequenced (Chen et al. 2007) and possessed only the G143A SNP, which is the most frequent, we would suggest that the G143A mutation does not induce a resistance cost in *P. viticola*. However, Fisher et al. (2004) have investigated the impact of the introduction of G143A SNP on bc complex activity of *Saccharomyces cerevisiae* and shown that it has little effect on the enzyme activity. In contrast, the association of the G143A with F129L, or other mutations, led to various deleterious effects on enzyme activity. It would be useful to assess the resistance cost of different mutations (*i.e.* G143A and F129L) in the laboratory, to understand why the

F129L SNP was less frequent in *P. viticola* populations than in other pathogen populations.

The management of pesticides in the vineyard requires to a better understanding of how the resistance takes hold, spreads and is maintained. Important questions about the adaptation of QoI resistance are: how does QoI resistance develop in the vineyard? What is the probability of obtaining various combinations of SNPs, and what is the consequence on population fitness? In view of the importance of the development of QoI-resistance over the past decade, today these products are never applied alone and are little present in current formulations or, if they are, they are then associated with other fungicides. It would be interesting to follow the development of QoI resistance under various selection pressures and survey the development of the different SNPs in the vineyard and to study their fitness in laboratory conditions. Understanding the development of QoI resistance in the vineyard is not easy. Data obtained in vineyard in Brazil on a trial site exhibiting 100% frequency of G143A in *P. viticola* populations showed that after stopping QoI treatment for 3 years, the frequency returned to 1% in 2002, suggesting a multiannual instability of the resistance. In contrast, five subsequent applications of QoI fungicide sprays increased the frequency again to 100% within one season, immediately producing a resistant population (Sierotzki et al. 2008). In addition, a survey of the development of QoI resistance in different sites in French vineyards with few resistant isolates (less than 10%) showed the maintenance of QoI resistance in sites without QoI treatment during the growing season (Corio-Costet et al. 2006, 2008). These studies suggest that the resistant isolates possess rather good one fitness during the growing season (asexual cycle), but that the fitness of QoI resistant populations could be less good during the sexual phase and/or without selection pressure. Therefore, the contribution of sexual reproduction to global fitness might play an important role in maintaining QoI resistance. Since the cytochrome *b* gene is coded by the mitochondrial genome, it most likely depends on mitochondrial inheritance. After sexual reproduction, the level of resistant allele might decrease, depending on the mating type carrying the resistant allele, and/or the sexual fitness, and could explain contrasted results obtained in fields after sexual or asexual cycles.

The possibilities of combating QoI resistance in *P. viticola* by mixing QoIs with other fungicides, or stopping the use of QoI fungicides for a few years in order to diminish resistance after each annual sexual cycle are currently solutions used by wine growers. The application of a QoI mixed with other fungicides does not seem to eliminate selection pressure, but delays it compared to the application of QoI fungicide alone (Toffolatti et al. 2007). From a practical point of view, care should be taken not to use QoI fungicides again on plots of land having presented QoI resistance during the growing season, or then to preserve products containing QoIs for rare or selective applications during very strong epidemic pressure, and by alternating treatments with fungicides possessing different mode of action.

In conclusion, the mitochondrial haplotype II is little present in *P. viticola* populations and does not seem to have repercussions on resistant populations. Resistance can be very variable according to the vineyard under consideration and the selection pressure exerted. During the experiment under controlled conditions, we did not detect a significant difference of fitness either between 23 resistant and sensitive isolates, or between haplotypes I and II. The composite fitness index used here may help us to classify isolates according to different phenotypic and genotypic traits (*i.e.* fungicide resistance, mating-type, agressiveness) and to have them compete according to several parameters of fitness, or used to carry out studies on the efficacy of alternative methods (*i.g.* elicitors, breeding, biological fight) on specific population or isolates. It would also be useful to estimate the survival potential of resistant isolates exhibiting single, double or triple resistance to fungicides (*i.e.* Qoi, CAA and/or phenylamide fungicides), already encountered in the vineyard; but that would require including parameters of overwintering, and of capacity of sexual reproduction, to be used under field conditions. A study of fitness of the resistant isolates during the sexual cycle would be necessary to complete this work and to acquire better understanding of QoI resistance evolution under various pressures of selection.

Acknowledgements We thank D. Gobbin and colleagues from IFV for their help in sampling of the collection and Aquitaine Region Government for their financial support. We thank S. Gambier, S. Richart-Cervera, and L. Douence for technical support. We thank J-L. Genet from Dupont Protection for the gift of famoxadone.

References

Abadie, P. (2007). Adaptation du mildiou de la vigne à la pression fongicide: étude du fitness. MSc thesis of Genetic and Plant Development. Bordeaux University. pp. 20.

Anderson, J. B. (2005). Evolution of antifungal-drug resistance: mechanisms and pathogen fitness. *Nature Reviews Microbiology, 7*, 547–556.

Antonovics, J., & Alexander, H. M. (1989). The concept of fitness in plant-fungal pathogen systems. In W. E. Fry & K. J. Leonard (Eds.), *Plant disease epidemiology* (pp. 185–214). New-York: Mc Graw-Hill.

Bartlett, D. W., Clough, J. M., Godwin, J. R., Hall, A. A., Hamer, J., & Parr-Bobrzanski, B. (2002). The strobilurin fungicides. *Pest Management Science, 58*, 649–662.

Baudoin, A. I., Olaya, G., Delmotte, F., Colcol, J. F., & Sierotzki, H. (2008). QoI resistance of *Plasmopara viticola* and *Erysiphe necator* in the Mid-Atlantic United States. *Plant Management Network. Plant Health Progress.* doi:10.1094/PHP-2008-0211-02-RS.

Chen, W.-J., Delmotte, F., Richard Cervera, S., Douence, L., Greif, C., & Corio-Costet, M.-F. (2007). At least two origins of fungicide resistance in grapevine downy mildew populations. *Applied Environmental Microbiology, 73*, 5162–5172.

Collina, M., Landi, L., Guerinin, P., Branzanti, M. B., & Brunelli, A. (2005). QoI resistance of *Plasmopara viticola* in Italy: biological and quantitative real-Time PCR approaches. In H. W. Dehne, U. Gisi, K. H. Kuck, P. E. Russell, & H. Lyr (Eds.), *Modern fungicides and antifungal compounds IV* (pp. 81–88). Hampshire: BCPC.

Corio-Costet, M. -F., Delmotte, F., Martinez, F., Giresse, X., Raynal, M., Richart-Cervera, S., et al. (2006). Resistance of *Plasmopara viticola* to QoI fungicides: origin and diversity. Paper presented at the *8th Int. Conf on Pest and Diseases 2006*, pp 612-620, AFPP Eds, CD-Rom.

Corio-Costet, M. -F., Martinez, F., Delmotte, F., Douence, L., Richart-Cervera, S., & Chen, W. -J. (2008). Resistance of *Plasmopara viticola* to QoI fungicides: Origin and diversity. In H. W. Dehne, U. Gisi, K. H. Kuck, P. E. Russell, & H. Lyr (Eds.), *Modern fungicides and Antifungal compounds V* (pp. 107–112). Brauschweig: DPG Selbstverlag.

Cowen, L. E., Kohn, L. M., & Anderson, J. B. (2001). Divergence in fitness and evolution of drug resistance in experimental populations of *Candida albicans*. *Journal of Bacteriology, 183*, 2971–2978.

Fisher, N., Brown, A. C., Sexton, G., Cook, A., Windass, J., & Meunier, B. (2004). Modeling the Q_o site of crop pathogens in *Saccharomyces cerevisiae* cytochrome *b*. *European Journal of Biochemistry, 271*, 2264–2271. doi:10.1111/j.1432-1033.2004.04169.x.

Furuya, S., Suzuki, S., Kobayashi, H., Saito, S., & Takayanagi, T. (2009). Rapid method for detecting resistance to QoI fungicides in *Plasmopara viticola* populations. *Pest Management Science, 65*, 840–843.

Galet, P. (1977). Mildiou. In P. Galet (Ed.), *Les maladies et les parasites de la vigne* (pp. 89–222). Montpellier: Paysan du midi.

Genet, J.-L., Steva, H., Vincent, O., & Cazenave, C. (1997). A method for measuring the level of sensitivity of *Plasmopara viticola* populations to cymoxanil. *EPPO bulletin, 27*, 217–225.

Genet, J.-L., Jaworska, G., & Deparis, F. (2006). Effect of dose rate and mixture of fungicides on selection for QoI resistance in populations of *Plasmopara viticola. Pest Management Science, 62*, 188–194. doi:10-1002/ps.1146.

Gisi, U., Sierotzki, H., Cook, H., & McCaffery, A. (2002). Mechanisms influencing the evolution of resistance to Qo inhibitor fungicides. *Pest Management Science, 58*, 859–867.

Grasso, V., Palermo, S., Sierotzki, H., Garibaldi, A., & Gisi, U. (2006). Cytochrome b structure and consequences for resistance to Qo inhibitor fungicides in plant pathogens. *Pest Management Science, 62*, 465–472.

Heaney, S. P., Hall, A. A., Davies, S. A., & Olaya G. (2000). Resistance to fungicides in the QoI-STAR cross-resistance group: current perspectives. In *Proc. Brighton Conf. Protec Conf. Pest and diseases* (pp. 755–764). Farham: BCPC.

Hu, J. H., Hong, C. X., Stromberg, E. L., & Moorman, G. W. (2008). Mefenoxam sensitivity and fitness analysis of *Phytophthora nicotianae* isolates from nurseries in Virginia, USA. *Plant Pathology, 57*, 728–736. doi:10.111/j.1365-3059.2008.01831x.

Jordan, D. B., Livingston, R. S., Bisaha, J. J., Duncan, K. E., Pember, S. O., Picollelli, M. A., et al. (1999). Mode of action of famoxadone. *Pesticide Science, 55*, 105–108.

Kadish, D., & Cohen, Y. (1988). Competition between metalaxyl-sensitive and metalaxyl-resistant isolates of *Phytophthora infestans* in the absence of metalaxyl. *Plant Pathology, 37*, 558–564.

Magnien, C., Micoud, A., Glain, M., & Remuson, F. (2003). QoI resistance of downy mildew-monitoring and tests 2002. *Paper presented at the 7th Int. Conf. on Pest and Diseases 2003,* 8 pages. AFPP Eds, CD-Rom.

Montarry, J., Glais, R., Corbiere, R., & Andrivon, D. (2008). Adaptation to the most abundant host genotype in an agricultural plant-pathogen system-potato late blight. *Journal of Evolution Biology, 21*, 1397–1407. doi:10.1111/j.1420-9101-2008.01557.x.

Piganeau, B., & Clerjeau, M. (1985). Influence différentielle de la température sur la germination de sporocyste et la sporulation des souches de *Plasmopara viticola* sensible et résistantes aux phénylamides. In *Fungicides for Crop Protection. BCPC monograph, 31*, 327–330.

Pringle, A., & Taylor, J. W. (2002). The fitness of filamentous fungi. *Trends in Microbiology, 10*, 474–481.

Rosenberg, S. M. (2001). Evolving responsively: adaptive mutation. *Nature Review Genetic, 2*, 504–515.

Sierotzki, H., Kraus, N., Assemat, P., Stanger, C., Cleere, C., Windass, J., et al. (2005). Evolution of resistance to QoI fungicides in *Plasmopara viticola* populations in Europe. In H. W. Dehne, U. Gisi, K. H. Kuck, P. E. Russell, & H. Lyr (Eds.), *Modern fungicides and Antifungal compounds IV* (pp. 73–80). Hampshire: BCPC.

Sierotzki, H., Kraus, N., Pepin, S., Ferandes, N., & Gisi, H. (2008). Dynamics of QoI resistance in *Plasmopara viticola*. In H. W. Dehne, U. Gisi, K. H. Kuck, P. E. Russell, & H. Lyr (Eds.), *Modern fungicides and antifungal compounds V* (pp. 151–157). Brauschweig: DPG Selbstverlag.

Sirven, C., & Beffa, R. (2003). Resistance to fenamidone: monitoring by real-time quantitative PCR on *Plasmo-*

para viticola. *Pflanzenschutz-Nachrichten Bayer, 56*, 523–5332.

Toffolatti, S. T., Serrati, L., Sierotzki, H., Gisi, U., & Vercesi, A. (2007). Assessment of QoI resistance in *Plasmopara viticola* oospores. *Pest Management Science, 63*, 194–201.

Tooley, P. W., Sweigard, J. A., & Fry, W. E. (1986). Fitness and virulence of *Phytophthora infestans* from sexual and asexual populations. *Phytopathology, 76*, 1209–1212.

Vanderplank, J. E. (1982). *Host-pathogen Interactions in plant disease* (p. 27). New-York: Academic.

Wong, F. P., & Wilcox, W. F. (2000). Distribution of baseline sensitivities to azoxystrobin among isolates of *Plasmopara viticola*. *Plant disease, 84*, 275–281.

Zolan, M. E., & Pukkila, P. J. (1986). Inheritance of DNA methylation in *Coprinus cinereus*. *Molecular Cell Biology, 6*, 195–200.

Eur J Plant Pathol (2011) 129:199–206
DOI 10.1007/s10658-010-9677-y

Evolution of QoI resistance in *Plasmopara viticola* oospores

Silvia Laura Toffolatti · Marisol Prandato · Luca Serrati · Helge Sierotzki · Ulrich Gisi · Annamaria Vercesi

Accepted: 23 August 2010 / Published online: 9 September 2010

Abstract QoI resistance in *P. viticola* was first detected in France and Italy in 1999. Molecular and biological assays have been carried out since 2000 in order to provide reliable methods of detecting and quantifying resistance. Oospores were collected in vineyards located in northern and southern Italy. QoI resistance was evaluated by the germination rate of oospores on azoxystrobin amended medium and the frequency of mutant alleles in the DNA extracted from oospores. Both methods correlated to each other and were used side by side to test QoI resistance. Due to the spontaneous occurrence of the G143A mutation in wild type populations and the immigration from surrounding vineyards, resistance frequencies up to 10% were found in samples collected from vineyards never treated with QoIs. Particularly high values, about 90%, were associated with the application of five to six QoI treatments within the same season, while lower percentages, about 30%, were detected in vineyards treated with QoI used in mixture with fungicides belonging to a different resistance group. A progressive decrease of resistance frequency was observed when QoI applications were reduced in number or completely suspended for at least one season. Therefore, a full recovery of sensitivity may be achieved even in vineyards characterized by high levels of resistance, if particular care is taken during disease control by using QoIs only in mixtures and reducing the number of QoI treatments.

Keywords Downy mildew · Fungicide resistance · Grapevine · Resistance management · Sensitivity assay

S. L. Toffolatti (✉) · M. Prandato · A. Vercesi
Facoltà di Agraria, Di.Pro.Ve.
Sezione di Patologia Vegetale,
Università degli Studi di Milano,
via Celoria 2,
20133 Milano, Italy
e-mail: silvia.toffolatti@unimi.it

L. Serrati
Development Center South East Europe,
Syngenta Crop Protection,
via Gallarate 139,
20151 Milano, Italy

H. Sierotzki · U. Gisi
Research Biology, Syngenta Crop Protection,
WST-540, Schaffhauserstrasse,
4332 Stein, Switzerland

Introduction

Grapevine downy mildew caused by *Plasmopara viticola* (Berk. *et* Curt.) Berl. and De Toni is a serious disease in climates characterized by moderate temperatures and frequent rainfall in spring and summer (Gilles 2004; Jeger and Pautasso 2008). These conditions occur in different viticultural area and years leading to severe epidemics, as it was observed in the past few years first in southern and then in northern Italy. Chemical control is currently the most effective measure to protect the plants from disease. Numerous treatments with fungicides are carried out in vineyards during the grapevine growing season to prevent severe epidemics that can reduce both yield and quality of grapes. Due to its biological features (Burruano 2000; Lafon et al. 1988), *P. viticola* is considered as a high risk pathogen for resistance developing to fungicides (FRAC pathogen risk list, www.frac.info). It is therefore necessary to avoid high selection pressure on populations, which may favour the appearance of resistant strains. In order to prevent failures in disease control and to preserve the performance of products, it is necessary to optimize the use of all fungicide classes available against grapevine downy mildew.

Among the fungicides most frequently used in vineyards, QoI fungicides such as azoxystrobin, famoxadone and fenamidone are important (Gisi and Sierotzki 2008). All QoIs are sharing the same target site since they all inhibit mitochondrial respiration by preventing binding and subsequent oxidization of hydroquinone at the Quinol outer (Qo) site of cytochrome bc1 complex (Link et al. 2003; Kuck and Gisi 2007). Because of the single-site mode of action of QoIs and their extensive use, resistant strains in *P. viticola* population were detected already in 1999, first in field trials in Italy and France and subsequently in commercial vineyards (Bartlett et al. 2002). Monitoring methods are commonly used to evaluate the frequency of resistant strains and the effect of different strategies of fungicide application on pathogen populations (Brent and Hollomon 1998).

Since the discovery that QoI resistance in *P. viticola* correlates with a single nucleotide polymorphism in the cytochrome b gene leading to an amino acid change at position 143 (G143A) (Sierotzki et al. 2005; Chen et al. 2007), molecular methods allow a fast and accurate detection of QoI resistance by an allele-specific, real-time PCR assay (Sierotzki and Gisi 2003). The possibility of using this highly sensitive method improved the possibility of monitoring the spread of resistant strains and studying the evolution of fungicide resistance at the population level (Ma and Michailides 2005; Sierotzki et al. 2005). In general, QoI sensitivity can be monitored by collecting *P. viticola* strains in vineyards and evaluating their sporulation on leaf discs treated with different fungicide concentrations (www.frac.info) or by molecular methods (Sierotzki and Gisi 2003).

Biological and molecular assays have also been carried out by using oospores (Vercesi et al. 2002; Toffolatti et al. 2007). The sexual structures of the pathogen are differentiated in the infected leaves at the end of the grapevine growing season. They overwinter and germinate the following season, providing inoculum for primary infections (Vercesi et al. 2000). QoI resistance is inherited maternally (Gisi et al. 2002); therefore, QoI resistance can be studied in oospores collected at the end of the grapevine growing season, reflecting the result of the selection imposed by the fungicide applications made during the past season, and indicating the potential frequency of resistance in the primary inoculum for the following season.

In this paper, the evolution of resistance was investigated in vineyards which received either no QoI treatments or were treated with solo formulations or mixtures with non-cross resistant fungicides. The results support a more rational rather than routine use of these fungicides.

Materials and methods

Sampling

Leaves showing downy mildew symptoms were randomly collected during the seasons 2004 to 2007 in 33 vineyards located mainly in southern Italy (Puglia) but also in northern Italy (Lombardia, Veneto, Piemonte and Emilia Romagna). In particular, 7 vineyards were monitored for at least 2 consecutive years in order to follow the evolution of QoI resistance after QoI use suspension (Tables 1, 3): in the two vineyards located in Emilia Romagna, two adjacent plots, respectively untreated (3P and SG) and treated with QoI fungicides (3PT and SGT) were sampled. A single leaf fragment (0.5 cm^2), containing a high number of oospores (as determined microscopically with Leitz Orthoplan), was cut from each leaf. Fifty leaf fragments per replicate were transferred to nylon bags (pore diameter 100 μm) and overwintered at 5°C in the dark laying on a sand substrate kept water-saturated. Four replicates were prepared per sampling location.

Biological and molecular assays

The biological assay, based on the germination rates of oospores on azoxystrobin amended medium at the

Table 1 Origin of samples and average rates of RO and MA

Sample	Region	Year of sampling	QoI treatment	RO (%)	MA (%)
BAE	Puglia	2004	solo	81	95
BAE	Puglia	2005	solo	36	48
BAE	Puglia	2006	solo	85	76
BAA	Puglia	2004	solo	12	19
BAD	Puglia	2004	solo	97	100
BAS	Puglia	2004	solo	94	98
BAG	Puglia	2004	solo	74	76
BAF	Puglia	2004	solo	94	98
BAF	Puglia	2006	solo	70	93
BAZ	Puglia	2004	solo	31	18
BAB	Puglia	2004	solo	79	85
BAN	Puglia	2004	solo	1	10
BAP	Puglia	2004	solo	14	70
BAU	Puglia	2004	solo	52	83
BAO	Puglia	2004	solo	87	98
BAM	Puglia	2004	solo	20	24
BBC	Puglia	2004	solo	0,4	1
BBI	Puglia	2005	solo	22	29
BBL	Puglia	2005	solo	28	39
BBL	Puglia	2006	solo	66	43
MO	Piemonte	2005	solo	13	11
MO	Piemonte	2006	solo	0	0.5
BAH	Puglia	2004	mixture	0	0
BAB	Puglia	2005	mixture	21	31
BBH	Puglia	2005	mixture	72	71
BBH	Puglia	2006	mixture	55	98
BBM	Puglia	2006	mixture	10	15
BAT	Puglia	2004	mixture	16	22
BAR	Puglia	2004	mixture	0	0
BBN	Puglia	2006	mixture	24	47
SO	Lombardia	2006	mixture	42	36
BAF	Puglia	2005	no QoI	10	8
BAM	Puglia	2005	no QoI	25	25
BAI	Puglia	2005	no QoI	0	1
BAI	Puglia	2006	no QoI	11	16
BAI	Puglia	2007	no QoI	0	9
BBI	Puglia	2006	no QoI	2,6	5
BBD	Puglia	2004	no QoI	3,1	7
MT	Veneto	2004	no QoI	5	9
MT	Veneto	2005	no QoI	0	10
MT	Veneto	2006	no QoI	13	20
MS	Veneto	2005	no QoI	6	8
MS	Veneto	2006	no QoI	4	11
BE	Lombardia	2005	no QoI	12	0.5
BE	Lombardia	2006	no QoI	2	3

discriminatory dose of 10 mg l^{-1}, was carried out every year at the end of January as described by Toffolatti et al. (2007). Oospore germination (1200 oospores per sample) was assessed on 1% water agar (Agar Noble, Difco) with and without 10 mg l^{-1} of azoxystrobin (technical grade) at 20°C. Percentage of oospore germination (G) was determined by counting the number of germinated structures, which was checked daily over a period of 14 days under the microscope (Leica Wild M10). The rate of resistant oospores (RO) was calculated as Gaz×100/Gwa, where Gaz and Gaw are the germination rates of the oospores on water agar amended with azoxystrobin and on water agar.

The molecular assay was carried out according to Toffolatti et al. (2007, 2008) by allele specific detection and quantification in real-time PCR of the SNP leading to G143A amino acid substitution in cytochrome b protein. Quantification of both mutated and wild type alleles in the DNA extracted from the oospores (CTAB method) was performed on ABI prism 7000 thermal cycler (Applied Biosystems, USA) by using ARMS-Scorpion or SYBR Green detection methods. Double distilled water and DNA extracted from the leaves were used as controls in order to detect possible unspecific amplifications. The efficiency of the PCR assay was estimated from standard curves based on serial dilutions of purified plasmids containing the target sequences.

The frequency of mutated alleles ([1/(r+1)]×100) was calculated from the ratio (r) of both alleles as $r=2^{\Delta Ct}$, where $\Delta C_t = C_t\text{mut} - C_t\text{wt}$ and Ct are the threshold cycle values (C_t) for G143 (wt) and 143A (mut) alleles.

Statistical analysis

Shapiro-Wilk's test for normality was used to evaluate the distribution of the mean frequency of resistant oospores (RO) and mutated allele 143A (MA). Non-parametric correlation between the two variables was assessed by Spearman's R and Kendall's Tau coefficients. Differences between samples collected from vineyards with different use strategies of QoI's (solo, in mixture with partner belonging to another resistance group, and not applied at all) were visualized by box-plots. Box-plots visualise the variability existing among populations even if the underlying statistical distributions are unknown. They show the shape, characteristic values and variability of a distribution by using the maximum and minimum values ("whiskers"), the interquartile range (i.e. the difference between the third and first interquartile), IQR (box), and the median (central line) of the data set. One-way ANOVA and *a priori* contrasts, tested with t statistics, were performed on transformed RO and MA (arcsin $\sqrt{x/100}$) with the aim of determining differences in the mean values of different treated samples. Statistical analysis was carried out with the SPSS Statistics 17.0 software.

Results

Frequency of resistant oospores and mutated allele

Since the hypothesis of a normal distribution was rejected following the results of Shapiro-Wilk's test on the mean frequency of RO (W=0.83, p=0.000) and MA (W=0.84,

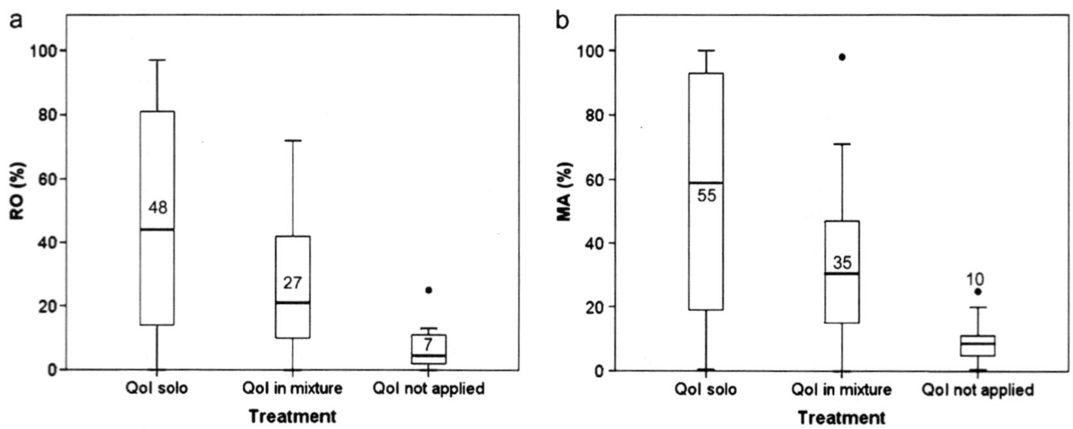

Fig. 1 Boxplot analysis of RO (**a**) and MA (**b**) following different strategies of QoI treatment. Numbers indicate mean values, *black circles* represent mild outliers

Table 2 *A priori* sets and values of contrasts on transformed values of RO and MA and results of *t*-test

	Contrast	Value of contrast	Standard error	*t*	df[a]	Significance
RO	Solo+mixtures against untreated	0.7791	0.23536	3.310	42	0.002
	Solo against untreated	0.5254	0.12099	4.342	42	0.000
	Mixture against untreated	0.2538	0.15120	1.678	42	0.101
	Solo against mixture	0.2716	0.14003	1.940	42	0.059
MA	Solo+mixtures against untreated	0.8591	0.25974	3.308	42	0.002
	Solo against untreated	0.5647	0.13353	4.229	42	0.000
	Mixture against untreated	0.2945	0.16687	1.765	42	0.085
	Solo against mixture	0.2702	0.15454	1.749	42	0.088

[a] Degrees of freedom

p=0.000), Spearman's and Kendall's coefficients for non-parametric correlation between variables were calculated. A strong linear and positive correlation between the values of MA and RO (R=0.93, Tau= 0.80) was found, showing that both parameters can be used to estimate QoI resistance in *P. viticola* oospores.

Effect of QoI treatments

The application of QoIs as solo formulations resulted in high variability in the rates of both RO and MA, which ranged from 0 to 100% with an IRQ of 70% and a non-normal distribution of the samples (W=0.89, p=0.019

Table 3 Origin of samples collected from vineyards where treatments with QoI fungicides were suspended and average RO

Sample	Region	Year of sampling	QoI treatment	RO (%)
3P	Emilia Romagna	2001	no QoI	98
3P	Emilia Romagna	2002	no QoI	18
3P	Emilia Romagna	2003	no QoI	33
3P	Emilia Romagna	2004	no QoI	17
3P	Emilia Romagna	2005	no QoI	20
3P	Emilia Romagna	2006	no QoI	27
3PT	Emilia Romagna	2001	solo	93
3PT	Emilia Romagna	2004	no QoI	17
3PT	Emilia Romagna	2005	no QoI	16
3PT	Emilia Romagna	2006	no QoI	4
SG	Emilia Romagna	2002	no QoI	72
SG	Emilia Romagna	2004	no QoI	69
SG	Emilia Romagna	2005	no QoI	17
SG	Emilia Romagna	2006	no QoI	17
SGT	Emilia Romagna	2002	mixture	86
SGT	Emilia Romagna	2004	no QoI	42
SGT	Emilia Romagna	2005	no QoI	14
SGT	Emilia Romagna	2006	no QoI	9
BAF	Puglia	2005	solo	90
BAF	Puglia	2006	no QoI	10
BAM	Puglia	2004	solo	22
BAM	Puglia	2005	no QoI	25
BAM	Puglia	2006	no QoI	27
BBI	Puglia	2005	mixture	22
BBI	Puglia	2006	no QoI	3

for RO, W=0.88, p=0.013 for MA). Despite the mean values of RO and MA (48 and 55%, respectively, Fig. 1), the samples can be divided into two main groups: the first and the larger one is characterized by low (0–50%) and the second by high (70–100%) rates of resistance (Table 1).

The box-plots showed reduced median resistance values and decreasing variation (IQR) in the samples collected from vineyards where QoIs were applied in mixture: for both indices measured, RO and MA, the 50% portion of the examined samples ranged from 10 to 40–45%, and the mean values were 27 and 35%, respectively (Fig. 1). Particularly low mean values and IQR, about 10%, were detected for oospores collected in vineyards which were not treated with QoIs at all (Fig. 1). In addition, the extreme values were very close to the median in the untreated samples, indicating a narrow distribution of the data.

Statistically significant differences were found between the means of samples with different treatments: F=9.558, $P(F)$=0.000 for RO and F=9, $P(F)$= 0.000 for MA. *A priori* contrasts showed differences between treated (solo and mixtures) and untreated samples and between solo treated and untreated samples, whereas samples treated in mixture did not differ from solo nor untreated (Table 2).

Effect of QoI use suspension

In seven vineyards samples were collected in subsequent years in order to assess the evolution of resistance following QoI use suspension (Table 3). QoI sprays in the first year of monitoring resulted in a mean RO of 63%, but 25% of the population showed particularly high rates, ranging between 85% and 95% (Fig. 2). In Emilia Romagna, similar rates of resistant oospores were found in the untreated plots, 3P and SG, located near the QoI treated ones. Despite the wide range of RO observed in the first year since QoI suspension, much smaller values were observed for IQR (15–55%) and the median (25%), indicating a shift in the pathogen population towards reduced resistance (Fig. 2). Two years after suspension, RO showed a positively skewed distribution with a reduced IQR, ranging between 15% and 30%, and a median value of 18%. At the end of the third year, the number of resistant oospores further decreased towards values similar to those observed in vineyards where QoI were not applied (mean 16%, median 13%).

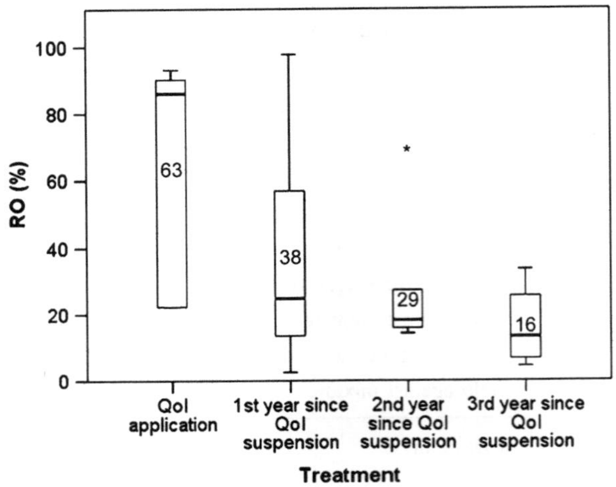

Fig. 2 Box-plot analysis of RO in samples collected from vineyards where QoI sprays were suspended. Numbers indicate mean values, asterisks extreme outliers

Discussion

Since a large number of oospores, each representing a single individual, can be easily collected in many vineyards, detection of *P. viticola* resistance to QoI fungicides using the oospore method is highly representative for entire populations and can be particularly useful in base-line sensitivity assays. Moreover, rapid and quantitative detection of resistant strains in pathogen populations enables the prediction of the resistance risk and may prevent control failure by using adapted fungicide use strategies (Ishii 2006; Macdonald 2008).

Since the discovery of resistance, numerous studies have been carried out on both the underlying mechanism and inheritance (Blum and Gisi 2008; Gisi et al. 2002; Grasso et al. 2006). Phylogenetic analysis of a large mitochondrial DNA fragment including cytochrome *b* gene sequences of different strains of *P. viticola* showed that QoI resistance emerged independently several times in Europe (Chen et al. 2007). The strong correlation between the rates of RO and MA confirms that QoI resistance is mainly based on the point mutation in the cytochrome b gene leading to the amino acid substitution G143A (Sierotzki et al. 2005; Toffolatti et al. 2007; Chen et al. 2007).

The effect of the different QoI treatments (singly and in mixture) on pathogen sensitivity was evaluated by analysing the distribution of RO and MA by ANOVA and *a priori* contrasts on the transformed data. The repeated application of QoI fungicides, especially as a solo formulation, which was done in past years, significantly increased the initially low proportion of resistance of *P. viticola* populations from 0% to 20% to variable but large values (0% to 100%), not only in the directly treated plots, but also in the untreated vineyards parts located nearby, due to the migration of the pathogen strains (Toffolatti et al. 2008). Moreover, solo treatments induced a separation of the samples in two subpopulations, characterized by low (about 15%) and high (about 80%) rates of resistance. Due to sexual recombination and migration of sensitive strains from surrounding vineyards, a high selection pressure does not always result in oospore populations with high resistance rates. The samples collected from vineyards following the utilization of QoI in mixture with fungicides belonging to different resistance groups, such as inhibitors of RNA synthesis (phenylamide fungicides) or multi-site inhibitors such as folpet, showed reduced frequencies of resistance, compared to solo applications, confirming the lower selection pressure of this use strategy (Russell 2003; Genet et al. 2006).

A shift towards low resistance levels similar to those in untreated vineyards was observed following the interruption of QoI sprays for at least 2 years. A substantial reduction in the distribution of resistance was found starting from the first year of interruption, followed by a further recovery of sensitivity starting from the second year onwards.

Studies carried out with microsatellite markers suggested that QoI resistant strains of *P. viticola* possess fitness properties as high as sensitive strains during asexual cycles *in vitro* and *in natura* (Corio-Costet et al. 2008; Lafarge et al. 2008). However, a decline in the frequency of resistant strains of *P. viticola* was observed after stopping QoI treatments (Gisi and Sierotzki 2008). Also in the present study, we demonstrated that resistance in oospore populations decreases when selection pressure is reduced. Therefore, a more sensitive primary inoculum can be expected for the next grapevine season, if fungicide mixtures instead of solo formulations are used or, in cases of high resistance frequency, if QoI treatments are completely suspended for at least 2 years.

The presented data suggest that usage of QoIs in mixture is a suitable strategy to manage resistance. The data suggest that while QoIs cannot be totally excluded, their usage in mixtures is a suitable strategy to manage resistance. However, it is necessary that the frequency of resistant strains decreases to a low level in the population and that the population is monitored continuously for changes.

The preservation of many fungicides is more and more critical since the application of the Directive 91/414/EEC which will be replaced by the recent regulation (EC) No 1107/2009 of 21 October 2009, concerning the placement of plant protection products on the market in the European Community, and of the Directive 99/45/EC, related to the classification, packaging and labelling of dangerous preparations. Therefore, the number of fungicides available on the market could be reduced. As a consequence, it is very important to maintain as many fungicide classes as possible through sound resistance monitoring and use strategies.

References

Bartlett, D. W., Clough, J. M., Godwin, J. R., Hall, A. A., Hamer, M., & Parr-Dobrzanski, B. (2002). The strobilurin fungicides. *Pest Management Science, 58*, 649–662.

Blum, M., & Gisi, U. (2008). Inheritance of fungicide resistance in *Plasmopara viticola*. In H. W. Dehne, H. B. Deising, U. Gisi, K. H. Kuck, P. E. Russel, & H. Lyr (Eds.), *Modern fungicides and antifungal compounds V* (pp. 101–104). Braunschweig: DPG Selbstverlag.

Brent, K. J., & Hollomon, D. W. (1998). *Fungicide resistance: The assessment of risk*. Brussels: Global Crop Protection Federation.

Burruano, S. (2000). The life-cycle of *Plasmopara viticola*, cause of downy mildew of vine. *Mycologist, 14*, 179–182.

Chen, W. J., Delmotte, F., Richard-Cervera, S., Douence, L., Greif, C., & Corio-Costet, M. F. (2007). At least two origins of fungicide resistance in grapevine downy mildew populations. *Applied and Environmental Microbiology, 73*, 5162–5172.

Corio-Costet, M. F., Martinez, F., Delmotte, F., Douence, L., Richart-Cervera, S., & Chen, W. J. (2008). Resistance of *Plasmopara viticola* to QoIs fungicides: Origin and diversity. In H. W. Dehne, H. B. Deising, U. Gisi, K. H. Kuck, P. E. Russel, & H. Lyr (Eds.), *Modern fungicides and antifungal compounds V* (pp. 107–112). Braunschweig: DPG Selbstverlag.

Genet, J. L., Jaworska, G., & Deparis, F. (2006). Effect of dose rate and mixtures of fungicides on selection for QoI resistance in populations of *Plasmopara viticola*. *Pest Management Science, 62*, 188–194.

Gilles, T. (2004). Forecasting downy mildew disease. In P. Spencer-Phillips & M. Jeger (Eds.), *Advances in downy mildew research* (Vol. 2, pp. 35–67). Dordrecht: Kluwer Academic.

Gisi, U., & Sierotzki, H. (2008). Fungicide modes of action and resistance in downy mildews. *European Journal of Plant Pathology, 122*, 157–167.

Gisi, U., Sierotzki, H., Cook, A., & McCaffery, A. (2002). Mechanisms influencing the evolution of resistance to Qo inhibitor fungicides. *Pest Management Science, 58*, 859–867.

Grasso, V., Palermo, S., Sierotzki, H., Garibaldi, A., & Gisi, U. (2006). Cytochrome b gene structure and consequences for resistance to QoI inhibitor fungicides in plant pathogens. *Pest Management Science, 62*, 465–472.

Ishii, H. (2006). Impact of fungicide resistance in plant pathogens on crop disease control and agricultural environment. *Japan Agricultural Research Quarterly, 40*, 205–211.

Jeger, M. J., & Pautasso, M. (2008). Comparative epidemiology of zoosporic plant pathogens. *European Journal of Plant Pathology, 122*, 111–126.

Kuck, K.-H., & Gisi, U. (2007). FRAC mode of action classification and resistance risk of fungicides. In W. Krämer & U. Schirmer (Eds.), *Modern crop protection compounds* (pp. 415–432). Weinheim: Wiley-VCH.

Lafarge, D., Cigna, J., Abadie, P., Douence, L., Dufour, M. C., & Corio-Costet, M.-F. (2008). Competitive fitness and adaptative value of QoI-resistant *Plasmopara viticola* strains. In H. W. Dehne, H. B. Deising, U. Gisi, K. H. Kuck, P. E. Russel, & H. Lyr (Eds.), *Modern fungicides and antifungal compounds V* (pp. 181–185). Braunschweig: DPG Selbstverlag.

Lafon, R., Clerjeau, M., Pearson, R. C., & Goheen, A. C. (1988). Downy mildew. In R. C. Pearson & A. C. Gohen (Eds.), *Compendium of grape diseases* (pp. 11–13). St. Paul: APS Press.

Link, T. A., Iwata, M., Björkman, J., van der Spoel, D., Stocker, A., & Iwata, S. (2003). Molecular modeling of inhibitors at Qi and Qo sites in cytochrome bc1 complex. In G. Voss & G. Ramos (Eds.), *Chemistry of crop protection* (pp. 110–127). Weinheim: Wiley.

Ma, Z., & Michailides, T. J. (2005). Advances in understanding molecular mechanisms of fungicide resistance and molecular detection of resistant genotypes in phytopathogenic fungi. *Crop Protection, 24*, 853–863.

Macdonald, O. C. (2008). Regulatory aspects of resistance management. In H. W. Dehne, U. Gisi, K. H. Kuck, P. E. Russell, & H. Lyr (Eds.), *Modern fungicides and antifungal compounds V* (pp. 113–120). Alton: BCPC.

Russell, P. (2003). Taking the path of least resistance. *Pesticide Outlook, 14*, 57–61.

Sierotzki, H., & Gisi, U. (2003). Molecular diagnostics for fungicide resistance in plant pathogens. In G. Voss & G. Ramos (Eds.), *Chemistry of crop protection* (pp. 71–88). Weinheim: Wiley.

Sierotzki, H., Kraus, N., Assemat, P., Stanger, C., Cleere, S., Windass, J., et al. (2005). Evolution of resistance to QoI fungicides in *Plasmopara viticola* populations in Europe. In H. W. Dehne, U. Gisi, K. H. Kuck, P. E. Russell, & H. Lyr (Eds.), *Modern fungicides and antifungal compounds IV* (pp. 73–80). Alton: BCPC.

Toffolatti, S. L., Serrati, L., Sierotzki, H., Gisi, U., & Vercesi, A. (2007). Assessment of QoI resistance in *Plasmopara viticola* oospores. *Pest Management Science, 63*, 194–201.

Toffolatti, S. L., Prandato, M., Serrati, L., Sierotzki, H., Gisi, U., & Vercesi, A. (2008). Monitoring QoI resistance in *Plasmopara viticola* oospore populations. In H. W. Dehne, H. B. Deising, U. Gisi, K. H. Kuck, P. E. Russel, & H. Lyr (Eds.), *Modern fungicides and antifungal compounds V* (pp. 159–165). Braunschweig: DPG Selbstverlag.

Vercesi, A., Sirtori, C., Vavassori, A., Setti, E., & Liberati, D. (2000). Estimating germinability of *Plasmopara viticola* oospores by means of neural networks. *Medical & Biological Engineering & Computing, 38*, 109–112.

Vercesi, A., Vavassori, A., Faoro, F., & Bisiach, M. (2002). Effect of azoxystrobin on the oospores of *Plasmopara viticola*. In P. T. N. Spencer-Phillips, U. Gisi, & A. Lebeda (Eds.), *Advances in downy mildew research* (pp. 195–199). Dordrecht: Kluwer Academic.

Eur J Plant Pathol (2011) 129:207–219
DOI 10.1007/s10658-010-9698-6

EMS and UV irradiation induce unstable resistance against CAA fungicides in *Bremia lactucae*

Avia E. Rubin · Ayelet-Chen Werdiger · Mathias Blum · Ulrich Gisi · Helge Sierotzki · Dietrich Hermann · Yigal Cohen

Accepted: 7 October 2010 / Published online: 23 October 2010

Abstract Wild type (WT) field isolates of *Bremia lactucae* failed to germinate in vitro or infect lettuce leaves in the presence of CAA (carboxylic acid amide) fungicides. Minimal inhibitory concentrations (MIC) for mandipropamid, dimethomorph, benthiavalicarb and iprovalicarb were 0.005, 0.5, 0.5 and 5 $\mu g ml^{-1}$, respectively. Mutagenesis experiments showed that spores exposed to EMS (ethyl methane sulphonate) or UV irradiation (254 nm) could infect lettuce leaves in the presence of up to 100 $\mu g ml^{-1}$ CAA. The proportion of infected leaves relative to the number of spores inoculated (infection frequency) was inversely related to the concentration of CAA used, ranging between 0 and 160 per 1×10^6 spores. Resistant mutants (RM) lost their resistance within 1–14 reproduction cycles on CAA-treated plants. Crosses were made between RMxWT isolates and RMxRM isolates with an attempt to obtain stable homozygous resistant off-springs. Such crosses yielded few resistant but unstable progeny isolates. Mutagenic treatments given to hybrid isolates also failed to produce stable resistance. Previous gene sequencing data showed that stable resistance to CAAs is based on a single SNP in the *cellulose synthase 3* (*CesA3*) gene of *Plasmopara viticola*. Therefore, we sequenced a 582 bp DNA fragment of *Ces3A* of WT, RM and hybrid isolates of *B.lactucae*. No mutation in this gene fragment was found. We conclude that mutagenic agents like EMS or UV may induce resistance to CAA in *Bremia lactucae* but this resistance is not stable and not linked to mutations in *CesA3* gene.

Keywords Chemical control · Lettuce downy mildew · Oomycetes

A. E. Rubin · A.-C. Werdiger · Y. Cohen (✉)
Faculty of Life Sciences, Bar Ilan University,
Ramat-Gan, Israel
e-mail: ycohen@mail.biu.ac.il

M. Blum · U. Gisi
University of Basel,
Basel, Switzerland

M. Blum · U. Gisi · H. Sierotzki · D. Hermann
Syngenta Crop Protection,
Stein, Switzerland

Abbreviations

BENT	Bentiavalicarb
BC	Back-cross
CAA	Carboxylic acid amides
Ces	Cellulose synthase
Dpi	Days post inoculation
DMM	Dimethomorph
EMS	Ethyl methanesulfonate
IPRO	Iprovalicarb
MIC	Minimal inhibitory concentration
MPD	Mandipropamide
RM	Resistant mutant
PCR	Polymerase chain reaction
UV	Ultra violet
WT	Wild type

Introduction

The Carboxylic Acid Amides (CAA), mandipropamide (MPD), dimethomorph (DMM), bentiavalicarb (BENT) and iprovalicarb (IPRO) are effective fungicides against oomycete plant pathogens including *Phytophthora infestans* (Cohen and Gisi 2007), *Plasmopara viticola* (Reuveni 2003; Gisi et al. 2007), *Pseudoperonospora cubensis* (Cohen et al. 1995) and *Bremia lactucae* (Cohen et al. 2008). Resistance against CAAs occurs in field populations of *P. viticola* (Gisi et al. 2007) and *P. cubensis* (Y.Cohen, unpublished data, CAA FRAC working group, www.frac.info) but not in *P. infestans* or *B. lactucae*. The reasons for the differential occurrence of field resistance in some but not other oomycetes are not clear. Major reasons may be that resistance is hardly fixed in the population because of its recessive nature controlled by one or two nuclear genes (Gisi et al. 2007) and that pathogen species differ in basic biological processes such as frequency of sexual reproduction and migration. With *P. infestans*, we were able to artificially mutate sporangia for stable resistance against the phenylamide fungicide mefenoxam but failed to select mutants with stable resistance against CAAs (Rubin et al. 2008), suggesting a low risk of resistance developing in this pathogen in field populations against CAAs.

The aim of the present study was to estimate the likelihood of *B. lactucae* to develop resistance against CAA fungicides. This was done by exposing spores to UV light or to the mutagenic agent EMS and examining their capability to infect lettuce leaves after being mixed with CAA, or to infect CAA-treated leaves. The information provided here is important for formulating resistance management strategies to be employed in the field. Preliminary reports were published recently (Cohen et al. 2009; Werdiger and Cohen 2008).

Materials and methods

Fungicides

Four CAA (carboxylic acid amide) fungicides were used: mandipropamid (MPD; Syngenta, mw=412), dimethomorph (DMM; BASF, mw=266), iprovalicarb (IPRO; Bayer, mw=320) and benthiavalicarb (BENT; Kumiai Chemicals, mw=339). Technical grade fungicides were dissolved in DMSO (10 mgml^{-1}) and diluted in double distilled water to the desired concentrations. Formulated fungicides used were: mandipropamid 250SC, dimethomorph (Forum) 50WP and Benthiavalicarb (5EC), a gift from Syngenta. All concentrations are presented in units of active ingredient (a.i.).

Fig. 1 **a**. The effect of CAA fungicides on germination of six wild type isolates of *Bremia lactucae* in vitro. Spores (about 200 in 20 μl droplets) were mixed with technical grade CAAs and placed in glass slide depressions. The proportion of germinating spores (producing a germ-tube of≥10 μm) was determined microscopically after incubation at 15°C in the dark for 20 h. Vertical bars represent standard deviation of the mean (n=3). **b**. One-way Anova analysis of the germination data. Anova unit=square root of arc-sin% germination. Horizontal diagonal=mean; vertical diagonal=standard error. In each panel, diagonals sharing the same letter are not significantly different (Tukey-Kramer test, α=0.05)

Pathogen

The following wild type field isolates of *B. lactucae* Regel were used: IL60 and BS (both from Israel); BL-18, BL-21, BL-24 and BL-25 (from A. Lebeda, Olomouc University, Czech Republic); and isolate CH (from Syngenta, Stein, Switzerland). Isolates were maintained by repeated inoculation on detached lettuce cotyledons in Petri dishes in a growth chamber (15°C, 12 h light/day).

Plants

The susceptible lettuce (*Lactuca sativa*) cv. Noga (cup type; Hazera Genetics, Mivhor, Israel) was used. Plants were grown from seeds in 250 ml pots containing peat/vermiculite mixture (1/1, v/v) to give 30–40 plants per pot. Plants were grown in a growth cabinet (20–22°C) and used 1 week after sowing, when two cotyledon leaves had developed. Also detached cotyledons laying on a wet filter paper in Petri dishes were used. In some experiments, plants were grown in 0.5-L pots and used at the 10th leaf stage.

Mutagenesis

Two mutagenic agents were used: EMS and UV irradiation. EMS (ethyl methane sulphonate) was purchased from Sigma. It was dissolved in water and supplied to spores of *B. lactucae* at the concentrations

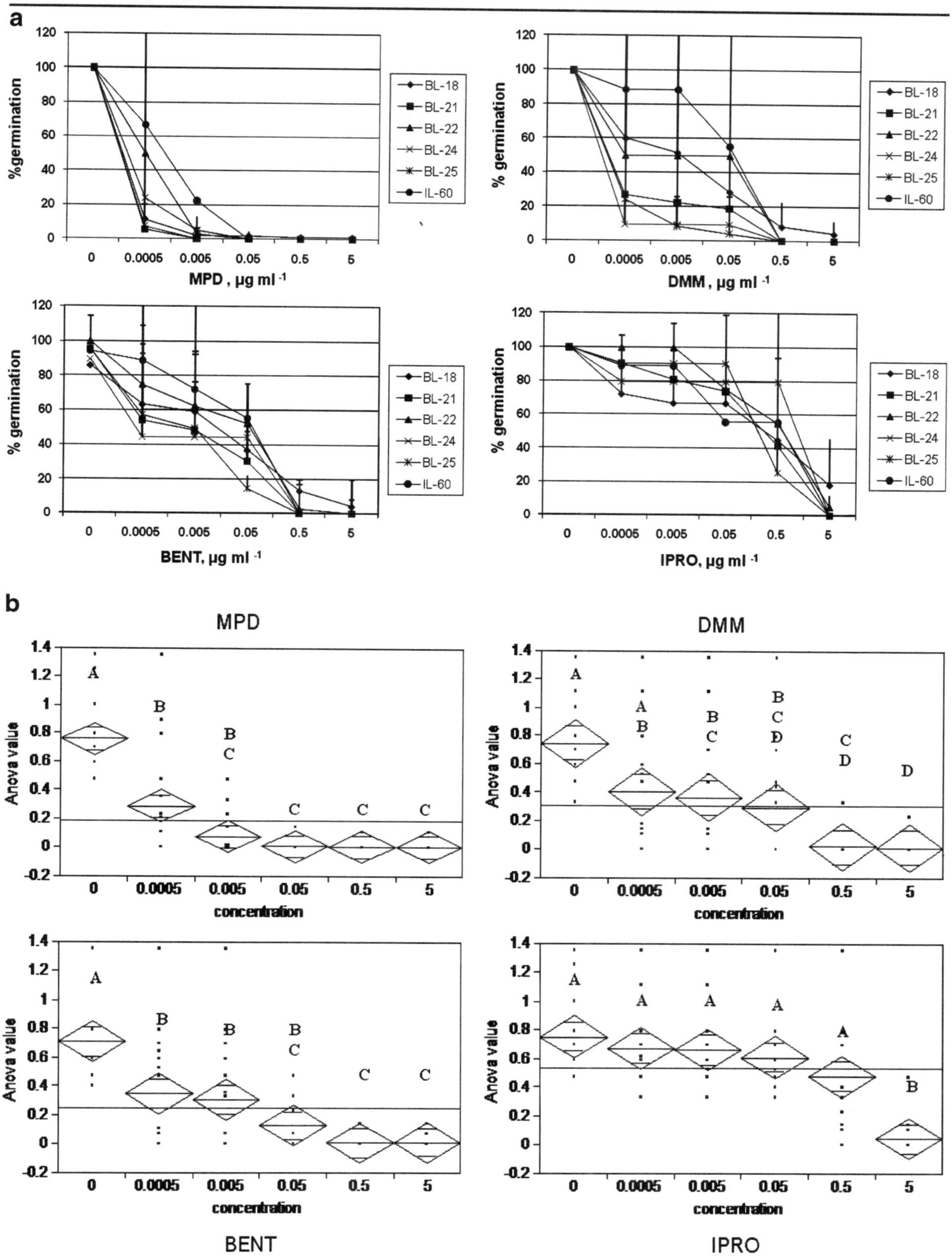
a
% germination
MPD, µg ml -1
DMM, µg ml -1
BENT, µg ml -1
IPRO, µg ml -1
BL-18
BL-21
BL-22
BL-24
BL-25
IL-60
b
MPD
DMM
BENT
IPRO
Anova value
concentration

indicated in Tables 2 and 3. UV light (254 nm) was supplied to spores suspended in ice water at the doses and time periods indicated in Tables 2 and 3.

Sensitivity to CAAs

Sensitivity of isolates to CAAs was tested at the germination stage and the infection stage as described before (Cohen et al. 2008). For germination, spores were mixed with technical CAA of 0.001 to 100 μg ml^{-1} (1:1, vol/vol) and applied to depressions in glass slides (20 μl/depression) and germination was examined at×40 magnification after 20 h of incubation at 15°C in the dark. For infection, spores (2–3× 10^4 ml^{-1}) were mixed (1:1, vol/vol) with technical grade CAA of 0.001 to 100 μgml^{-1} (×10 step dilutions) and inoculated onto the lower surface of detached cotyledons, 20 μl droplet/cotyledon, 4–12 cotyledons/dose/fungicide. The proportion of sporulating cotyledons and/or the sporulation intensity (0–3 visual scale) on each cotyledon was recorded at 7 dpi for wild type isolates and up to 21 dpi for mutant isolates. Sensitivity to CAAs of mutant isolates was also tested in intact cotyledons growing in pots. Plants were sprayed to run-off with formulated CAAs of 0.78 to 50 or 100 μg ml^{-1} (×2 step dilutions), one pot with 30–40 plants per dose, and thereafter spray-inoculated with spores of the mutant isolate. Inoculated plants were placed at 100% relative humidity at 18°C in the dark for 20 h to ensure infection, and then in a growth chamber at 15–18°C (12 h light/day, 100 μE $m^{-2}s^{-1}$).

Crosses

Crosses between isolates of opposite mating types were made in the manner described for *P. infestans* (Rubin and Cohen 2006). WT isolates were crossed with mutants expressing unstable resistance to CAAs to possibly obtain homozygous resistant progeny isolates. Crosses were also made between resistant mutants. To make a cross, spores (asexual) of the two parental isolates were mixed at 1:1 ratio and inoculated onto detached cotyledons or first true leaves in Petri dishes. At 7–10 dpi, when plenty of oospores were produced, leaves were homogenized in ice water, the homogenate was poured into 9 cm Petri dishes, 10 ml/plate, incubated in a ventilated hood to dry out, re-suspended in water and dried again to ensure the killing of the vegetative structures of the pathogen. Oospores in water suspension (1-5× 10^3 ml^{-1}) were inoculated onto intact cotyledons growing in pots, 20 μl droplet/cotyledon, and the pots sealed in 1 l transparent plastic containers incubated for 1–4 weeks in a 15°C growth chamber, until progeny spores were formed on the infected cotyledons (usually 1–5 sporulating cotyledons out of about 100 inoculated). Oospores of progeny isolates were collected and used for sensitivity tests as described above. In one cross, single spore isolates were produced, propagated in the absence of fungicides, and each was tested for sensitivity to CAAs.

Fig. 2 **a**. The effect of CAA fungicides on infection of lettuce cotyledons with six wild type isolates of *Bremia lactucae*. Spores (about 200 in 20 μl) were mixed with technical grade CAA and inoculated onto detached lettuce cotyledons in Petri dishes. Plates were incubated at 15°C for 20 h in the dark and then at 15°C under 12 h light/day. The number of sporulating cotyledons was determined at 7 dpi. Vertical bars represent standard deviation of the mean (n=4). **b**. One-way Anova analysis of the infection data. Anova unit=square root of arc-sin infected cotyledons. Horizontal diagonal=mean; vertical diagonal=standard error. In each panel, diagonals sharing the same letter are not significantly different (Tukey-Kramer test, α=0.05) ▶

DNA extraction

Samples for DNA extraction were obtained by collecting spores (of WT, RM, F1's and BC's isolates) from infected lettuce cotyledons or leaves. Samples were immediately frozen in liquid nitrogen and stored at −20°C until used. DNA extraction was done using MasterPure Yeast DNA Purification Kit (Epicentre Biotechnologies, Madison, Wisconsin, USA) following manufacturer instructions.

PCR manipulations and analysis of the *CesA3* gene fragment

PCR amplifications were performed in a TGradient Thermocycler (Biometra, Göttingen, Germany); primers were obtained from Syntezza (Jerusalem, Israel). PCR reactions were done in a total volume of 30 μl containing 50 ng of extracted gDNA, 1.25 U of GoTaq DNA polymerase (Promega), 0.2 mM dNTP and 0.2 μM primers (BlCesA3fw TGTGTGGCGTGCACAGCAGACG, BlCesA3rev CGTACATGGCTACACTTTGGCTACACTTCC) in the polymerase manufacturer's buffer. The PCR program was as follows: first 4 min initial denatur-

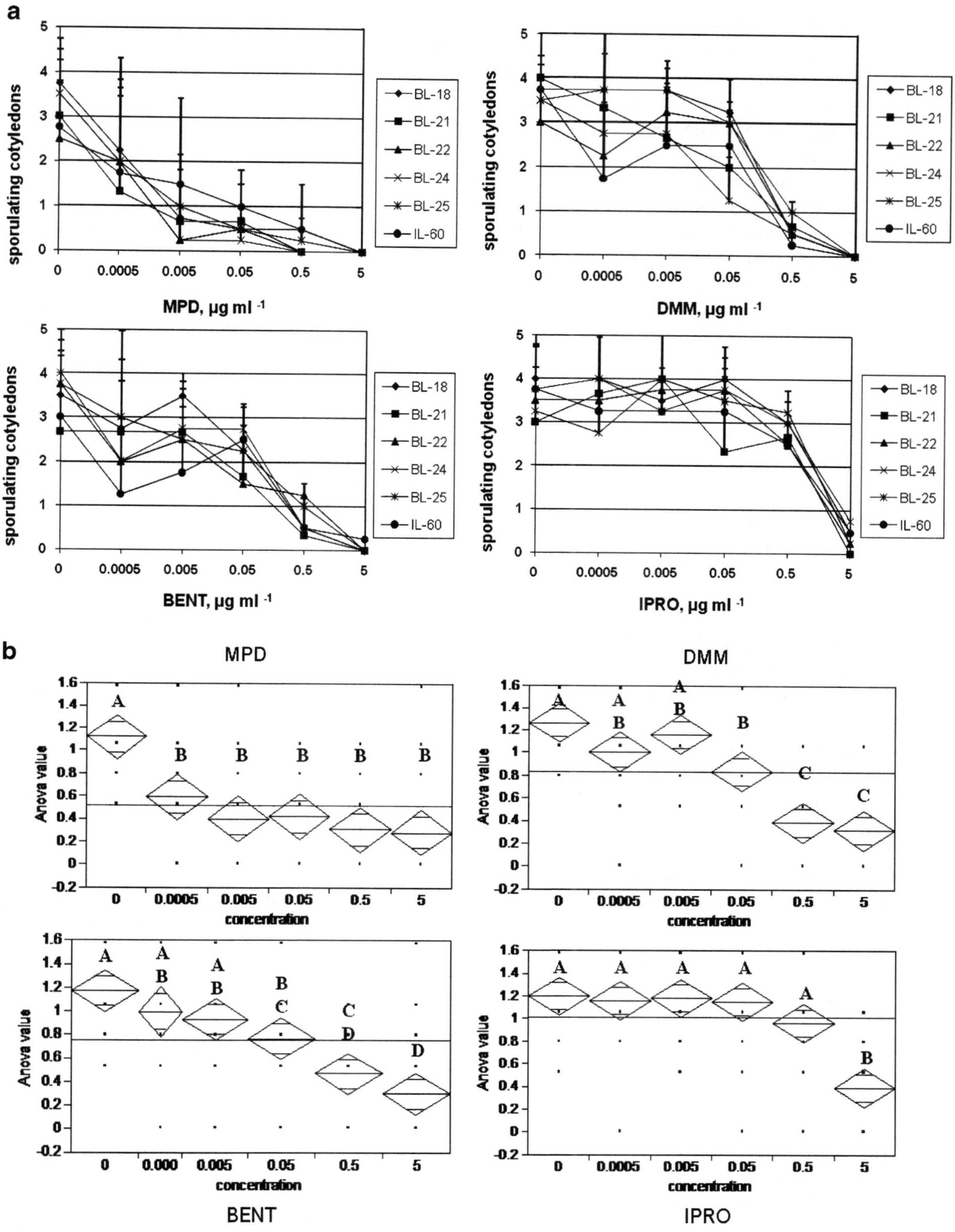
a
sporulating cotyledons
MPD, µg ml -1
DMM, µg ml -1
BENT, µg ml -1
IPRO, µg ml -1
0
0.0005
0.005
0.05
0.5
5
BL-18
BL-21
BL-22
BL-24
BL-25
IL-60
b
MPD
DMM
BENT
IPRO
Anova value
concentration

ation at 94°C, then 35 cycles of 30 s at 94°C, 30 s at 59°C, 45 s at 72°C followed by a final extension at 72°C for 5 min. Prior to sequencing, PCR products were analysed on 1.5% agarose gels and purified using the NucloeSpin Extract II kit (Macherey-Nagel).

Sequencing reactions were done with BigDye Terminator v3.1 Cycle Sequencing Kit (Applied Biosystems) according to manufacturer's instructions using sequencing primer (BlCesA3seq TATTCAGGCTCGTGCGA). Sequence analysis was performed on a 3130 Genetic Analyzer (Applied Biosystems) and further processed using Lasergene software (DNASTAR, Inc.).

Results

Sensitivity of wild-type isolates to CAAs

The wild-type isolates were highly sensitive to CAAs. When spores were incubated in the presence of technical CAAs in vitro, germination was strongly suppressed (Fig. 1a). Anova (Fig. 1b) revealed MIC values of 0.005, 0.5, 0.5 and 5 μgml^{-1} for MPD, DMM, BENT and IPRO, respectively. Similar results were obtained when spores were mixed with CAAs and inoculated onto cotyledon leaves (Fig. 2a and b). Two-way Anova analysis showed that inhibition of infection was significantly affected by the CAA, its dose and their combination but not by the isolate (Table 1). Due to its low efficacy, IPRO was not included in further experiments.

The effect of UV light or EMS on development of resistance in wild type isolates

Table 2 summarizes the results of 43 mutagenesis experiments conducted to induce resistance in wild type isolates of *B. lactucae* against CAAs. EMS applied to WT spores of *B. lactucae* at 0.15, 0.17, 0.2, 0.5 or 0.66% caused a reduction in infection frequency of about 50, 60, 70, 80 and 90%, respectively. Spores exposed to UV 254 irradiation at 100 $mJcm^{-2}$ for 5 or 10 min lost about 50 and 90% of their infection frequency, respectively.

Exposure to EMS or UV light enabled spores of *B. lactucae* to infect lettuce in the presence of 50 $\mu g\ ml^{-1}$ CAA at generation 0. Control experiments repeatedly showed that when mutagenic treatments were not employed, spores failed to infect in the presence of≥0.5–5 μgml^{-1} technical CAAs or≥ 0.78–3.12 μgml^{-1} formulated CAAs. The infection frequency at generation 0 of spores mixed with EMS and CAA was much higher for DMM or BENT as compared to MPD (Fig. 3a) and was inversely related to the concentration of MPD (Fig. 3b), DMM or BENT (see Table 2 for details).

The data presented in Table 2 show that mutagenesis did not result with the formation of stable resistance against CAAs, *vis.* isolates that can repeatedly infect plants in the presence of≥50 $\mu g\ ml^{-1}$ CAA. Nonetheless, in many experiments, resistant mutants (spores germinating at up to 12.5 μgml^{-1} CAA and infecting leaves treated with up to 50 $\mu g\ ml^{-1}$ CAA) did show up at generation 0 (using spores immediately after mutagenesis treatment). However, resistance diminished after one or a few inoculation transfers on CAA-treated lettuce leaves. Two examples are given in Fig.4. In the first (Fig. 4a), spores that were exposed to UV light and thereafter mixed with CAA were infective to detached cotyledons in the presence of up to 50 μgml^{-1} BENT (Table 2 Exp.1–9). Resistance to BENT (and viability) diminished after 5 transfer inoculations in the presence of CAA (Fig. 4a). In the second (Fig. 4b), spores that were mixed with 0.5% EMS

Table 1 Two-way Anova analysis of the single and combined effects of the compound (CAA), dose, isolate and their combinations on infection of lettuce cotyledons with wild type isolates of *Bremia lactucae*. The data in Fig. 2 were used for analysis

Source	Parameters	DF	Sum of squares	F ratio	Prob>F
Compound	3	3	9.234470	26.5434	<.0001*
Concentration	5	5	22.772119	39.2735	<.0007*
Compound X Concentration	15	15	4.628105	2.6606	0.0007*
Isolate	5	5	0.612663	1.0566	0.3840
Isolate x Compound	15	15	0.873199	0.5020	0.9395
Isolate x concentration	25	25	1.892276	0.6527	0.9009
Isolate x Compound X concentration	75	75	3.727254	0.4285	1.0000

Table 2 Description of the experiments conducted to induce resistance against carboxylic acid amides (CAAs) in seven wild type isolates of *Bremia lactucae*

Exp.	Isolate	Spores × 1000	Mutagen	Mutagen dose	CAA, μg ml^{-1}	Cotyledons inoculated	Cotyledons sporulating	dpi	Infection frequency	Resistant generations
1	BL-18	nd	UV	100 mJ/cm2, 5 min	MPD 1, 10	36	1	25	nd	4
2	BL-22	nd	UV	100 mJ/cm2, 5 min	MPD 1, 10	36	5	25	nd	4
3	BL-24	nd	UV	100 mJ/cm2, 5 min	MPD 1, 10	36	3	25	nd	4
4	BL-18	nd	UV	100 mJ/cm2, 5 min	DMM 5, 10	36	1	25	nd	4
5	BL-22	nd	UV	100 mJ/cm2, 5 min	DMM 5, 10	36	5	25	nd	4
6	BL-24	nd	UV	100 mJ/cm2, 5 min	DMM 5, 10	36	3	25	nd	4
7	BL-18	nd	UV	100 mJ/cm2, 5 min	BENT 5, 10	36	18	25	nd	4
8	BL-22	nd	UV	100 mJ/cm2, 5 min	BENT 5, 10	36	23	25	nd	4
9	BL-24	nd	UV	100 mJ/cm2, 5 min	BENT 5, 10	36	10	25	nd	4
10	BL-18	nd	UV	100 mJ/cm2, 5 min	DMM 5	300	34	11	nd	2
11	BL-21	nd	UV	100 mJ/cm2, 5 min	DMM 5	150	4	11	nd	2
12	BL-22	nd	UV	100 mJ/cm2, 5 min	DMM 5	150	3	15	nd	2
13	BL-24	nd	UV	100 mJ/cm2, 5 min	DMM 5	150	8	15	nd	2
14	BL-25	nd	UV	100 mJ/cm2, 5 min	DMM 5	150	9	11	nd	2
15	IL-60	nd	UV	100 mJ/cm2, 5 min	DMM 5	150	2	15	nd	2
16	BL-25	3250	EMS	0.15%	MPD 5	942	28	10	8.6E-6	2
17	BL-25	3250	EMS	0.15%	DMM 5	938	307	10	94.5E-6	1
18	BL-25	3250	EMS	0.15%	BENT 5	946	55	10	16.9E-6	1
19	BL-18	1690	EMS	0.15%	MPD 5	402	18	12	10.7E-6	1
20	BL-21	1250	EMS	0.15%	MPD 5	450	42	12	33.6E-6	2
21	BL-22	440	EMS	0.15%	MPD 5	456	5	12	11.4E-6	1
22	BL-24	352	EMS	0.15%	MPD 5	434	12	12	34.1E-6	2
23	BL-25	530	EMS	0.15%	MPD 5	402	7	12	13.2E-6	1
24	IL-60	424	EMS	0.15%	MPD 5	368	4	12	9.4E-6	1
25	CH	182	EMS	0.15%	MPD 5	408	2	12	11.0E-6	1
26	BL-18	3600	EMS	0.15%	DMM 5	786	236	11	65.6E-6	1
27	BL-21	4180	EMS	0.15%	DMM 5	804	380	11	90.9E-6	1
28	BL-22	3740	EMS	0.15%	DMM 5	876	366	11	97.9E-6	1
29	BL-24	3500	EMS	0.15%	DMM 5	902	309	11	88.3E-6	1
30	BL-25	2565	EMS	0.15%	DMM 5	802	256	11	99.8E-6	1
31	IL-60	2090	EMS	0.15%	DMM 5	814	264	11	126.3E-6	1
32	CH	1600	EMS	0.15%	DMM 5	860	190	11	118.8E-6	1
33	BL-18	664	EMS	0.15%	BENT 5	537	80	11	120.5E-6	1
34	BL-21	826	EMS	0.15%	BENT 5	552	84	11	101.7E-6	1

Table 2 (continued)

Exp.	Isolate	Spores × 1000	Mutagen	Mutagen dose	CAA, $\mu g ml^{-1}$	Cotyledons inoculated	Cotyledons sporulating	dpi	Infection frequency	Resistant generations
35	BL-22	1080	EMS	0.15%	BENT 5	852	97	11	89.8E-6	1
36	BL-24	1020	EMS	0.15%	BENT 5	819	83	11	81.4E-6	1
37	BL-25	675	EMS	0.15%	BENT 5	636	46	11	68.1E-6	1
38	IL-60	270	EMS	0.15%	BENT 5	531	30	11	111.1E-6	1
39	CH	560	EMS	0.15%	BENT 5	600	90	11	160.7E-6	1
40	CH	2000	EMS	0.50%	MPD 5	600	5	14	2.5E-6	7
41	CH	2000	EMS	0.50%	DMM 5	600	40	14	20.0E-6	9
42	CH	2000	EMS	0.50%	BENT 5	600	30	14	15.0E-6	10
43	Il60	8000	EMS	0.20%	MPD 12.5	800	4	15	500.0E-9	nd
43	Il60	8000	EMS	0.20%	MPD 25	800	2	15	250.0E-9	nd
43	Il60	8000	EMS	0.20%	MPD 50	800	1	15	125.0E-9	nd
43	Il60	8000	EMS	0.20%	MPD 100	800	0	15	000.0E+0	nd

Resistant generations=the number of successful transfer inoculations of detached cotyledons in the presence of 0.5–5 $\mu g ml^{-1}$ technical CAA or 3.12–25 $\mu g ml^{-1}$ formulated CAA in intact seedlings

and 5 $\mu g ml^{-1}$ CAA were infective to plants treated with up 25–50 $\mu g ml^{-1}$ CAAs but this resistance was gone after 7–10 transfer inoculations on CAA-treated plants (Exp.40–42, Table 2.).

Response of sexual progeny isolates to CAAs

The Israeli isolates IL60 and BS (Bet-Shaan) belonged to the same mating type as they produced no oospores when mixed-inoculated onto lettuce leaves. Similarly, all European isolates shared the same mating type as they all did not produce oospores when mixed-inoculated onto lettuce leaves in all possible combinations. However, the Israeli isolates did mate with any European isolate to produce oospores, indicating that IL and EU isolates belong to opposite mating types. We took advantage of this fact to cross mutants belonging to opposite mating types in order to possibly produce homozygous resistance to CAAs. Altogether, 42 crosses were made: 10 F1's between wild types IL60 or BS and resistant mutants (RM) derived from EU isolates (the mating type of isolates have not changed after mutagenic treatments); 3 BC's between F1's and IL60; 12 BC's between F1's and RM's; and 17 F2's between BC's. Bioassays showed that most progeny isolates were sensitive to CAAs (data not shown) but a few were resistant. For example, the BC x F1 and the BC x BC progenies presented in Fig. 5 were resistant to 6.25–

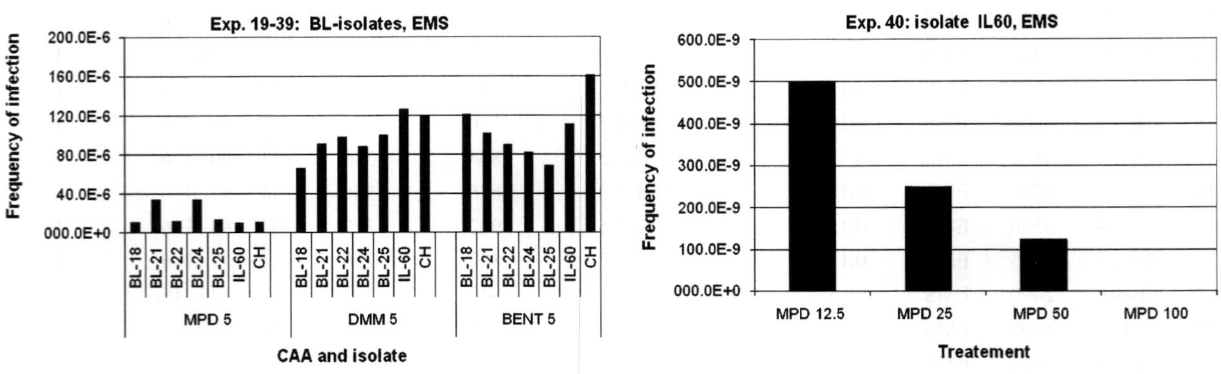

Fig. 3 Infection frequency of EMS-treated spores of *Bremia lactucae* in the presence of CAAs. **a**. Exposure of 7 WT isolates to 5 $\mu g ml^{-1}$ MPD, DMM or BENT. **b**. Exposure of isolate IL60 to increasing concentrations of MPD. More details in Table 2

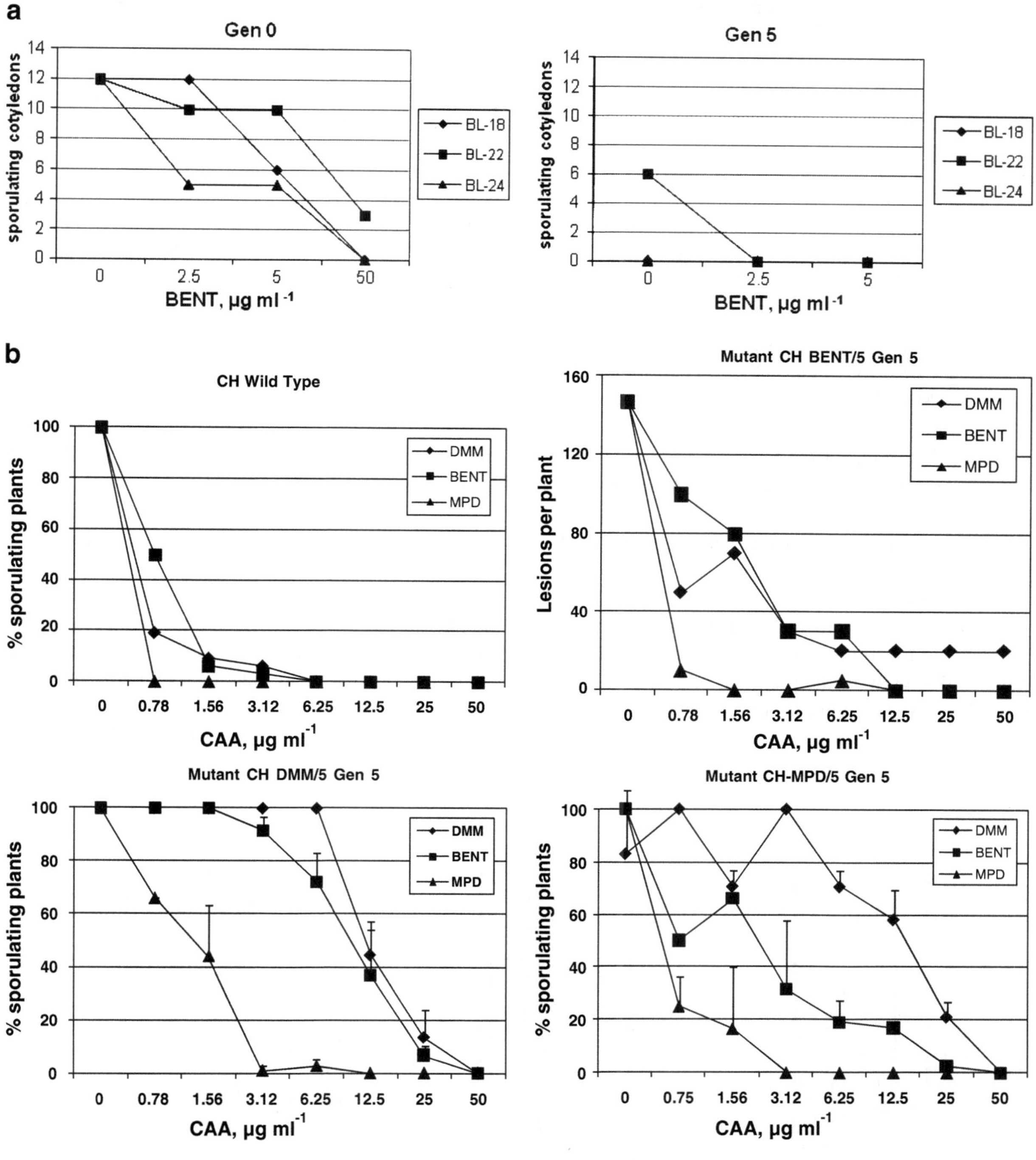

Fig. 4 **a**. Persistence of resistance against benthiavalicarb (BENT) induced by UV light (Exp.7–9, Table 2) during 5 generations in three wild type isolates of *Bremia lactucae*. Only Gen 0 and Gen 5 are shown. The spores produced at each infection cycle were all collected, mixed with benthiavalicarb at the indicated concentrations and inoculated onto detached cotyledons to produce the next generation. Note that two isolates lost infectivity at Gen 5. **b**. Efficacy of formulated CAAs against CH wild type and resistant isolates derived from it by EMS at Generation 5 (Exp.40–42, Table 2). The resistance of these mutants was diminished after 7–10 transfer inoculations on CAA-treated plants. See Table 2 for more details

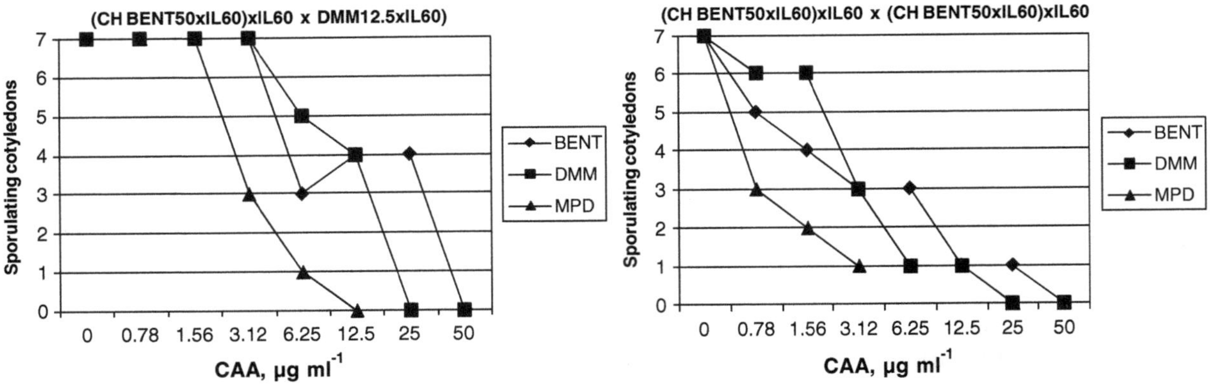

Fig. 5 Sensitivity of two sexual progeny isolates of *Bremia lactucae* to CAAs. Spores were mixed with formulated CAAs and drop-inoculated onto detached cotyledons. The number of sporulating cotyledons was counted at 18dpi. **a**. Cross 13. **b**. Cross 32. More details in Table 3

25 μgml^{-1} formulated CAA, but subsequent transfer inoculations of these progenies on CAA-treated plants revealed sensitive isolates only.

Single spore isolates were derived from an F1 progeny of a cross between the mutant CH BENT50 and the wild type BS, and each was tested for resistance to technical CAAs in detached cotyledons. Results are presented in Fig. 6. The five isolates that grew in the presence of 5 or 50 μgml^{-1} CAA lost resistance in the subsequent transfer cycle. Unfortunately, all single spore isolates belonged to the EU mating type (or were sterile) so that F2 progenies could not be generated.

Mutagenesis of RM and sexual progeny isolates

EMS or UV light was applied to resistant mutants or sexual progeny isolates with an attempt to bring resistance to the homozygous state. Data presented in Table 3 show that in 17 out 23 experiments, mutagenesis induced infection in cotyledons treated with 50 μg ml^{-1} MPD or BENT and 100 μgml^{-1} DMM. However,

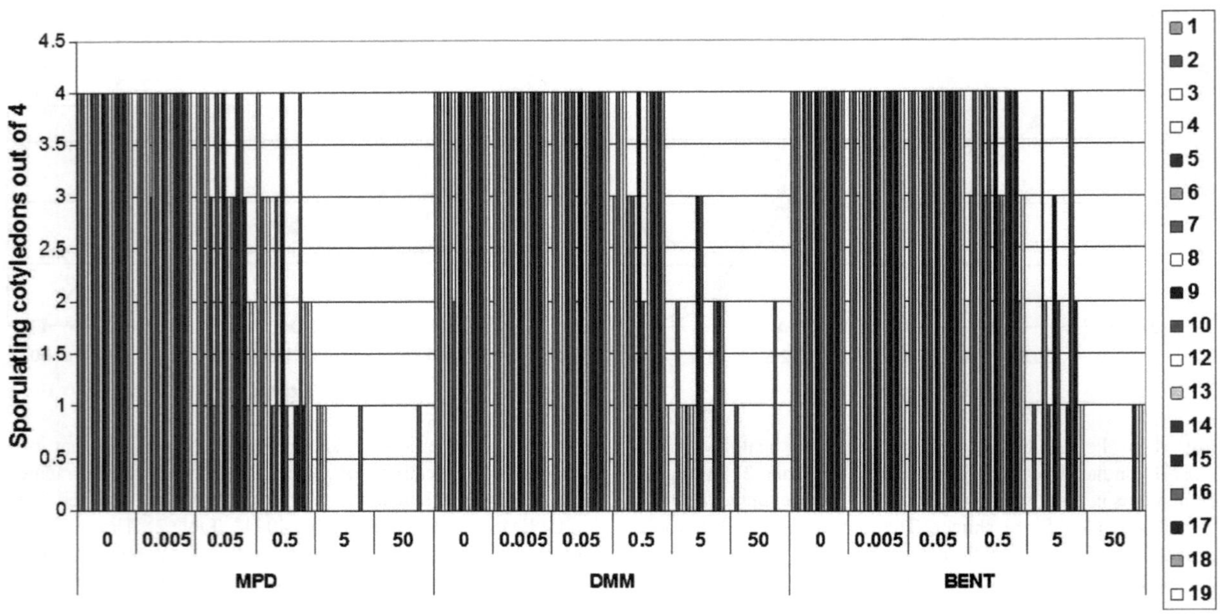

Fig. 6 Sensitivity to CAAs of 19 single-spore isolates derived from the F1 progeny of a cross between the resistant mutant CH BENT50 and the wild type isolate BS. Note that a few isolates could infect leaves in the presence of 50 μgml^{-1} CAA

Table 3 Description of the experiments conducted to induce resistance against carboxylic acid amides (CAAs) in mutants and hybrid isolates of *Bremia lactucae*

Exp.	Isolate(s)	Spores × 1000	Mutagen	Mutagen dose	CAA, μg ml^{-1}	Cotyledons inoculated	Cotyledons sporulating	dpi	Infection frequency	Resistant generations
1	CH BENT 50	3000	EMS	0.50%	DMM 40	1000	54	8	18.0E-6	14
2	CH BENT 5	5000	EMS	0.66%	BENT 50	3000	37	11	7.4E-6	5
3	mix hybrids	6000	EMS	0.20%	MPD 50	600	9	11	1.5E-6	3
4	mix hybrids	1200	EMS	0.20%	DMM 50	90	4	11	3.3E-6	3
5	mix hybrids	6000	EMS	0.20%	BENT 50	800	14	11	2.3E-6	3
6	mix hybrids	5000	EMS	0.20%	DMM 40	600	20	8	4.0E-6	2
7	mix hybrids	5000	EMS	0.20%	BENT 40	600	0	8	000.0E+0	0
8	mix hybrids	5000	EMS	0.20%	MPD 40	600	0	8	000.0E+0	0
9	hybrid 5	2000	EMS	0.17%	DMM 100	600	2	8	1.0E-6	2
10	hybrid 13	2000	EMS	0.17%	DMM 100	600	10	8	5.0E-6	2
11	hybrid 14	2000	EMS	0.17%	DMM 100	600	2	8	1.0E-6	2
12	hybrid 15	2000	EMS	0.17%	DMM 100	600	0	8	000.0E+0	0
13	hybrid 17	2000	EMS	0.17%	DMM 100	600	25	8	12.5E-6	2
14	hybrid 18	2000	EMS	0.17%	DMM 100	600	15	8	7.5E-6	2
15	hybrid 19	2000	EMS	0.17%	DMM 100	600	0	8	000.0E+0	
16	hybrid 32	2000	EMS	0.17%	DMM 100	600	25	8	12.5E-6	2
17	hybrid 5×7	2000	EMS	0.17%	DMM 100	600	5	8	2.5E-6	2
18	hybrid 8×8	2000	EMS	0.17%	DMM 100	600	0	8	000.0E+0	0
19	hybrid 18×7	2000	EMS	0.17%	DMM 100	600	3	8	1.5E-6	2
20	hybrid 8×8	1000	UV	100 mJ/cm2, 10 min	DMM 50	600	0	10	0	0
21	mix hybrids	1500	UV	100 mJ/cm2, 10 min	DMM 50	240	4	16	2.7E-6	1
22	mix hybrids	1500	UV	100 mJ/cm2, 10 min	BENT 50	240	7	16	4.7E-6	1
23	CH B50 Gen 11	2000	UV	100 mJ/cm2, 10 min	DMM 6.25	360	1	10	500.0E-9	3

Resistant generations=the number of successful transfer inoculations of detached cotyledons in the presence of 0.5–5 μg ml^{-1} technical CAA or 3.12–25 μg ml^{-1} formulated CAA in intact seedlings

here again, resistance diminished after 1–14 subsequent transfer inoculation cycles.

Sequencing of *CesA3* gene fragment in WT, RM and mixed hybrid isolates

DNA was extracted from spores of 26 isolates of *B. lactucae*: six WT, four RM, ten F1's and six BC's isolates. DNA of the 10 F1's was mixed before used. *CesA3* sequence analysis revealed no difference in the 582 bp fragment (Figure 7a.). No mutation was found at position 1105 that was shown to affect sensitivity to CAAs in *Plasmopara viticola* (Blum et al. 2010a) All isolates displayed Gly1105 encoded by codon GGC (as exemplified in Fig. 7b).

Discussion

This paper reports on three major findings. The first, EMS or UV treatments enabled spores of *B. lactucae* to infect lettuce leaves in the presence of up to 100 μg ml^{-1} DMM or 50 μg ml^{-1} BENT or MPD. For MPD, this is × 1000 the dose required to completely inhibit infection by WT isolates. The second, resistance to CAAs in *B. lactucae* was not linked to mutation in the

a

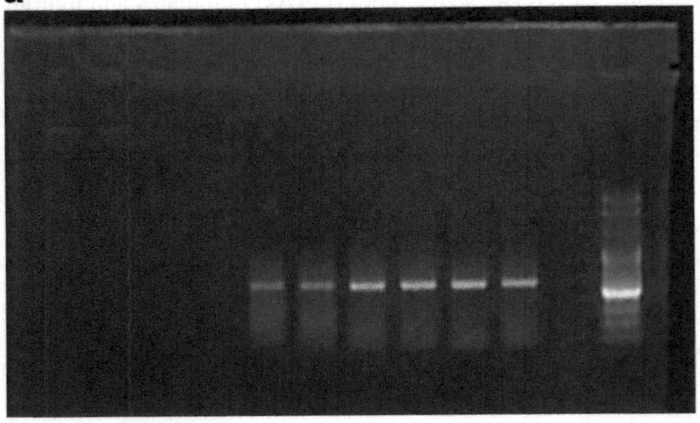

b

CH-WT-*CesA3* sequence

CGAAGGTTAGTCGTGGGCGAAACTGGTGTGGGCAGAAGAATCATG
GACGGAAATCCCTAATGTGCTGTTCTTTTTCCGCTTATGTTTAGCCA
AGTCGTGGCGCTCATTCGTTTCTTTGAGTACGAGAATGCAACGAAT
CCGTGGAATTACGTGTCAGCAATGTTTTTTGGCTTTTTCGTCATGAG
CCAATTCTATCCCATGGTCAAGATGAGTATCACGGAATATTGTGGAT
GGGACCACACGGCCGCCACGTTCACGGCCAATGTCTTC**GGC**TCC
TTGTTGGTCGTATACGTTGTCGTGTTCGTACAGCTATGGCAAGTGT
ATTACGAGGGCAACTTGTTAGTTGCTCAAGGTGCTGATGGCGGTGG
TGGCGCTGAGACTGCTACTGCGGTGTAACCGAGAAGCGTGTTCGG
ATCAATTGCTATGACGATTAGGCGGAGAGGTAAGGGTGCAACAAAA
ACGGAAGTGTAGCCAAAGTGTAGCATGGTACGACGTGTGGCGTCC
TCCGCAGACGACGGGGGAACCAAAATGTCGTCGTGTACGAAGTGT
GCCCCCCCACGTAACGTGTTGCGTCTACTGCACACCACGTTGGAG
CGGAGGTGG

M50-*CesA3* sequence

GCCGGCCACCAACACATCATGGACGGAAATCCCTAATGTGCTGTTC
TTTTTACGCTTATGTTTAGCCAAGTCGTGGCGCTCATTCGATTCTTT
GAGTACGAGAATGCAACGAATCCGTGGAATTACGTGTCAGCAATGT
TTTTTGGCTTTTTCGTCATGAGCCAATTCTATCCCATGGTCAAGATG
AGTATCACGGAATATTGTGGATGGGACCACACGGCCGCCACGTTC
ACGGCCAATGTCTTC**GGC**TCCTTGTTGGTCGTATACGTTGTCGTG
TTCGTACAGCTATGGCAAGTGTATTACGAGGGCAACTTGTTAGTTG
CTCAAGGTGCTGATGGCGGTGGTGGCGCTGAGACTGCTACTGCGG
TGTAACCGAGAAGCGTGTTCGGATCAATTGCTATGACGATTAGGCG
GAGAGGTAAGGGTGCAACAAAAACGGAAGTGTAGCCAAAGTGTAG
CCATGTACGAC

Mixed-hybrids-*CesA3* sequence

- GGAATGGTCAGTGGGCGAACACTGGTGCTGGGCAGAAGAATCATG
GACGGAAATCCCTAATGTGCTGTTCTTTTTCCGCTTATGTTTAGCCA
AGTCGTGGCGCTCATTCGTTTCTTTGAGTACGAGAATGCAACGAAT
CCGTGGAATTACGTGTCAGCAATGTTTTTTGGCTTTTTCGTCATGAG
CCAATTCTATCCCATGGTCAAGATGAGTATCACGGAATATTGTGGAT
GGGACCACACGGCCGCCACGTTCACGGCCAATGTCTTC**GGC**TCC
TTGTTGGTCGTATACGTTGTCGTGTTCGTACAGCTATGGCAAGTGT
ATTACGAGGGCAACTTGTTAGTTGCTCAAGGTGCTGATGGCGGTGG
TGGCGCTGAGACTGCTACTGCGGTGTAACCGAGAAGCGTGTTCGG
ATCAATTGCTATGACGATTAGGCGGAGAGGTAAGGGTGCAACAAAA
ACGGAAGTGTAGCCAAAGTGTAGCAT

Fig. 7 **a**. PCR product of a 582 bp DNA fragment of *CesA3* of *Bremia lactucae* derived from spores of the wild type isolate CH (2 lanes), the resistant mutant M50 at generation 0 (2 lanes), and mixed hybrid isolates (2 lanes). Each DNA sample was loaded twice in order. **b**. Sequence analysis of the 582 bp DNA fragment of *CesA3* of *Bremia lactucae*. Triplett at position 1105 (underlined, bold) is identical in CH WT, the resistant mutant M50, and the mixed hybrid isolates

cellulose synthase 3 (*CesA3*) gene. In *P. viticola* and *P. cubensis*, resistance to CAAs was associated with a mutation at position 1105 of this gene (Blum et al. 2010a, b). The third, resistance to CAAs in *B. lactucae* was unstable, diminishing after one or several transfer inoculations on CAA-treated plants. Attempts to stabilize resistance by producing sexual progenies have failed, as did repeated mutagenesis treatments applied to mutants or hybrid isolates. These findings indicated that the likelihood of *B. lactucae* to develop stable resistance against CAA fungicides is rather small.

We assume that the failure of *B. lactucae* to continue growing on CAA-treated plants at subsequent generations after mutagenesis, as well as the failure to bring resistance to homozygosity, resulted from the lack of mutation in the *CesA3* gene. Cellulose synthesis is crucial for germ-tube development in oomycetes and hence for infection (Blum et al. 2010c). CAA fungicides were shown to inhibit cellulose biosynthesis, interfering with cellulose synthase, and avoiding germ-tube development and infection (Blum et al. 2010c). A mutation at SNP 1105 of *CesA3* alters the protein making it insensitive to CAAs, thus allowing spore germination and infection in the presence of CAAs (Blum et al. 2010a, b).

While we may understand the reason for the failure of *B. lactucae* to continue growing on CAA-treated plants at subsequent generations after mutagenesis (lack of mutation), we do not understand the mechanism which governs the appearance of resistant mutants in the first generation after mutagenic treatments (Generation 0). In fact, this resistance could not be inherited by sexual crosses as it was observed for resistance in *P. viticola* (Gisi et al. 2007) and may therefore be related to mechanisms other than mutations at the target site *(CesA3 gene)* of CAA fungicides. One explanation might be that the stress imposed by the mutagenic agents temporarily interferes with the binding of CAA to the CesA3 protein. This is probably a rare phenomenon, which may have happened in only a few spores. Another, remote, possibility is that the stress imposed by EMS or UV have activated the expression of CAA-insensitive *Ces* genes, probably via promoter methylation. Because these cellulose synthase enzymes are not as efficient as Ces3A during spore germination and infection, a re-set mechanism reactivates *Ces3A* during progressive generations, restoring sensitivity to CAAs.

The data presented here are similar to those we published recently for *P.infestans* (Rubin et al. 2008). There we showed that UV or chemical mutagens applied to sporangia induced the appearance of mutants resistant to CAAs, but all showed erratic, unstable resistance in planta, diminishing after 1 to 8 asexual infection cycles, and failed to grow on CAA-amended medium. However, mutants with stable resistance to mefenoxam or a shift in mating type were obtained, suggesting that *P. infestans* could be artificially mutated for resistance against mefenoxam and a shift in mating type, but not against CAAs (Rubin et al. 2008). Indeed, based on those and other findings, mandipropamid was released to the market as a solo product *Revus* for controlling late blight in potatoes.

Our failure to produce stable CAA resistant mutants of *B. lactucae* in the laboratory, together with our failure to enforce selection of resistant isolates by heavy and continuous sprays of CAAs onto downy mildew-infected lettuce plants growing in closed plastic tunnels (Y.Cohen, unpublished data) suggest that CAA-resistant mutants of *B. lactucae* are unlikely to appear soon in nature. Although CAA fungicides (especially dimethomorph) have been applied since many years to control downy mildew in lettuce under commercial conditions, no resistant isolates have been detected (Syngenta internal information, and CAA FRAC working group). In contrast, mefenoxam resistance is known to be present in *B. lactucae* populations in several countries since many years.

More research is required to understand the mechanisms responsible for the temporary appearance of resistance to CAAs in *B.lactucae* after mutagenic treatments.

Acknowledgement The assistance of B.H. Ben-Daniel and L.Tsror of Volcani Center for Agricultural Research, Gilat, Isreal, and M. Galperin of Bar-Ilan University is gratefully acknowledged.

References

Blum, M., Waldner, M., & Gisi, U. (2010a). A single point mutation in the novel *PvCesA3* gene confers resistance to the carboxylic acid amide fungicide mandipropamid in *Plasmopara viticola. Fungal Genetics and Biology, 47*, 499–510.

Blum, M., Sierotzki, H., & Gisi, U. (2010b). Comparison of cellulose synthase 3 (*CesA3*) gene structure in different oomycetes. In H. W. Dehne, H. B. Deising, U. Gisi, K. H. Kuck, P. E. Russell & H. Lyr (Eds.), *Modern fungicides and antifungal compounds VI.* DPG, Braunschweig, Germany, (submitted).

Blum, M., Boehler, M., Randall, E., Young, V., Csukai, M., Kraus, S., et al. (2010c). Mandipropamid targets the cellulose synthase-like PiCesA3 to inhibit cell wall biosynthesis in the oomycete plant pathogen *Phytophthora infestans. Molecular Plant Pathology, 11*, 227–243.

Cohen, Y., & Gisi, U. (2007). Differential activity of carboxylic acid amide fungicides against various developmental stages of *Phytophthora infestans. Phytopathology, 97*, 1274–1283.

Cohen, Y., Baider, A., & Cohen, B. H. (1995). Dimethomorph activity against oomycete fungal plant pathogens. *Phytopathology, 85*, 1500–1506.

Cohen, Y., Rubin, A. E., & Gotlieb, D. (2008). Activity of carboxylic acid amide (CAA) fungicides against *Bremia lactucae. European Journal of Plant Pathology, 122*, 169–183.

Cohen, Y., Rubin, A., Gisi, U., Sierotzki, H., & Hermann, D. (2009). Activity of CAA fungicides against *Phytophthora infestans* and *Bremia lactucae. Phytopathology, 99*, S169. Abstract.

FRAC, CAA working group, www.frac.info.

Gisi, U., Waldner, M., Kraus, N., Dubuis, P. H., & Sierotzki, H. (2007). Inheritance of resistance to carboxylic acid amide (CAA) fungicides in *Plasmopara viticola. Plant Pathology, 56*, 199–208.

Reuveni, M. (2003). Activity of the new fungicide benthiavalicarb against *Plasmopara viticola* and its efficacy in controlling downy mildew in grapevines. *European Journal of Plant Pathology, 109*, 243–251.

Rubin, E., & Cohen, Y. (2006). An improved method for infecting tomato leaves or seedlings with oospores of *Phytophthora infestans* used to investigate F1 progeny. *Plant Disease, 90*, 741–749.

Rubin, A. E., Gotlieb, D., Gisi, U., & Cohen, Y. (2008). Mutagenesis of *Phytophthora infestans* for resistance against carboxylic acid amide and phenylamide fungicides. *Plant Disease, 92*, 675–683.

Werdiger, A. C., & Cohen, Y. (2008). Mutagenesis of *Bremia lactucae* for resistance against CAA fungicides. *Phytoparasitica, 36*, 10. Abstract.

Eur J Plant Pathol (2011) 129:221–227
DOI 10.1007/s10658-010-9679-9

Influence of dosage, storage time and temperature on efficacy of metalaxyl-treated seed for the control of pearl millet downy mildew

Ram P. Thakur · Veeranki P. Rao · Rajan Sharma

Accepted: 23 August 2010 / Published online: 9 September 2010

Abstract Metalaxyl (Apron 35WS) as a seed treatment has been used extensively to control downy mildew (caused by *Sclerospora graminicola*) in pearl millet in India. However, the extent of disease control has varied across cultivars, years and locations. We investigated the effects of fungicide dosage, storage time and storage temperature of metalaxyl-treated seed on disease incidence in four pearl millet lines having varying levels of resistance. A linear relationship was found between fungicide dosage (0.5, 1.5 and 2 g a.i. kg^{-1} seed) and reduction in disease incidence up to 40 days after emergence in all the lines. The normal fungicide dose (2 g a.i. kg^{-1} seed) protected the crop for up to 20, 40 and 50 days after emergence in highly susceptible (7042S), moderately susceptible (4042R), and moderately resistant (ICMP 451) lines, respectively. However, the quarter and half the normal dosage of fungicide provided protection only up to 20 days after emergence in 7042R and 40 days after emergence in ICMP 451. Storage duration of metalaxyl-treated seed (2 g a.i. kg^{-1}) up to 9 months at 25±2°C did not affect fungicide efficacy. Storage temperatures (5, 25 and 40°C) and duration (30, 60 and 90 days) of metalaxyl-treated seed (2 g a.i. kg^{-1}) showed differential effects in two pearl millet lines 7042S and 843B with downy mildew incidence being significantly lower in 7042S than in 843B. Metalaxyl-treated seed of 7042S and 843B stored at 40°C for different durations showed phytotoxic effects and it was more pronounced in 843B stored for 60 and 90 days where seed germination was inhibited in pot soil.

Keywords Metalaxyl · Pearl millet · *Sclerospora graminicola* · Storage time · Temperature

R. P. Thakur (✉) · V. P. Rao · R. Sharma
International Crops Research Institute
for the Semi-Arid Tropics,
Patancheru 502324 Andhra Pradesh, India
e-mail: r.thakur@cgiar.org

Introduction

Downy mildew disease of pearl millet [*Pennisetum glaucum* (L.) R. Br.] caused by the obligate parasite *Sclerospora graminicola* (Sacc.) Schroet is widespread and economically important in India and several countries in Africa. The disease can be best managed through host-plant resistance. Several breeding lines, hybrids and open-pollinated varieties (OPVs) having downy mildew resistance have been developed (Talukdar et al. 1999; Hash et al. 1999; Thakur et al. 2001). In India, the disease is particularly serious on single-cross F_1 hybrids than on OPVs. This is due to a narrow genetic base and uniformity of the hybrids than those of OPVs that are highly heterogeneous. During the past 25 years there has been steady growth in the number of hybrids and area under hybrid cultivation in India. Currently, about 50% of the 9 million ha under pearl millet cultivation is grown with over 70 different single-cross hybrids in India (Rai et al. 2006). Downy mildew incidence has been quite variable on different hybrids and more than 90% incidence has been recorded on some hybrids in farmers' fields (Thakur et al. 2003; Rao et al. 2007). The estimated annual grain yield loss due to downy mildew is approximately 20 to 40% (Singh 1995; Hash et al. 1999; Hess et al. 2002) but this could be much higher under favorable conditions of disease development (Singh 1995; Thakur 1998, 2008). In addition to host-plant resistance, seed treatment with metalaxyl-based fungicide (Apron 35WS) has been found effective in managing downy mildews in pearl millet (Williams and Singh 1981; Singh et al. 1984; Singh and Shetty 1990) and other crops (Gisi and Sierotzki 2008; Thind 2008). Metalaxyl (Apron®), a phenylamide class of fungicide that inhibits ribosomal RNA synthesis, has been formulated as 35% WS for application as a dust or slurry to seed to control downy mildews in many crops, including pearl millet, sorghum and maize (CIBA-GEIGY Technical Product Information-3, 1989). Metalaxyl penetrates the seed coat and is systemically translocated to both shoots and roots during germination (Thind 2008). Seed treatment by metalaxyl protects the young plants from systemic primary infection and secondary infection during the early growth period for up to four weeks (Williams and Singh 1981; Singh and Shetty 1990). This normally helps prevent epidemics of downy mildews in graminaceous crops.

In India, most seed companies treat pearl millet hybrid seed with metalaxyl before marketing to farmers (Thakur et al. 2003; Rao et al. 2007). Private seed companies spend about US $1.8 million per year to treat pearl millet seed with metalaxyl (pers. Communication). This level of phenylamide fungicide use is insignificant compared to that used for controlling diseases caused by oomycetes on grapes, cucurbits, lettuce, onion and tobacco (Gisi and Sierotzki 2008). However, this treatment is effective only with moderately resistant hybrids in certain environments. The fungicide is not effective with susceptible hybrids, which are protected only up to 40 days after emergence and the disease appears on nodal tillers and as "green ear" at the later stages of crop growth. The cost of treated seed is about 10% higher than untreated seed and farmers have to pay the additional price without any assurance of protection from the disease. Also, currently no such regulation exists in India under which the concerned seed companies can compensate farmers for crop loss. This chemical approach to downy mildew management may also lead to the emergence of fungicide resistant biotypes.

In India, with increasing land area under hybrid cultivation since the 1970s downy mildew has become more severe and more widespread (Thakur et al. 2006). The most cost-effective management of the disease can be obtained by breeding downy mildew-resistant pearl millet hybrids. There has been considerable success in breeding for downy mildew resistance using conventional pedigree breeding, and a large number of disease resistant hybrids have been developed and deployed (Khairwal et al. 2004). This has contributed to management of widespread downy mildew epidemics since the 1990s (Thakur et al. 2006). Marker-assisted backcross breeding has further enhanced the ability and efficiency of downy mildew resistance breeding (Hospital et al. 1992; Hash et al. 1999; Hash and Witcombe 2002). However, in view of the increasing severity of the disease and evolution of new more virulent pathotypes (Thakur et al. 2004) there is a need to develop a long-term strategy for downy mildew management. In order to optimize the use of metalaxyl we investigated the effects of different dosages, storage duration and storage temperature of treated seed on downy mildew incidence in pearl millet lines.

Materials and methods

Metalaxyl concentration

Seed of a highly susceptible pearl millet line 7042S and two moderately resistant lines 7042R and ICMP451 were treated with four concentrations of metalaxyl (Apron 35 WS) (0.5, 1.0, 1.5 and 2.0 g a.i. kg^{-1} seed). Treated seed was sown in the downy mildew plot infested with *S. graminicola* oospores along with the untreated control. In addition to oospore inoculum in the infested plot, infector rows of the susceptible line 7042S were grown after every four rows as inoculum donor though asexual spores (Singh et al. 1993). The experiment was conducted in a randomized complete block design (RCBD) with 2 rows of 4 m length and 3 replications. Disease incidence was recorded at 10, 20, 30, 40, 50 and 60 days after emergence.

Storage time

Seed of three lines (7042S, 7042R and ICMP 451) were treated with metalaxyl (2 g a.i. kg^{-1} seed) and stored for 0 (fresh seed treatment) to up to 9 months at laboratory temperature (25±2°C). The experiment was conducted in a RCBD with 9 storage periods as treatments and 6 replications, 4 rows of 2 m long/replication. Downy mildew incidence was recorded at 10 day-intervals after seedling emergence up to 70 days.

Storage temperature

Metalaxyl-treated seed (2 g a.i. kg^{-1}) lots of two susceptible lines 7042S and 843B were stored in incubators at 5, 25 and 40°C for 3 months in darkness. An experiment with six treatments (2 lines × 3 storage temperatures) with three replications using a completely randomized design (CRD) was conducted in the greenhouse. Seed were sown in pots filled with autoclaved potting mixture soil, sand and farm yard manure (2:1:1 by volume) and 30–35 seedlings were

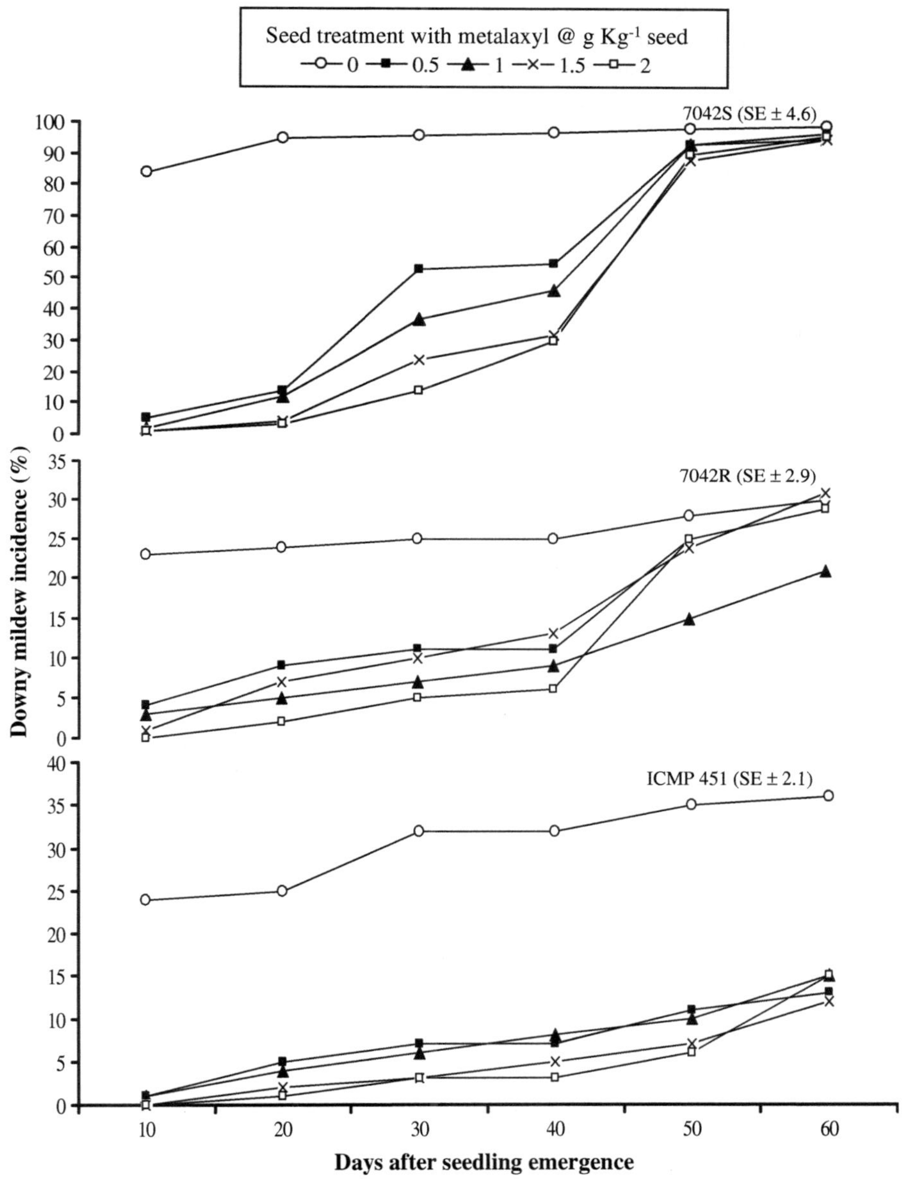

Fig. 1 Effect of different dosages of metalaxyl (Apron 35WS) treatment on downy mildew incidence in three pearl millet lines (7042S, 7042R and ICMP 451) in a field experiment

maintained in each pot. Seedlings at the 2-leaf stage were spray inoculated with a sporangial suspension (1×10^5 sporangia ml^{-1}) of Patancheru isolate of *S. graminicola* (Sg 409). Downy mildew incidence was recorded two weeks after inoculation.

Results

Effect of metalaxyl concentration

Metalaxyl seed treatment provided protection from downy mildew (<10% incidence) up to 10 days after emergence to the highly susceptible line 7042S at all four concentrations compared to 84% incidence in the control (no fungicide treatment). At concentrations of 1.5 and 2.0 g a.i. metalaxyl kg^{-1} seed protection duration increased up to 25 days, but by 60 days disease incidence increased to above 80% in all concentrations indicating that fungicide was not effective 25 days after emergence in 7042S (Fig. 1). In moderately resistant lines 7042R and ICMP 451 protection from downy mildew occurred up to 40 and 50 days after seedling emergence, respectively, in treated seed at all concentrations. Downy mildew incidence increased rapidly after 40 days in 7042R to reach 25% while it increased slowly in ICMP 451 and reached 12% by 60 days. No significant differences were observed for downy mildew incidence in ICMP 451 with metalaxyl concentrations. Therefore, reducing the fungicide concentration to half (1.0 g a.i. kg^{-1} seed) the recommended dose (2 g a.i. kg^{-1} seed) may effectively control the disease in moderately resistant cultivars.

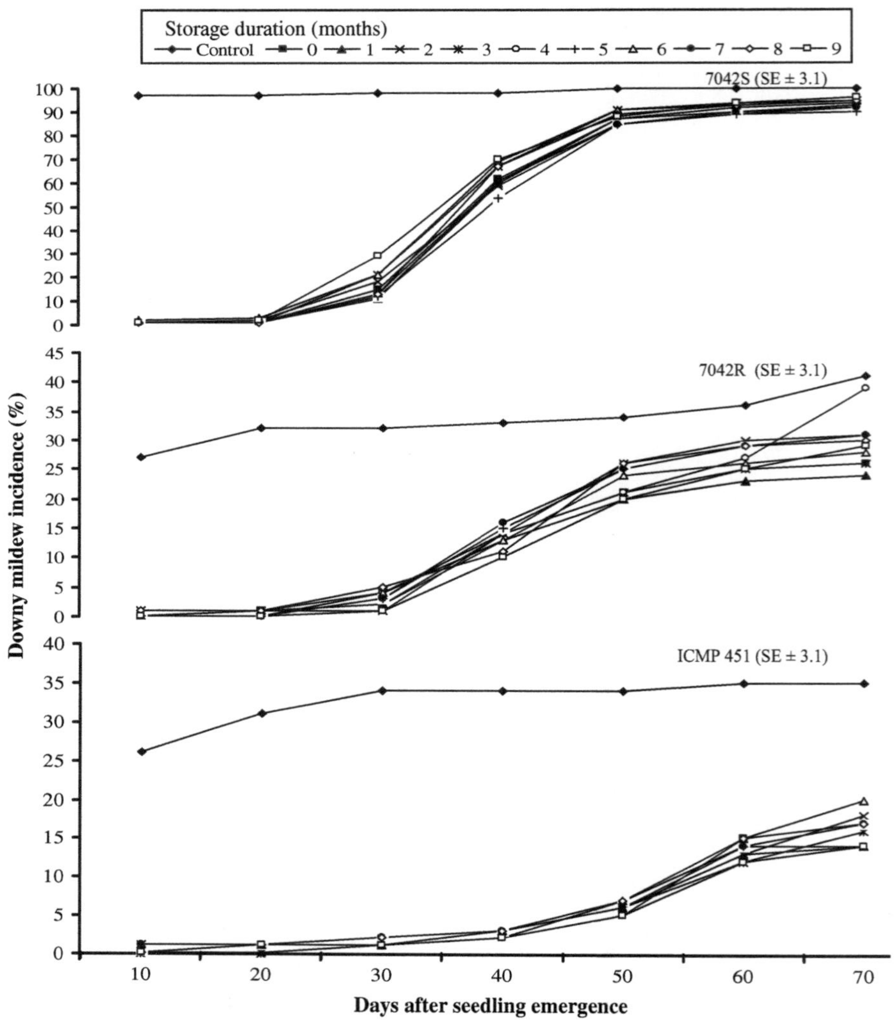

Fig. 2 Effect of storage duration of metalaxyl (Apron 35 WS)-treated seed on downy mildew incidence of three pearl millet lines (7042S, 7042R and ICMP 451) in a field experiment

Effect of storage duration

Metalaxyl treatment protected seedlings from downy mildew (<10% incidence) for up to 20 days in 7042S, 30 days in 7042R and 50 days in ICMP 451, irrespective of storage durations of treated seed (Fig. 2). The results indicate that storage duration of treated seed up to 9 months does not affect the efficacy of metalaxyl seed treatment. However, protection was longer in the moderately resistant line than in the highly susceptible line as reported above.

Effect of storage temperature

Storage temperatures of treated seed did not influence effectiveness of fungicide in reducing downy mildew incidence. Downy mildew incidence ranged from 0 to 38% in fungicide-treated seed compared to 60 to 100% in untreated controls in two pearl millet lines 7042S and 843B up to 90 days of crop growth (Fig. 3). In 843B, however, metalaxyl-treated seed stored at 40°C for 60 and 90 days did not emerge in pot soil indicating a phytotoxic effect on seed germination (Fig. 3). There was genotype-specific response of storage temperatures on the efficacy of metalaxyl seed treatment on seed germination in the two pearl millet lines. To test this, we conducted a blotter test (details not reported here) wherein metalaxyl-treated seed of 843B stored at 40°C for 30, 60 and 90 days recorded 16, 8 and 6% germination, respectively, compared to 77, 77 and 90% germination in untreated seed stored at the same temperature, while no such adverse effects were recorded for metalaxyl-treated seed of 7042S stored at 40°C for different time periods.

Discussion

With the increasing area under pearl millet hybrid cultivation in India, the prevalence and severity of

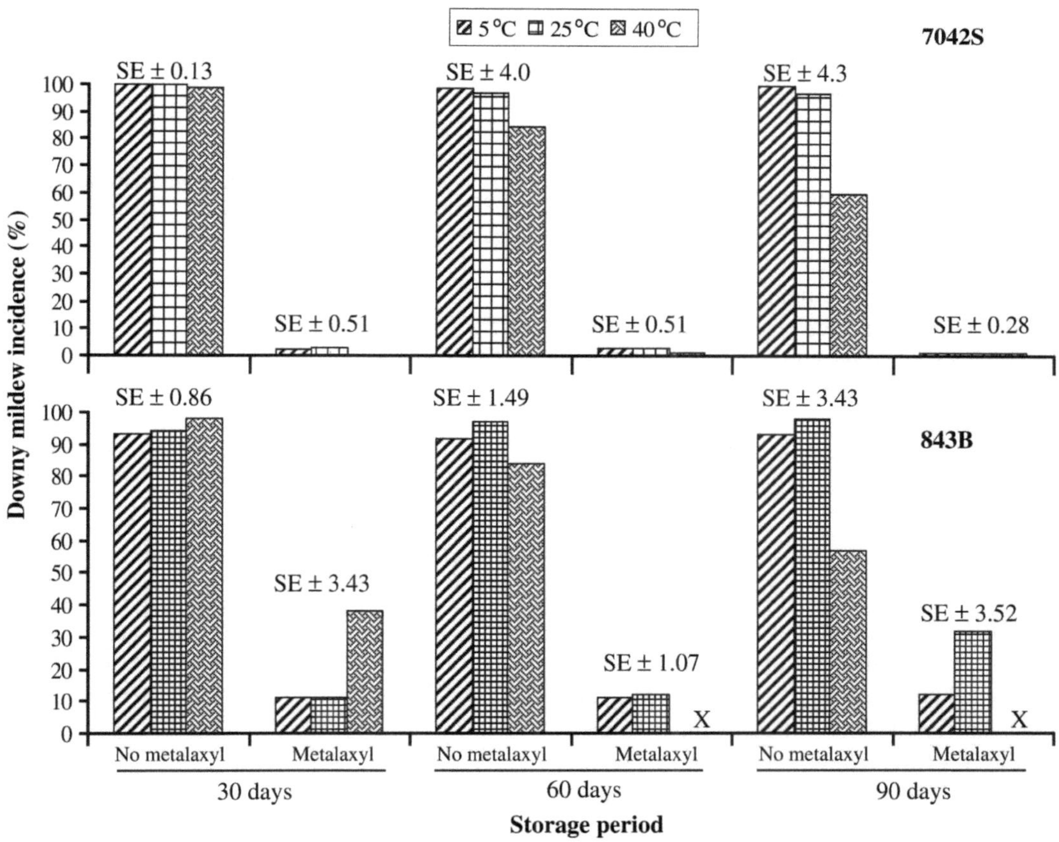

Fig. 3 Effect of storage temperature and duration of metalaxyl (Apron 35WS)-treated seed on downy mildew incidence of two pearl millet lines (7042S and 843B) in a greenhouse experiment. (Note that metalaxyl-treated seed of 843B stored at 40°C for 60 or 90 days failed to emerge (x) in pots indicating phytotoxic effect of metalaxyl)

downy mildew has increased over the years (Thakur et al. 2006). This has necessitated that private seed companies, the major player in promoting and marketing hybrid cultivars, treat the seed with metalaxyl prior to marketing through dealers. In many hybrids where genetic resistance was not adequate they succumbed to the disease even after metalaxyl seed treatment. We have encountered questions frequently from private seed companies regarding reducing the dosages of fungicide, effectiveness of fungicide after seed treatment, and the failure of fungicide to protect the crop from the disease. The results of this study have provided some useful information to the above queries. Reduced dosages of metalaxyl seed treatment in case of moderately resistant lines 7042R and ICMP 451 significantly reduced downy mildew incidence compared to that in the susceptible line 7042S (Fig. 1). Reduction in disease incidence was more pronounced even at 60 days in ICMP 451 than in 7042R most likely due to cultivar differences. This information is very useful for the seed companies to effectively and economically protect the crop and increase the commercial life of the hybrid through reduced fungicide dosages.

Metalaxyl-treated seed, under normal field conditions (adequate soil moisture and temperature, and low oospore inoculum load) has been reported to protect the crop from downy mildew only up to 35 days after emergence (Singh and Shetty 1990; Singh 1995). Our results have shown that fungicide protection lasted up to 20 days in the case of a highly susceptible line 7042S and up to 50 days in a moderately resistant line ICMP 451 and the disease progress up to 70 days (crop maturity) was much slower in ICMP 451 than in 7042S. Thus fungicide seed treatment in the case of a susceptible cultivar cannot be effective in protecting the crop.

Under tropical climatic conditions, where most pearl millet is grown, proper storage of treated seed is critical. In India, fungicide-treated seed of pearl millet hybrids are stored in small godowns for 2–3 months during hot summer months (April–June) when ambient temperatures reach 45°C (48°C in tin-shade go-downs) before selling to farmers after the onset of monsoon rains in June–July. Under these conditions, fungicide-treated seed lots are exposed to high temperatures in poorly-ventilated small godowns, resulting in reduced seed germination and even making the fungicide ineffective. Normally, for a fungicide to be effective the seed treatment should be done just before sowing, but that is not the case in India with pearl millet where seed treatment is done by private seed companies and given to dealers 3–4 months ahead of sowing time for selling to farmers. Results of this study have indicated little effect of storage duration and storage temperatures on effectiveness of fungicide seed treatment (Figs. 2 and 3); although in one line (843B), where fungicide-treated seed was stored at 40°C for 60 and 90 days, germination was adversely affected in blotter test and seedling emergence was inhibited in the potting mix. This clearly indicated variable phytotoxic effect of fungicide to different pearl millet lines. From these results, we propose that the long-term strategy of managing downy mildew in pearl millet should integrate the use of a disease-resistant cultivar (Thakur et al. 2008) with judicious use of fungicide seed treatment. This strategy, in addition to providing adequate protection to the crop will help reduce the chances of evolving fungicide-insensitive strains of the pathogen, which has been a major problem with several oomycete pathogens (Gisi and Sierotzki 2008; Thind 2008).

Acknowledgements Critical review and useful comments by our colleague Dr. Suresh Pande is gratefully acknowledged and contribution of Mr. SM Ahmed in refining the figures is appreciated.

References

CIBA-GEIGY (1989). *Technical Product Information-3.*

Gisi, U., & Sierotzki, H. (2008). Fungicide modes of action and resistance in downy mildews. *European Journal of Plant Pathology, 122*, 157–167.

Hash, C. T., Singh, S. D., Thakur, R. P., & Talukdar, B. S. (1999). Breeding for disease resistance. In I. S. Khairwal, K. N. Rai, D. J. Andrews, & G. Harinarayana (Eds.), *Pearl millet breeding* (pp. 337–379). New Delhi: Oxford & IBH.

Hash, C. T., & Witcombe, J. R. (2002). Gene management and breeding for downy mildew resistance. In J. E. Leslie (Ed.), *Sorghum and millets diseases* (pp. 27–36). Ames: Iowa State Press.

Hess, D. E., Thakur, R. P., Hash, C. T., Sérémé, P., & Magill, C. W. (2002). Pearl millet downy mildew: Problems and control strategies for a new millennium. In J. E. Leslie (Ed.), *Sorghum and millets diseases* (pp. 37–42). Ames: Iowa State Press.

Hospital, F., Chevalet, C., & Mulsant, P. (1992). Using markers in gene introgression breeding programs. *Genetics, 132*, 1199–1210.

Khairwal, I. S., Rai, K. N., Yadav, O. P., & Bhatnagar, S. K. (2004). Pearl millet cultivars. *All India Coordinated Pearl Millet Improvement Project*, Indian Council of Agricultural Research, Mandor, Jodhpur, India, 22p.

Rai, K. N., Kulkarni, V. N., Thakur, R. P., Houssmann, B. I. G., & Mgonja, M. A. (2006). Pearl millet hybrid parents research: approaches and achievements. In C. L. L. Gowda, K. N. Rai, Belum V. S. Reddy, & K. B. Saxena (Eds.), *Hybrid parents research at ICRISAT* (pp. 11–73). Patancheru, India: International Crops Research Institute for the Semi-Arid Tropics.

Rao, V. P., Kadwani, D. L., Sharma, Y. K., Sharma, R., & Thakur, R. P. (2007). Prevalence of pearl millet downy mildew, *Sclerospora graminicola* in Gujarat and pathogenic characterization of its isolates. *Indian Journal of Plant Protection, 35*, 291–295.

Singh, S. D. (1995). Downy mildew of pearl millet. *Plant Disease, 79*, 545–550.

Singh, S. D., & Shetty, H. S. (1990). Efficacy of systemic fungicide metalaxyl for the control of downy mildew (*Sclerospora graminicola*)of pearl millet (*Pennisetum glaucum*). *Indian Journal of Agricultural Sciences, 60*, 575–581.

Singh, S. D., Gopinath, R., Luther, K. D. M., Reddy, P. M., & Pawar, M. N. (1984). Systemic remissive property of metalaxyl against downy mildew in pearl millet. *Plant Disease, 68*, 668–670.

Singh, S. D., King, S. B., & Werder, J. (1993). Downy mildew disease of pearl millet. *Information Bulletin no. 37*. Patancheru, India: International Crops Research Institute for the Semi-Arid Tropics.

Talukdar, B. S., Khairwal, I. S., & Singh, R. (1999). Hybrid breeding. In I. S. Khairwal, K. N. Rai, D. J. Andrews, & G. Harinarayana (Eds.), *Pearl millet breeding* (pp. 269–301). New Delhi: Oxford & IBH.

Thakur, R. P. (1998). Disease management in pearl millet. In T. S. Thind (Ed.), *Diseases of field crops and their management* (pp. 53–76). Ludhiana: National Agricultural Technology Information Center.

Thakur, R. P. (2008). Pearl millet. In S. Lodha, R. Mawar, & B. S. Rathore (Eds.), *Disease management in Arid land Crops* (pp. 21–41). Jodhpur: Scientific Publishers.

Thakur, R. P., Rao, V. P., Amrutesh, K. N., Shetty, H. S., & Datar, V. V. (2003). Field surveys of pearl millet downy mildew—effects of hybrids, fungicide and cropping sequence. *Journal of Mycology and Plant Pathology, 33*, 387–394.

Thakur, R. P., Rao, V. P., Wu, B. M., Subbarao, K. V., Shetty, H. S., Singh, G., et al. (2004). Host resistance stability to downy mildew in pearl millet and pathogenic variability in *Sclerospora graminicola*. *Crop Protection, 23*, 901–908.

Thakur, R. P., Shetty, H. S., & Khairwal, I. S. (2006). Pearl millet downy mildew research in India: progress and perspectives. *International Sorghum and Millets Newsletter, 47*, 125–130.

Thakur, R. P., Rai, K. N., Khairwal, I. S., & Mahala, R. S. (2008). Strategy for downy mildew resistance breeding in pearl millet. *e-Journal of SAT Agricultural Research, 6*.

Thind, T. S. (2008). Fungicide resistance: a perpetual challenge in disease control. *Journal of Mycology and Plant Pathology, 38*, 407–418.

Williams, R. J., & Singh, S. D. (1981). Control of pearl millet downy mildew by seed treatment with metalaxyl. *Annals of Applied Biology, 97*, 263–268.

MIX
Papier aus verantwortungsvollen Quellen
Paper from responsible sources
FSC® C105338

If you have any concerns about our products,
you can contact us on
ProductSafety@springernature.com

In case Publisher is established outside the EU,
the EU authorized representative is:
Springer Nature Customer Service Center GmbH
Europaplatz 3, 69115 Heidelberg, Germany

Printed by Libri Plureos GmbH
in Hamburg, Germany